Good Laboratory Practice

Springer
Berlin
Heidelberg
New York
Barcelona
Hong Kong
London
Milan
Paris
Singapore
Tokyo

Jürg P. Seiler

Good Laboratory Practice

The Why and the How

With 34 Figures and 1 Table

 Springer

Dr. Jürg P. Seiler
Hoelzlistrasse
CH-3475 Riedtwil

ISBN 3-540-67938-3 Springer-Verlag Berlin Heidelberg New York

CIP data applied for

This work is subject to copyright. All rights are reserved, whether the whole or part of the material is concerned, specifically the rights of translation, reprinting, reuse of illustrations, recitation, broadcasting, reproduction on microfilm or in any other way, and storage in data banks. Duplication of this publication or parts thereof is permitted only under the provisions of the German Copyright Law of September 9, 1965, in its current version, and permission for use must always be obtained from Springer-Verlag. Violations are liable for prosecution under the German Copyright Law.

Springer-Verlag Berlin Heidelberg New York
a member of BertelsmannSpringer Science+Business Media GmbH

© Springer-Verlag Berlin Heidelberg 2001
Printed in Germany

The use of general descriptive names, registered names, trademarks, etc. in this publication does not imply, even in the absence of a specific statement, that such names are exempt from the relevant protective laws and regulations and therefore free for general use.

Product liability: The publishers cannot guarantee the accuracy of any information about the application of operative techniques and medications contained in this book. In every individual case the user must check such information by consulting the relevant literature.

Typesetting: Camera ready by the author
Cover design: design & production GmbH, Heidelberg
Printed on acid-free paper   SPIN 10748105   14/3130/ag   5 4 3 2 1 0

Preface

After more than twenty years of use Good Laboratory Practice, or GLP, has attained a secure place in the world of testing chemicals and other "test items" with regard to their safety for humans and the environment. Gone are the days when the GLP regulations were hotly debated amongst scientists in academia and industry and were accused of stifling flexibility in, imaginative approaches to, and science-based conduct of, all kinds of studies concerned with toxic effects and other parameters important for the evaluation and assessment of products submitted for registration and permission to market. The GLP regulations have developed from rules on how to exactly document the planning, conduct and reporting of toxicity studies to a quality system for the management of a multitude of study types, from the simple determination of a physical/chemical parameter to the most complex field tests or ecotoxicology studies. At the same time the term "Good Laboratory Practice" has become somewhat of a slogan with the aim to characterise any reliably conducted laboratory work.

The 1997 revision of the OECD Principles of Good Laboratory Practice has provided the reason to write this book and to present my views on GLP, to explain the changes that GLP has undergone, and to put into perspective the various possible interpretations of GLP requirements. The book is written not only with the Study Director or the Quality Assurance Manager in a regulatory environment as the target population in mind, but it is intended for, and directed to, all quality-minded scientists, less so for lecturing them with the exact interpretation of the strict requirements to be followed - as they have to be rigourously obeyed in a test facility mandatorily working under GLP - but to familiarise them with the intentions of GLP, to explain to them the real idea behind these three letters. It is the opinion of the author that the application of the GLP rules in other forms, adapted to specific situations - the PhD student working on his or her thesis, the research group in academia or industry scratching away at the frontier of science, or the central clinical-chemical laboratory doing routine determinations in the context of a clinical study - could help to increase the transparency, the quality and the integrity of any scientific investigation. Certainly, there are other quality systems which may be more suitable for some of these situations, but the idea behind the regulations on Good Laboratory Practice, namely to ensure the complete traceability of data and the full reconstructability of a study, would be applicable with high rewards to a number of situations outside the "legal realm" of GLP. In this sense, too, the book represents the very personal opinions of the author. My colleagues will certainly and easily recognise those areas, where I have been carried away by my favourite subjects and themes.

This book could not have been written, if I had not had colleagues from industry and from compliance monitoring authorities, too numerous to mention them individually, with whom I could discuss any number of questions with respect to the actual and exact interpretation of the GLP Principles. The Quality Assurance people and the Study Directors from Industry who kept pestering us with the most complex and intricate questions about special situations and the interpretations of the GLP Principles with regard to these, the colleagues from the other Swiss Compliance Monitoring Authorities with whom many heated discusssions arose around the industry questions, and the colleagues from the OECD Working Group on GLP, formerly the GLP Panel and its secretariat, headed by Dian Turnheim: To all of them I owe my sincerest thanks. Some of the material in this book was generously provided by Stan Woollen, Günther Menne, Rolf Vogel and Andreas Edelmann. Another important prerequisite for the successful completion of this book was the understanding of my wife, who constantly encouraged me and who, in consequence, bore patiently the innumerable evenings and weekends in the sole company of our cats, her favourite books (and the TV, if applicable), while her husband unsociably sat typing at the computer, and who, in the end, also helped greatly with the layout of the whole book.

Riedtwil, June 2000   Jürg P. Seiler

Table of Contents

Preface i

I. **What is Good Laboratory Practice All About ?** 1

   1. Introduction 1
   2. The History of GLP 6
   3. The Idea Behind GLP 14
   4. The Areas of Application 19
   5. The Pillars of Good Laboratory Practice 26
   6. Where Can GLP be Profitably Applied ? 31
   7. GLP and Other Laboratory Quality Systems 45

II. **How is Good Laboratory Practice Regulated ?** 51

   1. Introduction 51
   2. Definitions in GLP 53
      - 2.1 *Good Laboratory Practice* 53
      - 2.2 *Management* 54
      - 2.3 *Study Director and Principal Investigator* 57
      - 2.4 *Test Facility and Test Site* 61
      - 2.5 *The Study* 64
      - 2.6 *Short-Term Studies* 70
      - 2.7 *Initiation, Starting and Completion Dates* 76
      - 2.8 *Study Plan, Amendments and Deviations* 80
      - 2.9 *Raw Data* 85
      - 2.10 *The Phases of a Study* 88

| | 2.11 | The Master Schedule | 91 |

3. Responsibilities in Good Laboratory Practice 94
	3.1	Management	94
	3.2	Study Director and Principal Investigator	102
	3.3	Study Personnel	114
	3.4	The Sponsor	116

4. The Quality Assurance Programme 121
	4.1	General Considerations	123
	4.2	Quality Assurance Inspections	128
	4.3	Quality Assurance Inspection Reports	142
	4.4	Audits of Raw Data and of Final Reports	146
	4.5	The Quality Assurance Statement	150

5. Facilities 156
	5.1	General Requirements	157
	5.2	Test System Facilities	160
	5.3	Facilities for Handling Test and Reference Items	162
	5.4	Archive Facilities	165

6. Apparatus, Materials and Reagents 167

7. Computerised Systems 172
	7.1	Introduction	172
	7.2	Basic Considerations	177
	7.3	Data considerations	180
	7.4	Prospective Validation	182
	7.5	Retrospective Evaluation	188
	7.6	Maintenance and Change Control	190
	7.7	Security	192
	7.8	Levels of Complexity	194

8. Test Systems 196
| | 8.1 | Physical/Chemical Test Systems | 197 |
| | 8.2 | Biological Test Systems | 198 |

9. Test and Reference Items 202
- 9.1 Handling and Documentation 203
- 9.2 Characterisation 208
- 9.3 Expiry Dates 218
- 9.4 Sample Retention 220

10. Standard Operating Procedures 223
- 10.1 Introduction 223
- 10.2 The Format 225
- 10.3 Issue, Approval and Distribution 228
- 10.4 On-line SOPs 231
- 10.5 The Content 232
- 10.6 Where are SOPs required? 237

11. Study Performance and Reporting 240
- 11.1 The Study Plan 240
- 11.2 Study Conduct 246
- 11.3 The Final Report 249
- 11.4 Re-opening and Amending a Study 255

12. The Archives 259
- 12.1 Storage Period 259
- 12.2 Indexing and Retrieval 264
- 12.3 Security 265
- 12.4 Archive location, merging and dissolution 269

Appendix II.I 272

Appendix II.II 290

Appendix II.III 307

Appendix II.IV 330

III. How can Good Laboratory Practice be Introduced in a Test Facility? ... 337

- 1. Introduction 337

2. General Aspects 338

3. A General Way to Implementation 341
   3.1 The preliminaries 341
   3.2 The organisation 342
   3.3 Separation and distribution of facilities and equipment 342
   3.4 Interlude: Personnel documentation 345
   3.5 Distributing Responsibilities 346
   3.6 The Major Task: Standard Operating Procedures 347
   3.7 Second Interlude: Quality Assurance and IT 351
   3.8 The Personnel: Education and Training 351
   3.9 Study Plans 353
   3.10 Test and Reference Item Issues 354
   3.11 Study Conduct 354

IV. How is Compliance with Good   Laboratory Practice Monitored ? 357

1. Introduction 357

2. National Monitoring Authorities 358

3. MOUs, MRAs, and MJVs 361

Appendix IV.I 365

Appendix IV.II 373

References 389

Useful Internet Addresses 392

Subject Index 393

I. What is Good Laboratory Practice All About ?

1. Introduction

Good Laboratory Practice belongs to the ever increasing number of "Good Practices", starting with the Good Agricultural Practice and ending (probably) with the Good Zoological Practice. In many of these instances, the term "Good Practice" denotes nothing else than the established way of doing something, the way generally recognised as being the proper one. In many of these instances of "Good Practices" there are no "strings attached", i.e. no specific rules are strictly to be followed in order to comply with the respective Good Practice. The area, where Good Laboratory Practice is employed, however, is of such importance, that the conduct of activities under its terms has to follow stricter rules than are recommended or prescribed in other areas of Good Practices. The difference to many of these other Good Practices, that most clearly separates Good Laboratory Practice from them, is that compliance to the rules of Good Laboratory Practice is of fundamental importance and necessity in the area of investigations into the safety of commercial (chemical) products. Another difference is that compliance to the Principles of Good Laboratory Practice, in company of only a few other "Good Practices", notably those strongly connected with the field of pharmaceutical manufacturing and testing, is monitored by governmental, regulatory bodies. In contrast to this, Good Agricultural Practice, e.g., is just a notion used to set pesticide residue limits. It only describes the recommended farming practice, and, if a farmer is adhering in his spraying programme to the recommended dosage, to the recommended spraying intervals and to the recommended pre-harvest interval, he is using "GAP", even if there were the possibility of using less active ingredient per area, or of using longer intervals between sprayings than recommended. The pesticide levels arrived at in this fashion in field trials will also subsequently be considered as residue limits. GAP can also be used the other way around: If it is accepted practice in one country to "clean" lettuce from lice by spraying an insecticide one or two days before harvest, then the residue limits for this pesticide will have certainly to be set at a higher level than if insecticide treatment is allowed by the "good practice" in another country to be applied ten days before harvest at the latest. Thus, Good Agricultural Practice is not a universal set of rules, as is the Good Laboratory Practice, but changes with the habits, the necessities and possibly also the environmental consciousness from one country to the next.

Chemical substances as well as other items and devices are introduced into industrial, therapeutic, agricultural or household use by virtue of specific properties which are judged "positive" by all, or at least some, humans. Besides these useful, and thus economically exploited, properties, such items and devices, especially chemical substances will, however, exhibit in a lesser or greater degree also some unwanted, even dangerous properties, and these hazards may affect human health and the environment. Even with controlled use, but more so with uncontrolled spreading, chemical (and/or biological) substances will, therefore exhibit some inherent risks. In order to minimise or altogether prevent such risks wherever possible, control legislation has been introduced in most countries world-wide; these control measures generally call for testing and assessing these items to determine their potential hazards. The leading principle of such legislation is that safety test data on chemicals or other items to be put into use have to be generated for, and submitted to, a national Regulatory Authority. The competent authority will then scrutinise and evaluate this information, and determine whether all, or part of, these safety aspects have been addressed and resolved in a satisfactory way before any such item may be placed on the market, or be used in any other way. Furthermore, the requirement that these assessments be based on safety test data of sufficient quality, rigour and reproducibility is another one of the basic principles in such legislation. The issue of the necessary scientific rigour of safety testing has been taken care of by the development of internationally agreed guidelines for the conduct of such studies, as well as for the format and content of the respective submission packages. Although it could be argued that, by its intrinsic virtue, any scientifically conducted safety study would meet also stringent quality criteria, not only the historical facts but also a number of recent occurrences in academic as well as in commercial settings have shown a different picture. In the historical context it had been observed that studies submitted to Regulatory Authorities were of mediocre quality with respect to study design, study conduct and data reporting. Not only this lack of quality, but the detection of outright fraud in such "studies" finally led to the development of formal Principles of Good Laboratory Practice.

In order to be able to recognise what Good Laboratory Practice is all about, what this set of rules intends to achieve, and in what advantages the application of these principles should result, it might be useful to look at the interpretations of this term from different angles. The term "Good Laboratory Practice" may invoke three different notions, two of which are inaccurate if not even completely false:

To some people – they might mainly be characterised as "the administrative person at the sponsor's" – Good Laboratory Practice, or GLP, is the general and everyday practice of a good laboratory; they, figuratively speaking, do not write it with Capital Letters as "GLP", but only as plain "good practices used in the laboratory". They therefore expect any laboratory which claims to be "good" to be in possession of an official certificate attesting its conformity to GLP standards, even if the laboratory in question, and the activities performed in it, would not require such formal recognition. These people are not concerning themselves with the scope and the real meaning and intentions of GLP and they would thus expect GLP to be applied in all, even the most inappropriate settings. Typically, in the pharmaceutical field, this is the case with laboratory investigations of a routine nature which are part of, and performed within, clinical studies on humans, where the clinical Study Monitor would ask that the participating haematology laboratory to produce a GLP Certificate to demonstrate its prowess (Dent, 1994; Fox et al., 1995), even though GLP is clearly defined as applicable for non-clinical, i.e. in the pharmaceutical context mainly animal toxicology, studies only.

Other people - some at the lab bench, or some in charge of corporate finances - are looking at GLP as an administrative burden, imposed by bureaucrats, who have no idea about laboratory work and are therefore imposing futile requirements the adherence to which would be involving too much time and labour which could be spent more profitably in the actual laboratory activities. This feeling has found its expression in the early translation of the acronym GLP into "Gimme Lots'a Paper", which of course harbours a grain of truth. This opinion has been voiced especially by research-minded people, who tended to claim that the GLP requirements, especially the necessary strict adherence to protocols and standard operating procedures, would be stifling innovation in the conduct of scientific investigations. They used to maintain that truly scientific investigations into biological and toxicological properties of test compounds or other test items could only then be performed, when the investigator had the full freedom to let him- or herself lead where the data would be pointing; at every moment of an investigation, an unexpected result might thus necessitate the alteration of the pre-conceived way of conducting the study, and this flexibility should not be hindered by demands for a strict application of Standard Operating Procedures and by the requirement to have a fully developed and strictly formulated study plan ready before the start of the investigation. Although already from the outset of formulating GLP principles it had been stressed that these regulations were a management tool with the objective of promoting and maintaining the quality of safety test data, and that therefore they would not interfere with the exercise of scientific knowledge or practice, but would rather complement the scientific side of safety testing, this opinion of GLP as a hindrance to a really scientific conduct of studies was very widespread.

While the latter notion has lost some of its original importance, the former one has gained weight in recent years. However, Good Laboratory Practice, as it has been conceived and as it is being used, has a completely different meaning.

Good Laboratory Practice it is a quality system which intends to ensure, through careful and accurate documentation, covering all aspects of a study and of its environment, the quality, integrity and reliability of safety data. Certainly, there is a not-to-be-underestimated amount of administrative work connected with the conduct of studies under the conditions of GLP and, regarded in this way, GLP may indeed be looked at as somehow hemming-in scientific creativity. This "setting of guiding rails" should not, however, be confounded with having to perform studies within a rigid framework from which no deviations could be possible. On the contrary: Every scientist would certainly and wholeheartedly agree that any experiment should be conducted according to a well conceived plan, that the data should be recorded faithfully and completely, and that finally the results should be presented in a way truthfully reflecting the actual data, since only then could the conclusions from the study be really trusted and utilised to prove (or refute) the starting hypothesis. What GLP then does is to formalise these "common sense issues" in a way that would ensure their general application in order to make studies conducted under these principles of comparable trustworthiness.

Thus, the Principles of Good Laboratory Practice (GLP) have been developed to promote the quality and validity of test data used for determining the safety of chemicals and chemicals products. It is primarily a managerial concept covering the organisational process and the conditions under which laboratory studies are planned, performed, monitored, recorded and reported. Its principles are required to be followed by test facilities carrying out studies to be submitted to national authorities for the purposes of assessment of chemicals and other uses relating to the protection of man and the environment. They can, however, be seen as far more outreaching ideas possibly also influencing the conduct of studies which do not fall into the restricted area of "human health and environmental safety studies". Indeed it has been widely acknowledged that some of the principles underlying GLP should be observed in one or another form in a number of additional areas, too. A later section of this part will be dealing with some of the issues emanating from this notion.

This book now intends not only to introduce the reader to the principles of GLP as they are laid down in national and international regulations, the most important ones being the revised Principles of Good Laboratory Practice of the Organisation for Economic Cooperation and Development (OECD); its intentions go beyond a purely descriptive and explanatory approach. It intends to be educational rather than being a training textbook with ready-to-use recipes for the creation of QA programmes or the writing of study plans. There are a number of good textbooks which fulfil this training role and which, e.g., may give practical advice on how to write Standard Operating Procedures or which may present Check Lists for the preparation of laboratory inspections. Certainly, this book, too, will mainly deal with the various ways of interpreting the GLP guidelines and with the practical aspects of implementing the GLP principles in a laboratory. In this respect it is directed at those people in test facilities who have to work under GLP conditions, who have to deal with the various possibilities of interpreting the rules, and who have sometimes to adapt the principles to their individual, special problem. However, this book is also directed at people not directly involved with GLP, but who are working in an environment, where the principles expressed in the GLP regulations could (and should) be put to good use, since also activities not covered directly by the scope of GLP could certainly profit from the application of these principles. Given, e.g., today's atmosphere of great competitiveness in research, it might be important one day for a researcher or a laboratory to be able to unequivocally and convincingly demonstrate that their data were the result of a well planned experiment, the conduct of which, and all circumstances surrounding it, had been well controlled and documented in an impeccable way. It is thus not only for the "GLP professional" that these principles are valuable - for him/her, they are of course indispensable - but the quality of any study could benefit from the application of the basic tenets of Good Laboratory Practice. In this sense, the first of the "inaccurate or false notions about GLP" may be regarded as being not so wrong indeed (as it has to be looked at from a regulatory point of view). It should indeed be the customary hallmark of any "good" laboratory to have, in a general way and in some measure, introduced these Principles of GLP in the conduct of their daily activities.

2. The History of GLP

While the term "good laboratory practice" might have been used colloquially already for some time in many laboratories around the world, its first official use can be found in the 1972 New Zealand Testing Laboratory Act, where the "(promotion of) the development and maintenance of good laboratory practice in testing" had been made a task of the Council of Testing Laboratory Registration. In this rather general statement, constituting a prerequisite for the registration of any testing laboratories, the term "good laboratory practice" did not yet denote the defined regulations of the present times, but was rather an indication of a good quality level of the work conducted in such laboratories. To develop this general term into the one written with Capital Letters as "Good Laboratory Practice", consisting of a stringent set of rules and a defined area of application, an external impulse was necessary, which was given by an accumulation of occurrences, yielding negative headlines, in the area of toxicity testing, especially in the United States.

The end of the 1960s and the beginning of the 1970s had been characterised by a surge in the invention, production and use of chemicals for a variety of purposes. At the same time, regulatory requirements especially for safety testing of pharmaceuticals tightened in the wake of the thalidomide disaster. Safety testing of a rapidly increasing number of chemical substances was thus partly exceeding the capacities of chemical companies, and they increasingly turned to specialised testing laboratories ("Contract Research Organisations", CROs) for the conduct of the safety studies required by Regulatory Authorities. There are, of course, three different kinds of danger lurking in such a constellation, two of which are connected directly with the business of the CRO. One is the "wish to please", i.e. to deliver results that come as near as possible to the intentions of the sponsor, as the sponsor, naturally, has not much interest in an outcome of a safety test that could endanger the fate of the product. The second one is the wish to conserve, or even increase, the market share of the testing facility; thus, possibly, more commissions might be taken in than the test facility could cope with, and therefore, the quality of the study conduct certainly would tend to suffer. On the other hand the sponsor could, of course, also exercise some pressure on the CRO to deliver results that would be agreeable to them, in view of the fact that the CRO also intended not only not to lose this sponsor but to increase its market share. Such "unholy alliances" might then result not only in the sloppy conduct of tests, but could ultimately lead, in the absence of strong controls, to outright fraud.

This was exactly the situation at the beginning of the 1970s. At least some pharmaceutical companies had obviously supported their New Drug Applications (NADs) with data, that had been generated through studies designed to minimise negative findings and thus to procure a favourable outcome. When discrepancies between the data and conclusions submitted to the US Food and Drug Administration (FDA) with other data published in the open literature became obvious, the US Senate started, in the middle of 1975 and under the chairmanship of Senator Kennedy, a series of hearings on the "Preclinical and Clinical Testing by the Pharmaceutical Industry", the so-called "Kennedy Hearings" (Comm. on Public Labor and Welfare, 1975). In these hearings, FDA officials described a number of occasions, where they had detected deliberate changes, e.g. between the descriptions of tumours in the raw data and their description in the respective final report. In order to make the product look more innocuous, benign tumours that occurred in control animals had their description changed to "malignant", thus artificially boosting the incidence of malignancies in the controls, and minimising in this way the difference in tumour incidences between control and treated groups. There were also indications of fraudulent substitution of animals in order to cover up and negate positive findings. Several such examples were given at these hearings by Dr. Adrian Gross of FDA. He described for instance an incident in which the company in question went so far as to destroy evidence. In a carcinogenicity study on Metronidazole ("Flagyl"), the occurrence of an adenocarcinoma of the mammary gland in a control male was described. Since this is a very rare tumour in males, but relatively common in females, the occurrence in a control male would tend to decrease the significance of similar tumours found in treated males. In order to check, by chromosomal analysis, whether this tumour indeed had originated in a male animal, FDA inspectors tried to recover some tissue material from this animal, but the company first refused to hand over this material; subsequently the company rendered the whole tissue unanalyzable for this purpose, and the question of whether a fraudulent exchange of a female for a male had taken place could not be answered any more. Further examples of irregularities uncovered at some companies included the observation that there were entries made on necropsy protocols several months after the actual necropsy, and by a pathologist who could not have been present at the necropsy itself because his employment started only in the year after these necropsies had been performed. In other instances, whole autopsy reports from a number of animals in a study were outright missing. These irregularities at, and fraudulent behaviour of, some companies led then to a large investigation into the preclinical, toxicology testing at pharmaceutical companies and CROs.

Prominent amongst the CROs at this time was a company called Industrial Bio-Test Laboratories, or IBT for short. This company had developed from modest beginnings. Founded in 1953, IBT quickly grew into the largest contract toxicology facility of the United States, if not the whole world. During the 1950s and '60s it increased its business until, in the middle of the 1970s, it conducted an estimated 35 to 40 percent of all toxicology studies in the US. More than 22000 toxicology studies were performed by this company during its existence, and a high proportion of them served to support the registration of pesticides, cosmetics, pharmaceuticals and other chemical substances and were submitted to an agency for obtaining a marketing permission. IBT was reputed to do good quality work and, amongst sponsors, it was renowned for its moderate prices. Its scientific reputation was so good, that the name "IBT" on a study report practically guaranteed the full acceptance of its results and conclusions, without any questions being asked, by the authorities.

In 1976, however, in the wake of these Kennedy Hearings, investigators from the FDA uncovered at IBT what amounted to be the largest scientific fraud ever committed in the US, if not the whole world (Schneider, 1983).

In order on the one hand to deliver results that agreed with the wishes or the outright demands of the sponsors and on the other to cover up massive shortcomings in the design both of animal rooms and of studies alike, the company personnel either faked or suppressed scientific findings, invented data for activities that were never performed, and suppressed documentation on, or never documented, other activities, like the replacement of dead animals. The conditions at IBT must have been appalling: Mice and rats obviously could escape their cages nearly at will, test animals were given wrong doses or even wrong test compounds, dead animals were replaced by new ones without any notice. If such replacements were not done, and the mortalities, e.g. in control groups exceeded the limits for meaningful statistical analysis, then whole mortality tables were either made "internally consistent" (i.e. faked) by backward re-calculation, or data from a similar control group from another experiment were taken to fill the gap. If documentation on analytical results from haematology or urinalysis investigations could not be found anymore in the files, or if it had plainly been forgotten to perform these analyses, it did not matter: The data could very well be fabricated out of the existing ones, with, at the same time, favourable regard for the "most appropriate" result.

It was obvious that a better control of such safety data had to be instituted, from the planning of the studies all the way through the generation of results, the documentation of data and their retention, to the final report and its submission to the relevant authorities. Such controls should not only make fraud less easy to commit, but they should also do away with the sloppiness in study planning and study conduct, and in the handling of documentation connected with such studies. In this spirit the FDA published proposed regulations for Good Laboratory Practice in non-clinical laboratory studies in the Federal Register on November 19, 1976. The applicability of these rules was then tested by FDA in a pilot inspection program that started in December 1976 with its results being reported in 1977. After reviewing the written comments received and after considering the opinions expressed at public hearings, the FDA finalised these rules and published them in the Federal Register on December 22, 1978 under Title 21 of the Code of Federal Regulations as Part 58 (21 CFR 58), and these GLP regulations came into effect on June 20, 1979. In the introductory summary the FDA stated the following: "*(The issuing of GLP regulations) is based on investigatory findings by the agency that some studies submitted in support of the safety of regulated products have not been conducted in accord with acceptable practice, and that accordingly data from such studies have not always been of a quality and integrity to assure product safety ... Conformity with these rules is intended to assure the high quality of non-clinical laboratory testing required to evaluate the safety of regulated products.*" Due to the great importance of the quality and integrity of data for the assessment of safety the FDA published these principles of GLP as a Regulation rather than a mere Guideline, providing them with much higher legal weight. By a publication in the Federal Register of September 4, 1987, the FDA amended its GLP Regulations with the intent of reducing the regulatory burden on testing facilities, while avoiding to compromise the objectives of the GLP Regulations, i.e. to assure the quality and integrity of the safety data submitted to the agency.

The US Environmental Protection Agency, which had been as much touched by the scandal at IBT, or even more so, because practically the whole toxicity testing of new pesticides had been performed by this CRO, followed suit with its own Good Laboratory Practice standards which had to be placed under two different legislative umbrellas, i.e. the Federal Insecticide, Fungicide and Rodenticide Act (FIFRA), which deals with pesticides and their safety, and the Toxic Substances Control Act (TSCA), which is concerned with chemical substances in general. The proposed rulings were announced by EPA through publication in the Federal Register: On May 9, 1979, GLP standards for the investigation of health effects under TSCA were proposed, which were supplemented by proposed standards for physical, chemical, persistence and ecological effects testing on November 21, 1980; the respective proposal for regulation of GLP in toxicity testing under the pesticide program of FIFRA was separately published on April 18, 1980. The respective GLP regulations were finally published by EPA in 1983 in the Code of Federal Regulations (40 CFR 160 and 40 CFR 792, resp.).

Through a number of considerations that arose as a result of the application of EPA's two sets of GLP regulations the agency has recently attempted to combine the two into a single regulation. On December 29, 1999, the agency published in the Federal Register its proposal for a unified, modernised regulation which should not only take care of certain developments in the area of GLP, but should also facilitate the application of GLP in certain instances. The final rules will ultimately be published in the Code of Federal Regulations under the title of 40 CFR 806 and will thus replace both of the 1980 versions (160 and 792, respectively).

With the United States having taken the lead, the need for an international harmonisation of these standards became rapidly obvious. Non-clinical laboratory investigations included in submissions to US Regulatory Agencies had to comply with these GLP standards, and thus it was feared that studies not conducted according to these principles might have to be repeated in order to be accepted by the respective US Agencies. Such duplication of toxicology studies was considered not only to be wasteful of resources but also to be contrary to animal protection ideas. Therefore the Organisation for Economic Cooperation and Development (OECD) began, more or less at the same time as the FDA rules were published, with the task of an international harmonisation of these standards.

Such international harmonisation was urgently needed, since the issue of data quality had (and still has!) an important international dimension. Trade in chemical substances, pesticides and pharmaceuticals was large and ever increasing. Chemicals control legislation, including legislation on the control of pharmaceuticals, food and feed additives, cosmetics, pesticides and industrial chemicals relied on safety test data for registration of the respective compounds, and data quality played an important role in the proper functioning of all such legislation. While the globalisation of the international trade in chemicals and chemically-based commodities was being facilitated through alleviation or abolishment of tariff hurdles, the development of national, non-harmonised regulations on the proper conduct of safety studies threatened to establish new, non-tariff barriers to trade. However, if Regulatory Authorities in one country could rely on safety test data that had been developed in test facilities of another country, duplicative testing could be avoided. Not only would by this development savings in monetary costs and human resources be achieved, but at the same time a goal of animal protection, i.e. reduction of animal numbers used in toxicity testing, would be approached. Moreover, common principles for GLP would facilitate the exchange of information and prevent the feared emergence of additional, new non-tariff barriers to trade, while pro-actively contributing to the protection of human health and the environment.

The Principles of Good Laboratory Practice of the OECD were first developed by an Expert Group on GLP established in 1978 under the Special Programme on the Control of Chemicals. The expert group started by identifying three essential elements upon which such mutual acceptance of data could be based. Besides the utilisation of the OECD Guidelines for the Testing of Chemicals, they cited the application of GLP Principles and the establishment of harmonised national GLP compliance monitoring programmes as essential parts of the mutual acceptability of data. The Principles that were elaborated by this expert group were set out as an integral part of the Council Decision on Mutual Acceptance of Data in the Assessment of Chemicals, which states that *"data generated in the testing of chemicals in an OECD Member country in accordance with OECD Test Guidelines and OECD Principles of Good Laboratory Practice shall be accepted in other Member countries for purposes of assessment and other uses relating to the protection of man and the environment"*. The resulting Principles of Good Laboratory Practice were published and formally recommended for use in Member countries by the OECD Council in 1981 [C(81)30(Final)]. The working group of experts who had developed the OECD Principles of Good Laboratory Practice also proceeded to formulate and publish guidance for the Monitoring Authorities with regard to the introduction of procedures necessary for the monitoring of industry's compliance with these Principles, as well as guidance with respect to the actual conduct of the necessary control activities, i.e. laboratory inspections and study audits. This guidance was already incorporated in the final report of the expert group on GLP, but was subsequently published by OECD as separate documents in the OECD Series on GLP (OECD GLP Series Nos. 2 and 3, 1991, revised 1995).

In response to these developments, other countries began to develop Good Laboratory Practice standards, too, which in most cases were based on the GLP Principles of either FDA or OECD. Thus, during the late 1970s and early '80s, The Netherlands, Switzerland, UK and Japan formulated their national GLP standards. The European Community also, by the Council Directive 67/18/EEC of 18 December 1986, formally adopted the OECD Principles, including them within its framework of guidelines governing the submission of safety data for the marketing of chemical substances.

The GLP standards that had been formulated in these various publications were primarily based on the way toxicology studies were conducted at that time. Laboratory automation was not yet very much advanced, and most, if not all, data had to be recorded in handwriting. Thus, the GLP Principles called for faithfulness in the recording of data through requirements such as the one for the continued legibility of the original entry after correction, with the reason for the change being also recorded and acknowledged by dated signature (or initials). They also called for the creation of a single point of control in the person of the Study Director, who should be able to supervise the whole study conduct, and should thus also be able to bear the full responsibility for the quality, completeness, integrity, accuracy and faithful reporting of the data recorded in the study. However, the technical development through automation and computerisation, the fragmentation of studies into various parts and their out-sourcing to specialised laboratories, as well as the requirement for additional study types, other than animal toxicity testing, to be conducted under GLP, as well as some unresolved questions led to a continued need for adaptation of these Principles to new areas.

Such new developments were formally addressed by the OECD in a series of so-called Consensus Conferences, where experts from OECD member countries discussed single issues and developed guidance for the application of the GLP Principles to areas such as Field Tests, Short-Term Tests, Suppliers and Computer Validation. The documents originating from the deliberations of the respective groups were published as Consensus Documents, with the intent that the recommendations therein should help to arrive at a harmonised interpretation and implementation of the basic Principles. However, these documents have not the same formal status as the Principles themselves; therefore, they are not applicable, or enforceable, to the same extent as the Principles themselves, and they were thus not introduced to the same extent as the Principles themselves into legislative frameworks. On the other hand, several national GLP Monitoring Authorities independently addressed a number of these issues on an individual basis. The problem of the application of GLP Principles to computerised systems ranged foremost in such efforts to keep these standards up-to-date; the UK authority issued a guidance for the application of the GLP Principles to computerised systems in 1989, and the US EPA published its GALP ("Good Automated Laboratory Practices") guidance document in 1995.

These developments were then the reason why, after about fifteen years of use, OECD member countries considered that there was a need to review and update the Principles of GLP to account for scientific and technical progress in the field of safety testing as well as for the fact that safety testing was required in many more areas of testing than was the case at the end of the 1970's. On the proposal of the Joint Meeting of the Chemicals Group and Management Committee of the Special Programme on the Control of Chemicals, another Expert Group was therefore established in 1995 to develop a proposal to revise the Principles of GLP. The Revised OECD Principles of GLP were reviewed in the relevant policy bodies of the Organisation and were finally adopted by the OECD Council on 26th November, 1997. [C(97)186/Final].

Subsequently these revised Principles were again introduced into the legislatory framework of a number of OECD countries. As an example, the European Union issued the Commission Directive 1999/11/EC on 8 March 1999, therewith adapting the former regulation to the technical progress and the revised GLP Principles. At the same time the Commission Directive 1999/12/EC adapted the Annex to the Council Directive 88/320/EEC on the inspection and verification of compliance with Good Laboratory Practice to the new circumstances.

3. The Idea Behind GLP

Any claim for the presence of certain advantageous properties, or for the absence of noxious influences, in an item which is publicly available or widely distributed, be this a car, a food or a chemical substance, can either be accepted in mutual trust, or it has to be verified by some mechanism. In some cases the importance of being able to rely absolutely on the claim made by the manufacturer or the distributor may not be so great as to necessitate or justify such a verification. Whether it is true that a certain make of car could be able to run at a speed of 300 km/h will not be important to most drivers in their everyday situation, and they will therefore most probably not care whether this claim can be trusted to be true, or whether it is just a marketing exaggeration. In many other instances, however, mainly where safety aspects are at stake, it stands to reason that such claims should be not only trustworthy, but verifiable. It indeed belongs to the functions of the respective authorities to make sure that the safety of the products they are admitting to be marketed, used, and dissipated into the environment is proven and that these safety claims can be verified. Proof for the safety of chemical products to humans and the environment, or, to be more exact, evidence for the absence of noxious or dangerous properties of such products at the prospective doses or exposures under actual conditions of use, will be obtained mainly by experimental means, i.e. by testing these products in laboratories or in the field.

A Regulatory Authority may then obtain verification of such an experimental result by either of two ways: By repeating the experiment itself, or by the complete, step-by-step, reconstruction of all activities performed and circumstances encountered which had been leading to the result to be verified. While the former approach might be considered to give more confidence in the result, it is impractical, very costly, and even unethical, since it would entail, among other things, the endless repetition of numerous animal toxicity studies. On the other hand, although the latter approach will not give direct confirmation of the results themselves, it will implicitly lead to trust into them, because the planning and the conduct of the experiment, as well as the recording and the reporting of the data can be followed, and it can then be judged, whether the work in this test facility can indeed be considered trustworthy.

In the first instance, and based on the historical reasons for its development, Good Laboratory Practice may thus be seen as an instrument of mistrust. Certainly, the requirements of GLP have been developed for combating fraud in the generation and reporting of safety data. The idea behind GLP extends, however, much further than that. It is not only a control mechanism which enables Regulatory Authorities to judge the integrity of a study by obtaining information about the probability of whether it has been conducted in the way as it is being described in the study report, and whether it has really yielded the results submitted. Results, by the way, the authority has to rely on in the task of determining the safety of the product in question. The GLP Principles are designed as a tool enabling also the improvement of study and data quality by applying rigorous documentation requirements which allow for the possibility of reconstructing any activity all the way back to its inception.

The requirements laid down in the GLP Principles are addressing a number of issues, and they are directed at the various organisational levels of a facility performing safety studies. Many of the requirements within these issues can be regarded as constituting "common sense", which should be adhered to whether working under GLP circumstances or not. There are sound principles like having to work according to a previously agreed working protocol, from which deviations may be possible, but only on documented reasons. There are self-evident demands on the management, like the obligation to provide sufficient space, equipment and personnel for the tests, and to ensure the technical and scientific competence of the testing people. There are, however, of course also those formal requirements which do not yield readily to understanding and acceptance like having to date and initial a work sheet every time the floor of an animal room is mopped up.

All these requirements can be summarised in three issues that are central to the ideas behind GLP:

The first one is the possibility, for a third party, to reconstruct the whole course of a safety study, even years after it has been performed, and even in the absence of persons having been actively involved in the conduct of this specific study. This reconstructability is the reassurance needed by the Regulatory Authority that there have been no major flaws in the technical conduct of the study, that, e.g., all the animals have received the correct dose of the test item at all times, that the correct samples have been taken and analysed, and that the compilation of results faithfully reflects the actual data that had been collected. It provides reassurance that experiments have been performed in the exact way as they have been described in the report submitted to the Regulatory Authority. This is in some way related to, but not quite congruent with, the idea of "traceability". The term "traceability" is used in metrology to describe the mechanism by which any physical entity, e.g. of weight or length, can be traced back to the respective international standards. Thus, for the calibration of a balance, the manufacturer may provide a set of calibration weights, for which an unbroken chain of calibrations leads back to the original, international standard weight (to be more exact: the international prototype kilogram mass) as it is being kept at the Bureau International des Poids et Mesures at Sèvres near Paris. The fact that the calibration weight of any specific balance can be traced back to this international standard weight allows for reliance into the precision of the actual single weight determination, as well as for its comparability to other weight measurements. This specific notion of traceability has indeed also been taken up by the GLP Principles with the requirement that "*calibration should, where appropriate, be traceable to national or international standards of measurement.*" However, in GLP this term should be regarded in a more general sense, namely in its connection with the possibility of an exact reconstruction of all activities, events and decisions, which together make up the course of a study.

The second issue can be looked at as "accountability" and is very closely connected with the first one. The documentation needed in the GLP compliant conduct of a study will tell years later, who was doing what, and who could be held accountable for mistakes. On the other hand, if any question arises, it is also possible, because of this accountability, to call on the correct person, if still available at the test facility, the technician who actually did this piece of work, for clarification of this problem. Or, as another example, it makes it possible to look at the records accounting for amounts of test item received, used, and returned or destroyed at the end of the experiment in order to judge the plausibility of claims for doses given to animals or for concentrations used in field tests.

Thirdly, GLP increases awareness. Not only will it increase awareness for the greater issues: Of management for the never ending task to strive for optimal quality and transparency of the studies conducted at their test facilities; of Study Directors for the orderly performance of the studies they are to control. It also raises awareness for the small details that may or may not affect the fate of a study, or indeed of a test compound. Most importantly it raises awareness for the activities that are routine in a test procedure. Inherent in such routine activities that can be performed "in one's sleep" may be the danger that they would not receive the necessary attention and might therefore either become performed thoughtlessly, ending up with failing to observe a possibly important effect. Having to acknowledge the performance of such routine activities by dated signature (or initialling) can certainly become a routine activity in itself, but the knowledge that any omission or error, through inattention for the requirements of these activities, can be traced back to the responsible person, will increase the awareness for the correct performance even of such routine tasks.

All of these points necessitate, however, that one general principle be followed, namely that not only records will be generated for each and every activity, event and condition, but that these records should be retained in such a way as to allow an orderly retrieval of each single piece of information whenever that might be needed. A test facility, or a study, in which such orderly record keeping could not be guaranteed would never be considered as being GLP compliant. How important this way of thinking and acting is may be illustrated by a recent example of a clinical study, where the report had to admit that "source documentation for about 12 % of the patients is lacking", and that "for most of these, informed consent had been obtained, but the respective forms have been misplaced". The authorities receiving such study reports may choose to believe these assertions, or they may call them fairy tales and reject the study. Without documented proof for their data and information, the company seeking marketing approval for its product would certainly be on the losing end. A word of caution has to be inserted here, however. It is certainly not the intention here to pass the blame entirely on other "good practices". It has to be recognised that it will be much more difficult to control the "good documentary practice" in a clinical study involving scores of participating centres at hospitals and physicians in several countries, than in a GLP study involving one test facility or at most a few test sites. Furthermore, there may also be instances where test facilities operating under the conditions and rules of GLP may fail in some respect, as is demonstrated by the examples of test facilities, or GLP studies which had to be marked "non-compliant". But let us turn back to the main theme of this section.

While GLP does not formally address scientific issues in the choice of, or the necessity for, the tests performed, to follow these basic ideas behind the framework of GLP can nevertheless also give rise to better science and to a more rational study conduct. The necessity of having to prepare a study plan with full arguments about the reasons for the study and its proposed mode of conduct, could, in one or the other instance, lead to some sitting back and thinking about better ways to achieve a certain goal. It has to be acknowledged, that in everyday practice in many cases the study plans reside as general templates in the word-processing system of the test facility, and they are just printed out from this template, which is filled in with the specific information for the study to be planned, while the reasons given for the study itself ("regulatory requirement") and its mode of conduct ("the rat is an acknowledged animal species for this type of study") will stay more or less always the same. However, there are two points to be considered in this question: Since the study results will ultimately be judged by the Receiving Authorities, the relevant, rational and correct scientific principles have necessarily to be addressed in the study plan. Secondly, since a violation of the GLP Principles would occur, if the study would not be performed according to the study plan, the conduct of the study itself has therefore to follow the study plan with its built-in scientific principles. This would call for an *a priori* consideration of the science involved in the planned study. Also, insofar as scientific methodologies may affect the quality and integrity of studies, as for instance through the calibration of instruments, or the characterisation of the test item, such methodological points are subject to the GLP standards.

There is an old example about the differences in the way GLP and Science are looking at these studies. Let us consider the analytical determination of glucose in a number of biological samples. There are several methods with which the presence of glucose can be determined: There are elaborate, highly precise analytical technologies utilising HPLC or enzymatic reactions, there are some semi-quantitative means of roughly estimating glucose levels (e.g. the so-called "dip-sticks"), or one may simply wet the fingertip in the sample and try whether it "tastes sweet". Any of these methods may be performed in a GLP-compliant way, if it were ascertained by the presence of Standard Operating Procedures that the way in which these tests were performed would be standardised to an extent that would make it possible to reconstruct the whole way of obtaining the results. It is clear, however, that - according to the level of precision required by the scientific purpose of the study - the scientific evaluation, and finally the safety assessment, would reject any study performed with a methodology unable to produce results with the required degree of precision. There may be many more examples to be collected from all kinds of studies: erythema in an irritation study can be judged by the (experienced) technician by eye, or it can be determined through the use of a calibrated instrument measuring redness or skin temperature; the ripeness stage of a crop may be determined by some analytical method involving sophisticated equipment and looking for the presence or concentration of some lead chemical, or it may be guessed by the outer appearance or by some organoleptic trial. Any of these possible ways of assessing some property of the test system can be conducted to GLP standards, and the science behind the method is important only with regard to the purpose of the investigation and to the demands on the precision of the data to be obtained.

Thus, GLP is primarily intended to ensure data quality and integrity and is not concerned in a direct way with scientific issues; but in the application of GLP the scientific aspects of safety studies are indirectly addressed as well.

4. The Areas of Application

Many questions and problems about the application of the GLP Principles centre around, and originate from, an uncertainty about the real areas of applicability. Apart from the clear-cut, "classical" area of application, i.e. toxicology studies on pesticides and pharmaceutical ingredients, there may be investigations, where it could become difficult to judge whether or not these should be conducted under the strict regime of the GLP standards. There are, however, two firm determinants which can be used to define the necessity for application of the principles of GLP. They have to be applied in combination in order to determine whether or not for any specific study or study type strict adherence to GLP would be mandatory.

First of all, there is the scope of GLP as it is defined in the OECD Principles, and which states that GLP encompasses *"the non-clinical safety testing of test items contained in pharmaceutical products, pesticide products, cosmetic products, veterinary drugs as well as food additives, feed additives, and industrial chemicals. These test items are frequently synthetic chemicals, but may be of natural or biological origin and, in some circumstances, may be living organisms. The purpose of testing these test items is to obtain data on their properties and/or their safety with respect to human health and/or the environment."* (OECD, 1998). GLP is thus applicable to safety studies in two major areas: Effects on human health and on the environment. These two areas may share some types of studies that have to be conducted in order to test the safety of the respective test item, but other study types may exclusively be required for one or the other area.

The second point to be observed in the judgement on the necessity for GLP adherence is that these studies are not only conducted to *"obtain data on (the test item's) properties and/or its safety"*, but that they are *"intended for submission to appropriate regulatory authorities."* These are the two aspects that have to be considered when discussing the question of whether GLP should or could be deemed mandatory for any single study or type of study.

It is important, however, to stress that this restriction to tests on safety, and amongst these to such tests only, which are intended for submission to Regulatory Authorities, would not mean that the essential principles from the GLP regulations should not be applicable to any other study or type of study; on the contrary, we will see later on in this part (see section 6, page 31) that these principles can be profitably utilised in many areas where one would not, at first thought, consider applying them.

Let us therefore first dissect these prerequisites for the full, mandatory application of the GLP Principles.

The first delimitation of the areas of application of GLP is described as "non-clinical safety testing". While the OECD Principles leave it at this general expression, the FDA guidelines go somewhat further, in that they expressly exclude testing on humans and on human material from the requirement of GLP. In their definitions (21 CFR 58.3 d) it is stated that *"the term (non-clinical laboratory study) does not include studies utilising human subjects or clinical studies"*. Although it is commonly understood that "non-clinical testing" is only concerned with performing safety studies *in vivo* on animals and in *in vitro* systems, or with conducting environmental safety studies in the field, there are nevertheless attempts in the area of clinical testing of pharmaceuticals to introduce GLP into the laboratory parts of clinical studies on humans, too. Although, or because, the Principles of Good Clinical Practice (GCP) mostly do not deal expressly with the question, under which one of the available or applicable quality schemes the laboratory investigations (haematology, clinical chemistry, pharmacokinetics) within a clinical study should be conducted, the opinion that the (unspecified in, but nevertheless required by, GCP) quality system should be equal to the Good Laboratory Practice is finding therefore its advocates. Certainly, there are aspects in the rules of GLP which could profitably be employed also in the conduct of laboratory investigations within the context of clinical trials. However, to mandatorily require a laboratory, which is analysing samples from clinical trials, to be officially acknowledged as complying with GLP would lack the support of the wording and the intentions of the Principles of GLP.

1.4 Areas of Application

There is another side to this first delimitation: GLP is defined as *"a quality system concerned with the organisational process and the conditions under which non-clinical health and environmental safety studies are planned, performed, monitored, recorded, archived and reported."* It is not a system that guarantees either the scientific validity of the method used in any study, nor does it guarantee the ability of the test facility to generate accurate and precise measurements. It is only through the possibility of reconstructing each and every activity and process within a study, that the accuracy and precision of the data reported may be judged. This definition of "quality" distinguishes GLP from other quality systems which focus more strongly on the ability of the respective test facilities to reproducibly generate accurate and precise results and data. This aspect also makes it clear that GLP is not the instrument with which the conduct of a safety study according to high scientific standards can be warranted. Since it is mandatory in GLP, however, that the selection of the actual test system should be justified, this goal of the application of "good science" in a safety study can nevertheless be reached in an indirect way.

The applicability of GLP to "human health and environmental safety" testing spells out yet another facet of this delimitation which may give rise to a number of questions. Laboratory testing that is performed in a non safety-related way, e.g. only to ascertain the keeping of certain specification limits, or to analyse for the content of ingredients (e.g. nutrient content of food or feed), need not mandatorily be conducted under GLP, since they are purely quality control measurements, used to assure the conformity of the product, and therefore GLP does not apply to the work done to establish the specifications of a test item. Laboratory testing which is performed with the view of demonstrating the efficacy of the test item in its (prospective) use is also not a case for GLP, and neither would be the organoleptic evaluation of processed foods. Also, work done to develop chemical methods of analysis or the first validation trials for an analytical method are another case in point: If such validation trials are only conducted in a "pilot study" form to confirm the applicability of analytical methods used to determine, e.g. the concentration of test substance in animal tissues, or in drug dosage forms, then the GLP rules need not be strictly followed, insofar as there would be no inspections or audits by the QA. However, GLP has to be applied to the chemical procedures used to characterise the test item, to determine its stability, and to determine the homogeneity and concentration of its mixtures with any vehicle used in the application to the test system. Likewise, chemical procedures used to analyse specimens (e.g. clinical chemistry, urinalysis) have to be conducted under GLP.

The data to be generated under GLP have to have a connection to the assessment of safety for either of the two fields of human health or for the environment. In this regard there are apparent differences between these two fields which, however, are the logical extension of the differing safety aspects of certain data. The physical-chemical parameters of a pharmaceutical chemical substance might thus not be considered as related to an assessment of human health risks, and therefore the respective studies would not be mandatorily conducted under GLP. For a pesticide or another environmental substance, however, parameters like their vapour pressure, their acid-base equilibrium constant or their water/octanol partition coefficient, all of which will determine the nature and extent of environmental dissipation, compartmentalisation and accumulation, will certainly be safety-related data that have to be generated under GLP. In another context, the investigation of the pharmacodynamic properties of a chemical substance which may determine its efficacy in a therapeutic indication will not be conducted under GLP. On the other hand, the investigation of pharmacodynamic activities of a substance other than those related to its therapeutic efficacy, that is to say the investigation of the general or – *nomen est omen* – safety pharmacology, studies which are intended to generate data on possible adverse effects on human health, will have to be conducted under the requirements of GLP as far as possible.

Indeed, the second part of the sentence in the FDA regulations already mentioned above, that "...*(the term does not include) field trials in animals* ..." has given rise to some questions, since, in general, veterinary drugs have also to be safety-tested under GLP conditions. The term "field trials in animals" has, however, to be read in the context of "clinical trials" in animals, and it is certainly to be regarded as consequential that studies with the purpose of demonstrating efficacy have to be considered differently from those demonstrating safety. Therefore, clinical studies in humans as well as in animals do not have to be conducted under the stringent conditions of GLP.

With regard to the second requirement for the applicability of GLP, the submission of the data generated to a Regulatory Authority, an analogous dissection can be done from two angles: On the one hand according to the type of study, and on the other hand according to the nature and utilisation of the item to be studied.

In relation to this second point, the scope of the OECD GLP Principles mentions pharmaceutical products, pesticide products, cosmetic products, veterinary drugs, food and feed additives, and industrial chemicals as

1.4 Areas of Application

examples of items possibly subject to testing under GLP. Insofar as these products will have to be licensed, registered or approved for marketing by an appropriate Regulatory Authority, safety studies on these products and their ingredients have to be conducted under the strict rules of GLP. There may be differences in national policies and requirements, however: While, e.g., cosmetic products may need registration or licensing in some countries, they may be freely marketed without any need for registration (and thus without any need for the submission of safety data) in other countries. In the latter case, it will then not be necessary for manufacturers of cosmetic products to apply for a marketing permit; therefore, safety studies concerned with cosmetics would not need to be submitted to a regulatory agency, and there would thus be no legal requirement in these instances for conducting such studies (if any are indeed performed) under GLP. However, it has to be stressed that, notwithstanding the lack of a requirement for mandatory and documented adherence to GLP, the basic principles of GLP certainly represent a measure of good quality control, a goal that all testing facilities should strive to attain. Furthermore, in this world of ever increasing globalisation of industries and products, it would certainly be prudent to conduct any such safety studies under the conditions of the GLP Principles, since it can never be known with certainty whether any such study would, one day in the future or under different legal circumstances, have to be submitted to a Regulatory Authority.

On a second line, there are differences between the extent of mandatory GLP-compliant testing that may not be obvious at first sight. Two examples may illustrate this point. As mentioned in the scope of the GLP Principles, not only pharmaceutical products intended for treatment of human patients, but also the analogous products for animal use should be tested with regard to their safety under the conditions of GLP. In this latter case there are two different aspects to be considered: The first one is safety for the treated animal itself, but the second one is the human health aspect of possible consumption of products derived from treated animals. Thus, the GLP Principles should be applied to studies on animal health products dealing with the possible sequelae of overdosage in the target species, with the safety of the product in its intended, therapeutic application to the target species, as well as to tissue residue accumulation and depletion studies. On the other hand, so-called "field trials in animals" are considered to be similar to a human clinical trial with a pharmaceutical product, as they are conducted for the purpose of obtaining data on animal drug efficacy, and therefore these studies are excluded from coverage under the provisions of the GLP Principles.

In apparent deviation from this general principle, that efficacy studies are not required to be conducted under GLP conditions, the US EPA regulations state, that certain efficacy studies on pesticides have to be conducted under GLP, namely when they are considered as required studies under another heading of the US Federal Regulations (40 CFR 158.640). The EPA GLP regulations define a study as being *".. any experiment ... , in which a test substance is studied ... to determine or help predict its effects, metabolism,* **product performance** *... , environmental and chemical fate, persistence, or residue, or other characteristics ..."* GLP thus is mandated to apply to the conduct of all studies which support, or are intended to support, pesticide registrations, including studies on product performance, i.e. efficacy studies. The definition cited above does, however, exclude *"basic exploratory studies carried out to determine whether a test substance or a test method has any potential utility."* In a way this regulation may be seen in analogy to efficacy determinations of human drugs: While the basic, exploratory, non-clinical pharmacodynamic studies conducted to determine whether the substance might have any potential utility need not be performed under GLP, the ultimate efficacy studies, that is the clinical "product performance" studies on human subjects have to follow the rules of Good Clinical Practice. In the area of pesticide registration, the studies intended as final proof of the efficacy of the product would therefore also have to be of demonstrated adequacy, quality, integrity and validity. Similar to the human clinical efficacy trials, where under the rules of GCP the data should be traceable back to the single patients enrolled and investigated in these trials, study reconstruction must be possible also for such pesticide efficacy studies in order to allow the identification of any data of questionable integrity.

With this discussion the border between the viewpoint "nature of the test item" to the viewpoint "study type" has been crossed. The definition of study types that do, and those that don't fall under the provisions of GLP does again very much hinge on the aspects of safety and of regulatory submission. Looking at the second aspect of the applicability of GLP, namely the requirement that only such studies shall come under the rules of GLP which will be submitted to some Regulatory Authority, there are some interesting points to be made. First of all it will possibly not be definitely known at the time of study conduct whether the respective test item will really make it all the way through to a product submission. Thus, it will be uncertain whether the particular study in question will indeed be submitted to a Regulatory Authority. Therefore, it is really the possibility of submission which will determine whether a study has to be GLP compliant. It is therefore emphasised in the scope of the OECD Principles that GLP should be applied to such safety studies which are *"required by regulations for the purpose of registering or licensing ... "*; if the purpose of a study is to generate data that may be utilised in such a regulatory process, then GLP will apply. This point has also been emphasised in an EPA Advisory, where it is stated that *"Thus, at any time where it is known that study data are intended to be submitted to EPA under the scope and definition given in the regulation, that study must be performed according to GLP. However, we would advise that at any time that it is known that the data from a study may be submitted to EPA ..., that study should also be conducted according to GLP. ... The data submission may be rejected if the compliance statement indicates GLPs were not followed regardless of whether the data were intended for submission to EPA at the time that the study was performed."*

This interpretation has to be considered when judging the exemption from GLP formulated in the "study" definition of the EPA regulation. A fine but logical line separates exploratory studies which need not be conducted to the GLP standards from other such studies, where GLP would be regarded as mandatory. Let us assume as an example the various investigations on a chemical substance directed towards testing its properties as a prospective pesticide. Screening assays in the laboratory which lead to its characterisation as a potential fungicide would be considered as exploratory studies, and these would therefore be exempted from the strict application of GLP. The compound in question would, however, have to prove its fungicidal efficacy under actual conditions of use in the field, and a number of investigations with this product will be undertaken in the field. Since the purpose of these latter field trials is not confined to collecting data on the actual value of the product in the field, but may be extended to generating data for submission to the licensing authority, if the product would perform to expectations, these field trials would have to be conducted under GLP.

Yet another example may be described: In analytical chemistry, methods have to be developed for the determination of pesticides, their residues and metabolites. Again, studies performed entirely for internal use would not require compliance with GLP regulations, since such method development and validation studies could be regarded as exploratory. On the other hand, studies being performed because they form the required basis for the submission of a product to a Regulatory Authority should be regarded as being subject to the GLP rules. Here again, one may observe a difference between the requirements in a drug submission *versus* a pesticide submission. Pharmacokinetic studies, although being a mandatory part of the submission package for a (human or veterinary) drug, are exempted from the GLP requirement (the exception being the toxicokinetic investigations accompanying toxicology studies). As one part of these non-clinical pharmacokinetic investigations, metabolism and biotransformation studies of the test item are performed which are considered not to be safety related, since the safety of test item and its metabolites is being investigated together in the toxicity studies. Therefore, these metabolism and biotransformation studies do not need to conform to the GLP rules. In the pesticide field, however, it will be important, from a safety viewpoint, to gain knowledge about the biotransformation patterns of a compound in treated plants, in soil or in other environmental compartments, as the overall safety of a pesticide which is introduced into the environment will depend also on the activities, nature and fate of any metabolites formed.

These examples have been provided here to illustrate the point that the application of Good Laboratory Practice is not rigidly universal, in that one would be able to draw up a list of studies which are mandatorily subjected to the rules of GLP. Rather, the necessity to follow the strict regulations of GLP will be determined by the two basic principles of the scope of GLP – the safety-relatedness of the investigation and the foreseeable submission to an Regulatory Authority – the interpretation of which may, however, under different circumstances lead to different answers and conclusions.

5. The Pillars of Good Laboratory Practice

Any building needs a sound basis, on which to erect its visible structures. Thus, Good Laboratory Practice is based on four pillars which have to support the implementation and daily observance of its Principles:

- The Management;
- The Quality Assurance;
- The Study Director; and
- The National Compliance Monitoring Authority.

It is not by sheer coincidence that management would be mentioned here in the first place as one of the pillars of GLP. It is amply borne out by experience that GLP is only as well complied with as it is supported by test facility management's inner conviction. It is not sufficient to draft a nice declaration extolling the virtues of quality in general and of GLP in particular, when in everyday work the wrong cues are given to the test facility personnel with regard to the need for full adherence to the GLP Principles: When either the financial means for ensuring GLP conformity are severely curtailed, or when management is looking through the fingers or altogether the other way when a report of the Quality Assurance is asking for corrections that would necessitate some investments, then people will read between the lines of this statement and conclude that only appearances are important, but not the actual compliance, and they will behave accordingly.

On the other hand, a management which is convinced that GLP is a good thing in itself, and not just something that these silly bureaucrats in the government (who, in any case, do not have the slightest idea on how to run a business) are asking for and nagging about, but has merits of its own, that this system is really worth the efforts which have to be put into it, such a management will be rewarded with a smoothly running GLP system and with the delivery of real quality data and studies withstanding even the most detailed scrutiny by authorities. Therefore, even if the management of a test facility has nothing to do with the daily compliance with the GLP Principles, and has only to provide for the basic necessities to enable GLP to be implemented, its attitude towards this quality system, and its positive stance towards the efforts and expenses needed, will very much influence the way in which GLP will be observed within the test facility.

All this amounts to the requirement that it is the test facility management, who is ultimately responsible for ensuring full compliance with the GLP Principles throughout the facility as a whole. In order to deliver its responsibility, it will need some mechanism of continuous control. Therefore, an essential management responsibility is the appointment and effective **organisation of an adequate number of appropriately qualified and experienced staff throughout the facility, including those specifically required to perform QA functions.**

And this management responsibility brings us to the second pillar of Good Laboratory Practice.

The second pillar of this building, named GLP, is the Quality Assurance, an internal system for ensuring that the Principles of GLP are observed and that the studies which are conducted at the test facility are complying to the extent necessary with these Principles. The compliance with the GLP standards in the everyday work at a test facility can only be as good as the critical observational capability of the Quality Assurance inspector on the one hand, and also only as good as the ability of the Quality Assurance manager to succeed in carrying through any objections to the way GLP is handled by individual persons or by whole departments from receipt of the original message till their final resolution. For this end, the GLP Principles are regarding the independence of the Quality Assurance from the actual study conduct as a very important issue; from this requirement there can be not the slightest deviation. Any activities that are delegated to Quality Assurance must never compromise the independence of the Quality Assurance operation, and must not entail any involvement of Quality Assurance personnel in the conduct of the study other than in a monitoring role. On the other hand, it is also of utmost importance, that the person appointed to be responsible for Quality Assurance must have direct access to the different levels of management, particularly to top level management of the test facility. Quality Assurance has to be able to bring any deviations from the full observance of the GLP Principles detected in some part of the test facility to the immediate attention of test facility management, in order that corrective actions may be instituted at once (and before a Monitoring Authority inspects the facility and finds fault with the way GLP compliance is followed!). Quality Assurance may thus be regarded as the prolonged arm of management, which exercises its control over the GLP compliance within the test facility. However, it has also a bridging role between management and study personnel, in that failure to observe aspects of GLP may, e.g., be indicative of too great a workload in one particular part of the test facility, which could easily be remedied, if management just were made aware of it, and were, at the same time, willing to address these needs in an objective and adequate way.

The third pillar of the GLP system consists of one single person! The Study Director is the one single point of study control and the one single person on whom the whole study hinges from the beginning to the end. His prime responsibility is for the overall scientific conduct of the study and all duties and responsibilities as outlined in the GLP Principles stem from it. It is well known in all fields of human activities, e.g. in the military field, that a divided command will probably always lead to some smaller or greater disaster. This certainly holds true for the conduct of a study, where, based on this general knowledge, it is absolutely clear that there can be only one Study

Director at any given time. If this were not so, then personnel would be liable to receive conflicting instructions for the conduct of the study or for activities connected with it, which, ultimately, may lead to poor implementation of the study plan. In this regard, the Study Director serves to assure that the scientific, administrative and regulatory aspects of the study are fully controlled. This can only be accomplished by co-ordinating the inputs of management, scientific/technical staff and the Quality Assurance Programme. Certainly, some of the duties of the Study Director can be delegated to a Principal Investigator, as in the case of a subcontracted study, or to another "Responsible Scientist", as for the preparation and assessment of histopathological slides in the specialised laboratory within his test facility, but the ultimate responsibility of the Study Director as the single central point of control cannot be delegated. The Study Director has finally to acknowledge this by signing the GLP Statement in the final report of the study he has directed.

Another aspect comes to bear, too, in the person of the Study Director. The Study Director is usually the scientist responsible for study plan design and approval, as well as for overseeing data collection, analysis and reporting, and for drawing the final overall conclusions from the study. In this person, therefore, two worlds are meeting: On the one hand the issue of the formal study quality in terms of GLP and, on the other hand, the complex area of the scientific study quality in terms of design, data significance and assessment. In this regard the Study Director is an eminently important pillar in the whole structure of GLP.

What Quality Assurance is for the in-house control of adherence to the Principles of GLP, is the National Compliance Monitoring Authority for the international recognition and mutual acceptance of studies and test data. OECD has recognised the need for this further control instance and provided a framework for the institution of National Compliance Monitoring. The OECD Council Decision on Mutual Acceptance of Data (OECD, C(81)30(Final), 1981) therefore, logically, included an instruction for OECD to undertake activities *"to facilitate internationally-harmonised approaches to assuring compliance"* with the GLP Principles. Consequently, it recommended that member countries should institute such systems, and in order to promote the comparability in the different compliance monitoring procedures the Council further adopted the Recommendation concerning the Mutual Recognition of Compliance with GLP (OECD, C(83)95(Final), 1983). This Recommendation sets out the basic characteristics of the procedures for monitoring compliance with the GLP Principles, and following this Recommendation two guidance documents on "Compliance Monitoring Procedures for GLP" and on the "Conduct of Laboratory Inspections and Study Audits", both directed at the National Monitoring Authorities, were issued. These documents, which are also reproduced in this book (see Appendices IV.I and IV.II, pages 365 and 373, resp.), have been discussed and written mainly to develop common approaches to the technical and administrative problems related to GLP compliance and its monitoring. It was furthermore intended that by adherence to the procedures set out in these two documents, national approaches to GLP Compliance Monitoring should be harmonised with the final goal of arriving at a complete mutual recognition of the respective compliance monitoring procedures. It stands to reason that only in such a way the most important goal of these OECD Council Decisions, namely the mutual acceptance of safety test data among the OECD member countries, could be reached.

It had, however, also to be recognised that there would be a number of problems and difficulties on the way to attaining this goal. It was recognised that the OECD member countries would adopt the GLP Principles and establish compliance monitoring procedures, but that they would do so according to national legal and administrative practices, and according to priorities they would give to, e.g., the scope of initial and subsequent coverage concerning categories of chemicals and types of testing. Furthermore, and according to the legal framework for chemicals control in the individual countries, more than one GLP Monitoring Authority, and thus more than one GLP Compliance Programme could be established. One of the best known examples in this respect are the USA, where there is not only the division between the Food and Drug Administration (FDA) on the one hand, and the Environmental Protection Agency (EPA) on the other hand. There are even within the EPA two different sets of GLP regulations, which are based on two different laws: The Federal Insecticide, Fungicide and Rodenticide Act is the basis for the GLP standards published in the Code of Federal Regulations as 40 CFR 160, while the ones based on the Toxic Substances Control Act are published in 40 CFR 792; these should become replaced by the single reguletion under 40 CFR 806. The FDA GLP standards, on the other hand, are based on the Federal Food, Drug and Cosmetic Act as well as on the Public Health Service Act and are published under 21 CFR 58.

While in other countries all aspects of GLP monitoring can be assembled under the roof of one single GLP Monitoring Authority, there are still others who make use of the possibility of combining GLP Compliance Monitoring further with the monitoring activities in the area of other quality systems, like accreditation or ISO.

Whatever the structure and the function of such a Monitoring Authority, the most important aspect, from an international viewpoint, of this fourth pillar of GLP is the comparability of the monitoring procedures, and of the compliance assessments resulting from them, amongst the various countries and Authorities, since only then, mutual trust is achieved and the mutual acceptance of safety test data will be possible. How this comparability and equal functioning of Monitoring Authorities is assessed will be described in Section IV of this book.

> In summary, four pillars support the structure of Good Laboratory Practice. All of them serve important functions in the context of performing and monitoring safety studies, and all of them need to be based on the strong conviction that GLP is the one mean to achieve quality data. Certainly, there are other aspects and issues in GLP that may be seen as nearly equally important, and they will be dealt with extensively further on, but Test Facility Management, Quality Assurance, Study Director, and National Compliance Monitoring Authorities are the key positions where real adherence to the Principles of GLP, not only by the letter but by the spirit of them, is determined in the end.

6. Where Can GLP be Profitably Applied ?

As has been described above, GLP is a quality system which has found mandatory application in the safety testing of any items where the results of such testing will be assessed by some national Regulatory Authority for the purpose of registering or licensing this item. In these cases it is to be applied and followed to the full extent of its "letter and spirit". This does not mean that it is to be used exclusively within these defined and restricted boundaries. It is, furthermore, or luckily, not a "trade mark protected" term; any laboratory working according to these principles may claim adherence to them. It has to be emphasised, however, that only those test facilities which on the one hand are working in full compliance with the GLP Principles, and which on the other hand are included in a national monitoring system or are controlled by some national authority may claim official recognition of their GLP compliant status.

Nevertheless, there are a number of instances where test facilities could, or should, adhere to the principles of GLP. The most obvious case is the not so rare one of the test facility, where only very few studies are conducted according to GLP to the fullest extent; most of the studies performed there would either not qualify as safety studies, or they would not be conducted to GLP because the sponsor did not ask for a GLP compliant study. Where there is no obvious need for the application of the full requirements of GLP, e.g. no need for a formal Quality Assurance audit of the final report, it would seem to make no sense to apply these special regulations and thus to add to the administrative burden with no reward whatsoever. However, in such a test facility, it would be of the utmost importance that in all other respects studies would be performed as if they were conducted under GLP: Apparatus should be maintained and calibrated according to the respective SOPs, test systems should be properly located and identified, test items should be characterised and labelled, SOPs should be available for all activities performed at this test facility, the studies should be conducted to the applicable SOPs, the respective raw data should be treated in a manner analogous to those in a GLP study, and all these activities should be properly documented and recorded. Only if the personnel of this test facility were not allowed to apply two different standards in doing their work will it be ensured that a "real" GLP study, if one is to be performed, will truly be in compliance with the GLP Principles.

It has been stated at the beginning of this part that there are some misconceptions about the meaning of GLP. As one of these, it has been mentioned, that there is the wrong opinion that a laboratory, which views itself as working according to a good scientific or precision standard, should be able to apply for recognition as a test facility in compliance with GLP. Since the term "Good Laboratory Practice" is restricted to apply to such test facilities only, which are performing "human health and environmental safety studies", studies which furthermore have to be submitted to a Regulatory Agency for assessment in a registration or marketing permit procedure, it is evident that not every laboratory would fulfil these conditions. Therefore, formal recognition by a Compliance Monitoring Authority cannot be given to each and every laboratory that claims to work according to these GLP Principles.

However, this misconception may be, after all, not so wrong in its intentions. Other quality systems, such as those of the ISO series or the Accreditation schemes, which may be better suited for, and thus applicable in a more relevant way to the majority of, these cases, are also relying on similar measures as GLP for assuring the quality of the work performed. It goes without saying that also for a laboratory doing "only" routine tests of any kind, be it clinical chemistry determinations in blood samples from patients, determinations of microbial counts in food, surveillance of drinking water quality, measurements of environmental contamination, or analysis of chemical preparations for purity and content, it should be necessary to have an efficient Quality Assurance system in place. If such a laboratory should need an official document attesting to the quality of its work, then it should certainly apply for recognition according to the one standard that is best suited to its need and working area. Outside the field of "non-clinical human health and environmental safety" testing, where GLP has to apply, official recognition should be based on quality standards other than GLP, since GLP is absolutely confined to the said area.

However, if there is no need for submitting to a monitoring programme, e.g. when there is no pressure from sponsors to produce a document of official recognition, when the laboratory in question is one of research, or is based in a university, a hospital, or some branch of government, but the facility nevertheless wishes to institute some standard of quality, then the principles of GLP may provide good guidance to truly high quality in the standard of working. The necessity for the existence of a "good practice" system with an at least somewhat more than rudimentary kind of quality control is amply borne out by human nature itself and especially by the rise in importance of problems with scientific misconduct.

In recent years there have been a number of cases uncovered where not only scientific misconduct, but outright fraud in the context of clinical, university and industrial research has been suspected, suggested or proven (e.g., Humphrey, 1994; Law, 1999; Weiss et al., 2000). The Federal Register of the United States of 1999 contains 13 notices of cases, in which the Office of Research Integrity found evidence for scientific misconduct, i.e. studies where data were falsified or fabricated:

- The claim, that expression of wild type and mutant fibrinogen had been obtained in yeast cells, was falsified by "spiking" samples with mammalian fibrinogen before sending them to another laboratory for analysis;

- Monthly logs with patient data were falsified by re-dating and multiple submission;
- Instead of conducting the required interviews, patient signatures and responses to questionnaires were falsified;
- Data for entry criteria in clinical studies were falsified, reports on treatments and follow-up examinations were fabricated, and signatures on "informed consent" documents were forged;
- In a study which called for sending interview notes and the preliminary scores derived therefrom to an independent assessor, who should reassess the scores and send the results back by e-mail, the investigator failed to send the material and instead falsified electronic mail responses to indicate that the evaluation had been conducted;
- Autoradiographs of Northern blots from unrelated experiments (showing the effects of phorbol ester treatment on the expression of the myogenin gene) were re-labelled to make them appear to have come from different experiments; one of them was used in a publication purporting to demonstrate the effect of electrical activity on the expression of genes for subunits of the acetyl choline receptor;
- Another researcher fabricated data by cutting a scintillation counter printout from a former co-worker's notebook, pasting it into his own notebook, and representing it as his own results from a different experiment on the binding of angiotensin to transfected cells.

The reason for this apparent increase in scientific misconduct may have two roots: On the one hand the pressures to "publish or perish" certainly haven't lessened in basic research with increasing competition for (decreasing) grants and other resources as well as in its applications, especially in the biotechnology field, where large economic interests may be at stake. On the other hand, the competition may also encourage or drive the "whistle-blowers" to more readily come forward with their suspicions than in the past – though still with reservations (Rossiter, 1992). As in all instances, there is lesser or greater economic pressure behind such "scientific misbehaviour", which has also been at the bottom of a recent case of a patent dispute: There, the original patent had apparently been obtained at least in part on the basis of a key experiment which, however, had never been performed (Dalton, 1999). The same can be said of clinical studies with new medicaments, where cases have been uncovered in which data have been misrepresented, fraudulently altered or left unrecorded, or in which patients have even been invented – the catalogue could be long and would be reminiscent of the early days of preclinical studies before the advent of GLP.

Let us consider just one small point in the whole area of study conduct, and look at the example of data recording: The GLP Principles require that all original observations are immediately, clearly and legibly recorded. If there are observations that do not fall into the normal pattern, they nevertheless have to be recorded immediately. The Study Director may then certainly be asked about the significance of these observations, and if the Study Director decides, out of his scientific knowledge, that the observation could be just a spurious incident, he may declare it so by his dated signature under the reasoned explanation on why this fact should not be considered in the final assessment. But the fact that the respective observation has been made as originally recorded has to remain in the raw data in a clearly legible form. Even seemingly simple errors, e.g. in recording the date (who has never, in the first weeks of the year, written the figure for the "old" year instead of the "new" one, out of sheer habit?), spelling errors, or other "slips of the pen" are to be corrected in a way which does not obscure the original entry, and the corrections have to be dated, substantiated and initialled. However, if – as has happened in a clinical study – the nurse had instructions to notify the Study Physician, when it seemed to her that the blood pressure of a patient looked too high, before taking down the value measured, then one of the important principles of Good Practice (of whichever kind), namely truthfulness in recording the observations as they are made, is violated. It does not matter in this context whether the Study Physician would then try to cheat and instruct the nurse to write down a "normal" value, or whether he would tell her to record the one she measured: The principle of immediate recording of all observations is violated.

This principle is all the more important, as it sets a very stringent limit on the time within which original data have to be recorded. If any data are not immediately recorded, where would be the allowable time limit for this? Could the technician, the researcher, the nurse write it down after the coffee break? or at the end of the day? or even on the following morning, when it is still quiet in the office and nobody is disturbing this activity? It stands to reason that, the later such data or observations are recorded, the higher the chance that they will no longer reflect the actual observations: Did it already start to rain, when the field was still being treated, or was it only when the equipment was already safely stowed away and the technician started the car to drive back to the test facility? Had dog number 15 vomited after dosing, or was it dog number 16? Had the ELISA test been performed with the prescribed volumes pipetted in, or was it yesterday, or the day before yesterday that the Study Director advised to increase the volume of the test sera by 50%? Immediate recording would avoid the occurrence of such insecurities, and these questions would never have to be asked. In every study, be it in research or in development and safety testing, immediate recording of parameters, observations and events is tantamount to good quality of the single data and the whole study. Thus, this GLP requirement of immediate recording of events and observations can be regarded as a very simple example of where the Principles of GLP could be fruitfully applied and could be of value to every single person in a laboratory or test facility of any kind.

The importance of immediate and precise recording of data can best be illustrated with two details from one of the most illustrious cases of alleged fraud having given rise to a good many headline stories in scientific journals: The case of Tereza Imanishi-Kari the allegations in which, after ten years of investigations, were finally judged to be unfounded (Goodman, 1996). This case can be considered very illustrative as it can be demonstrated how the application of some simple rules of GLP could have obviated the need for or at least speeded up the respective investigations. Out of the nineteen counts of scientific misconduct with which the Office of Research Integrity (ORI) charged her, two may be taken as especially illustrative of the value of Good Laboratory Practice principles.

In the first instance, the investigators charged her with having fabricated the background radiation counts in her notebooks, as the hand-written figures deviated from the randomness to be expected with actual counts from natural background radiation. The explanation given by her for this deviation from randomness was that the actual figures had been rounded before transcription to the notebook; they did thus not constitute the original observations as recorded on the radiation counter tapes. Under GLP, such records in the laboratory notebook would not constitute original raw data. These should have been either filed as such, i.e. as the original counter tapes, or as "verified copies thereof". Even if rounded values would have been used for the calculations, these "secondary" data should have been retained only along with the original raw data. Had this been the case with the data in Imanishi-Kari's notebook, there would have been no question about the integrity of these background counts, since the reconstruction on how the rounded values had been obtained from the original raw data would have been clearly possible.

The second point to be made is also connected with the radiation counter records. Some of the tapes that had been attached to the notebook pages, which were claimed to represent data obtained at some crucial time point and which supported important aspects in the published paper, were investigated with very great efforts by the US FBI. The analyses of these paper data tapes compared colour of the tapes, ink and type fonts of the records with the respective parameters of tapes in the notebooks of other researchers and from other times than the claimed period in an attempt to establish whether these tapes could have been made at the time claimed. These analyses could find no matches for a series of tapes claimed to have been made in June 1985 and instead dated these tapes from before 1984 and as early as 1981, suggesting that data from old, unrelated experiments were used to fabricate the supportive results. Obviously, the counter did not print any date on the tapes, and the time of their production was suggested only by their place in the notebooks. In a laboratory working under GLP all these tapes would have been regarded as raw data and therefore, if no automatic dating system would have been present, they would have been dated and initialled by the responsible individual immediately at the time they were generated. They would furthermore have belonged to an earlier study and therefore they would have been archived at the conclusion of this study. Thus, no "old" tapes from unconnected experiments could have been used for the support of "new" data, or at least the "proof of innocence" would have been much easier, if the whole laboratory had been run under the rules of GLP.

Good Laboratory Practice can thus work both ways: It might be used to detect fraud, but it could also serve to protect the researcher from unfounded allegations. In this sense, the implementation of the basic rules of GLP could be of benefit even (or especially) to a research institution or laboratory.

Another example for the possibilities offered through the application of GLP in the research environment might be a somewhat more complex one, namely the Study Plan. It is certainly true that in academic research the scientific curiosity of the researcher should not be impeded by too strict directions, since the dictum is that a true research-minded academician should let him- or herself be led where the facts are pointing. Thus, for initial experimentation in research there may be really no place for any pre-defined study plan (although an experiment without any kind of planning would probably lead to nothing anyway). However, with every research there comes the stage, where the initial results will have to be reproduced, refined and extended. It is at this stage, where the experiments have to become a prospectively planned and well conducted affair. If there were no real planning, no adherence to preconceived procedures, and no records for the amendments and changes made to the experimental set-up and conduct, these experiments, thought to become the touchstone of the cherished hypothesis, would instead become meaningless waste and would certainly lose their argumentative power.

Having established the principle that also for research investigations some planning would be advisable, and that a formulated study plan would certainly be an asset, consideration can be given now to the question in which way the description of the study plan in the GLP rules could be applied. While the GLP Principles are enumerating the elements which should at the minimum be contained in a study plan in order to provide for full GLP compliance, not all of these elements might be needed in a study plan intended for the description of the activities in a research programme. It might for instance be considered unnecessary to include all of the more administrative information on the test facility and on the sponsor; it could also be considered that information on proposed timing of the experiments would be of minor importance, while the "approval" by dated signature of the Study Director would in any case be important information. On the other hand, a descriptive title, an exact description and characterisation of the test item, as well as detailed information on the experimental design (including a description of methods, materials and test conditions to be used, and type and frequency of measurements, observations and examinations to be performed) could certainly prove to be of benefit for the planning of such research activities. Also a justification for selection of the test system together with its characterisation, either by parameters like species, strain, source of supply, number, body weight range, sex, age and other pertinent information in the case of biological test systems, or by parameters such as type, modular composition, manufacturer and other pertinent instrument characteristics in the case of physical-chemical test systems would be of value in the possibility of judging – at a later time – the adequacy of the whole study and the quality of its results. The same holds for the description of any amendments to, and deviations from, the study plan. Presented in an orderly fashion with dated signature, this would constitute documentary evidence for the Study Director's train of thought in the development of the study, and for the intended or unintended changes in the course of the study which may or may not have influenced the ultimately obtained results.

The writing of some such kind of a study plan before initiating the respective investigations would not only improve the reconstructability of the studies conducted. Already the process itself of writing-up the various aspects of the premeditated study would serve the Study Director to clarify in his or her mind the relative value of the different ways to achieve the purpose of the research project and thus to set out in a logical manner all the possibilities for proceeding to this end.

When it comes to laboratory work of a routine nature, there are, of course, other considerations to be applied. When no distinct entities, which could be considered as being equivalent to studies in a GLP sense, are involved, then it would certainly become administrative nonsense to recommend or even require study plans to be formulated. In such instances, however, other requirements of GLP could very profitably be applied. The existence of Standard Operating Procedures for all routine activities on the one hand, and of maintenance and calibration documentation for instruments and apparatus on the other should be useful in more than one respect.

A further aspect of GLP may be considered as complementing the points that have been made in the foregoing with regard to the issue of adequate documentation of all steps and activities leading to a complete picture of the whole course of study conduct: In a number of research as well as routine laboratory activities other than those connected with safety studies *per se* the whole area of data recording, raw data and archiving would deserve as much attention as it gets in GLP. It would not call for major efforts nor for any great investments in time and money to exact a few principles for data recording and treatment of raw data; it would only be necessary to educate the individuals performing such studies, and possibly to monitor their adherence to these principles in one or another way. The important principles are not new, nor are they unknown; they can be summarised as follows:

- There should be a unique identification for the study and all of its parts.
- All original observations in a study should be immediately, clearly and legibly recorded.
- The recording should be indelible and corrections should be made so as not to obscure the original entry; for all corrections the respective reasons have to be provided.
- All records should be in the form of bound notebooks or on continuously numbered sheets.
- All entries and corrections to them should be dated and initialled.
- Records related to the test system itself (for biological test systems: acquisition, condition, suitability testing, etc.; for physical-chemical

test systems: specifications, manufacturer, model, etc.) should be gathered and retained.
- Specimens should be clearly identified so as to allow full traceability.
- At the end of a study, all raw data should be assembled, catalogued and archived.
- Archiving should provide for secure storage of all raw data, samples and specimens, together with any other documents such as study plan and study report (if any has been written).

The problems of scientific misconduct have certainly been noted and acknowledged by the scientific community as well as by the respective organisations and institutions, and a number of efforts have been made to address the problem through the formulation of guidelines for the ethical conduct of scientific investigations. Insofar as such guidelines are available, the issue of quality and integrity of research data might be considered as resolved. However, closer scrutiny of the available documents intended for the guidance of the researcher (e.g. DHHS, 1992; Clausen and Riis, 1997) reveals that the respective parts concerned with the possible means of establishing better data quality and integrity are rather vaguely worded. In contrast to the GLP Principles, they give only cursory guidance in the sense that they remain general and do not spell out precise requirements.

One might, for instance, compare the "Data Management" requirements of the "Guidelines for the Conduct of Research Within the Public Health Service" of the US Department of Health and Human Services (DHHS) with the analogous parts of the GLP Principles. The DHHS Guideline states the following:

"It is expected that the results of research will be carefully recorded in a form that will allow continuous and future access for analysis and review. Attention should be given to annotating and indexing notebooks and documenting computerised information to facilitate detailed review of data. All data, even from observations and experiments not directly leading to publication, should be annotated, indexed, and documented."

The GLP Principles are much more demanding in this respect, and in the section on the "Conduct of the Study", the following instructions are given for the recording of data:

1.6 Where can GLP be Profitably Applied?

"3. All data generated during the conduct of the study should be recorded directly, promptly, accurately, and legibly by the individual entering the data. These entries should be signed or initialled and dated."

"4. Any change in the raw data should be made so as not to obscure the previous entry, should indicate the reason for change and should be dated and signed or initialled by the individual making the change."

"5. Data generated as a direct computer input should be identified at the time of data input by the individual(s) responsible for direct data entries. Computerised system design should always provide for the retention of full audit trails to show all changes to the data without obscuring the original data. It should be possible to associate all changes to data with the persons having made those changes, for example, by use of timed and dated (electronic) signatures. Reason for changes should be given."

This comparison or confrontation of the two guidelines demonstrates clearly that they are neither exclusive nor redundant but that they should be considered as being rather complementary. In this way the GLP Principles could be used as the "executive arm" of the various existing research guidelines, providing the necessary detailed guidance on how to achieve the "*careful recording*" called for in the DHHS guideline.

An analogous situation can be seen in the requirements for data retention and storage, where the DHHS Guideline stipulates "*Similarly, research data, including the primary experimental results, should be retained for a sufficient period of time to allow analysis and repetition by others of published findings ... Retention time may vary under different circumstances. In some fields, five or seven years are specified as the minimum period of retention. A minimum of five years is required.*" Again, the GLP Principles are more precise in their requirements than just to mention "*data, including the primary experimental results*". They firmly require that not only all documentation directly related to the study should be archived ("*The study plan, raw data, samples of test and reference items, specimens, and the final report of each study*"), but that retention and storage should encompass a much wider range of documentation than just the one covering the experimental data. Again, the situation may be regarded as complementary, and the GLP Principles could be used to more precisely guide in the question of what and how to retain and store.

Finally, the custom to choose for publication only the most suitable results obtained in a series of investigations, to report only those values which are giving the best fit to the hypothesis, e.g. to "clean graphs" by omitting certain data points in order to make them look more convincing, may on the one hand represent the scientific judgement of the author and as such serve to clarify the situation, but the plain suppression of data might on the other hand be considered already to border on misconduct through intentional misleading. In the reporting of a GLP study, similar situations could arise, where the Study Director would have to exclude some data from the analysis and interpretation; in the case of a GLP study, this cannot, however, be done by simply suppressing these data, but the Study Director has to provide the scientific reasons for doing so. It could be argued that, in analogy, the authors of a paper submitted for publication should either present all the data in the manuscript, or that at least they should be submitted to the journal for peer review, together with the author's reasons for not including them into the final analysis.

There is one crucial question which will at this point be asked by anybody who does research or other work not mandatorily subjected to the requirements of the GLP Principles: "How would I, or my laboratory, profit from the implementation of, and adherence to, such a strict regimen of measures that are more administrative in nature than scientific?" The answer lies again in the possibility of a complete, one might say seamless, reconstruction of each and every detail in the various activities and study parts or whole studies which have been leading to the results and data eventually ending up in publication. It will add transparency to the science and will thus serve to increase the trustworthiness of the results. Any question about data integrity could be answered with more confidence, and any incrimination of scientific misconduct, if unfounded, could be rebutted much more easily. It would certainly not abolish totally the possibility of scientific misconduct, since, if anybody wants to forge data or invent results, this can also be done under GLP conditions, but it would be much more difficult to do so in a coherent way. Also, the obviously common practice of cutting out some data from an old notebook of a co-worker would be rendered more difficult by strict adherence to dating/initialling and archiving requirements, as well as to the application of strict rules for the control of archived material utilisation.

Even apart from questions about scientific misconduct, its proof or rebuttal made possible through adherence to the principles of GLP, there can be benefits for the research work itself, in that it can help to resolve issues and problems which might arise from seemingly contradictory or unexpected findings. To illustrate this point, an example can be cited, where GLP was instrumental in the explanation of a spurious result.

1.6 Where can GLP be Profitably Applied?

Within a safety testing programme of a new chemical substance, histopathological examination of brain sections was routinely performed. In one particular study, examination of such sections from treated groups, but not from control animals, showed a possibly dose-dependent increase in the occurrence and severity of neuronal vacuolation. This kind of lesion had not been observed in any other study, but with regard to the seriousness of the finding and the apparent dose-dependency of the lesion, the company scientists were hard put to find an explanation for this result. A peer review of the slides by another histopathologist could not resolve the issue, the study personnel and the Study Director had no obvious explanation to offer, and thus the possibility of a toxic influence on the brain had to be earnestly considered, even though there were no similar observations in other studies, as already mentioned. The fate of the compound hung in suspense.

It was the Quality Assurance who finally found the answer and saved the compound!

It was an artefact. The proof for this interpretation of the results came from the neat reconstruction of the study procedures through the use of the correctly recorded GLP documentation. In the first instance, study personnel, when asked whether there had been any unusual occurrences during the study, the necropsy and the following sectioning and slide preparation steps, could not think of anything out of the normal. They maintained that everything had been done as usual, nothing special had happened, and all procedures had been performed in the standard way. Scrutiny of all the raw data of this experiment, and especially of the protocol sheets of the slide preparation, by the Quality Assurance personnel revealed, however, an interesting fact, which provided the sought for explanation: Tissue sections had been cut on a Friday. Due to a minor delay in the further processing, only the slides from the control animal tissues could be stained on this day, and the other sections had to be stored unstained over the weekend. This storage was done in 70% ethanol, and on Monday, the staining of the remaining sections was resumed in a batch-wise manner, with the ones from the low dose group being stained first, followed by the ones of the mid- and high-dose groups. This storage in 70% ethanol, however, had effectively dissolved and washed out of some of the brain cells' lipids, producing an apparent vacuolation. Since the slides were processed in a "dose-dependent" manner, this wash-out process had been able to proceed for a longer time in the higher-dosed slides, thus giving the appearance of a dose-dependently higher incidence and severity of these "lesions", while the control slides, having been processed without this interim storage, could not, and indeed did not, show this effect at all.

This example provides a lesson for two different aspects of experimental work and its documentation. It is certainly important that in the conduct of a study the results obtained, and the observations made, are recorded completely, faithfully, with sufficient detail and in a clear-cut manner so as to allow an evaluation also at a later stage or time. It shows, however, that it is not only the results themselves, which can be of importance, but even the recording of such "unimportant" experimental details, such as the exact time and date of the performance of activities which, in research circles, is generally regarded as the "tedious bureaucracy of GLP", may in certain instances help in the interpretation of data and the resolution of apparent discrepancies. In the example presented above, it could not have been possible to resolve the question of a possible toxicity of the test substance to the central nervous system without performing a second similar study, i.e. without repeating the experiment. And even when the repeat experiment would have shown no such effects, a certain suspicion would have lingered on, while no obvious or indeed proven explanation could have been given for these results.

In this sense, it might be profitable for all scientists, wherever they are employed, whether in academia, in government or in industry, and whatever kind of work they perform, whether routine or research, to spend a thought or two on the Principles of Good Laboratory Practice: Firstly to look hard at the underlying ideas of ensuring data integrity and validity through adequate documentation allowing for the complete reconstructability of activities and processes, and secondly to determine the actual extent to which such principles could be implemented in their individual situation or workplace. In situations other than those concerned with "human health and environmental safety studies" the GLP Principles need not be implemented as the whole set of rules, but rather they could provide the starting point for improvements in the ways of planning, conducting, and documenting the activities connected with the work performed, and of storing and retrieving the respective records. While this would not guarantee an immediate improvement in the scientific quality of the work, it would certainly provide for enhanced transparency and data integrity, but it could also lead to improvements on the scientific side. The possibilities inherent in the applicable control mechanisms might subsequently provide the means for judging the shortcomings of the present activities and for improving the scientific quality of the future work.

7. GLP and Other Laboratory Quality Systems

One of the misconceptions about GLP which has already been mentioned earlier is simply connected with the terminology: The name, the "Good Laboratory Practice", implies that any laboratory capable of faultless operation and quality work must conduct its activities under the auspices of Good Laboratory Practice. The name seems thus to have completely usurped the quality field with respect to work conducted in laboratories. In reality GLP has a strictly defined area of application, which includes only some types of laboratories, and some types of studies, as has been described earlier (see section 4, page 19). Other quality systems do exist, however, which may apply to those laboratories and the work conducted therein, which are falling outside the area of GLP. They will be better tailored not only to the needs of sponsors and laboratories alike, but indeed to the type and nature of the work conducted in those types of laboratories.

Laboratory work may be of two different types.

- Either the result of the study will be an exact figure, and the sponsor expects this figure to be "true", and he will expect that the same figure would have been obtained in another laboratory. Thus the sponsor expects precision and reproducibility, while it would not matter too much, whether the exact proceedings of the study might be reconstructable later on.

- Or the result of the study will be information in a more general sense which will have to be interpreted, which will not be reproducible in the strict sense neither in this nor in another laboratory, and the sponsor will expect a scientifically sound result. The assessing authority will expect, however, that the study activities could be scrutinised and reconstructed, so that the authority could gain confidence in the way the study results had been obtained.

To illustrate these two points let us look at two situations:

When a physician, because of concerns about some risk factors for heart troubles, wants to have the blood sample of a patient analysed for the content of cholesterol, he or she will be interested in the precision of the result, because the decision to prescribe a lipid-lowering drug may critically depend on this information. Therefore the laboratory has to convince the physician of its technical expertise and of the precision and reproducibility with which it is able to determine this parameter. It will not be necessary to demonstrate to the physician, how the laboratory had organised the testing of the sample, who did the actual determination, and whether the procedure followed a pre-approved study plan. The importance of the precision of this determination lies in the fact that the physician will determine, guided by the generally accepted "cut-off" point, or range, of the cholesterol level in blood whether to treat the patient or not.

In a toxicology study, a similar clinical-chemical laboratory will determine the cholesterol level in blood samples of treated rats, possibly by identical instrumentation and methods. Contrary to the example above, precision of the measurement has not this very critical importance in this situation, as it has for the single patient. The purpose of the toxicology study is to arrive at an estimate of a dose level of the tested substance which may be interpreted as harmless for the animals and, by extrapolation, for man. This estimate can only be a crude one, since the spacing of the generally three dose levels to which the rats have been exposed will be wide enough to span the range of the completely innocuous to the distinctly toxic ones. Furthermore, for the assessment it may not matter, whether an individual rat exceeds to a slight degree the normal value, since it is the mean with the standard deviation which will determine the final judgement of the treatment effect on the cholesterol level.

In the first example, quality is determined in terms of precision and reproducibility of the result obtained, while in the second case, it is the reliability of the study that counts, because the results may not be challenged through a repetition of the study. In the first case, if the physician would have some reservations with regard of the precision of the reported values, he might send a second sample to another laboratory and compare the two sets of results. Either, they would correspond, in which case there would be no reason to mistrust the first laboratory, or they would not, in which case a third opinion might be sought. In the second example, as experience has shown, a repetition of the toxicology study will, in all probability, yield results which will quantitatively not be comparable to the results of the first one. Quite apart from considerations of animal protection which would anyway prohibit the repetition of studies just for the sake of corroboration, the repetition of such a study with the purpose of verification would therefore be scientifically objectionable.

It has been stated already that the primary purpose of GLP is not to guarantee primarily the scientific or technical quality of studies, but to provide transparency in enabling third parties to follow in retrospect the whole course of a study: to trace back all activities to procedural standards, to relate activities to the personnel that had performed them and decisions to the authorised individuals, in fact to reconstruct the whole study. While such a purpose of the quality system is valuable in situations where the outcome of a study may not be readily reproducible, and where repeating studies may be out of the question, other situations may require different approaches.

Thus, if "quality" is established in terms of precision and reproducibility of the results obtained in the "studies" (i.e. in the respective sets of measurements or experiments), the need to provide for each of the "studies" a study plan, approved by the head of the laboratory before the experiments or measurements can be started, will not be an important consideration. Certainly, Standard Operating Procedures will have to be observed, and the acknowledged methods will have to be followed, with any deviations to be described and justified. Since it is the quality of the result which counts for the determination of the test facility's "quality", and not the way on which it has been obtained, there is no need for a single point of study control in the person of the Study Director. Certainly, a laboratory head will have to be appointed, who has to ensure that the "quality" of the data obtained in the laboratory remains high, and who has to provide the necessary education and training for the technical personnel in order to enhance and update their technical expertise. If precision and reproducibility are the primary purpose of the test facility's quality concerns, then apparatus, instruments, equipment and computerised systems have to comply to the highest technical standards in terms of validation, maintenance and calibration.

It can be easily seen from this incomplete listing of rules and principles that are of importance in one or the other situation, that there are certain similarities between different quality systems, but that grave differences do also exist. There are still other differences which have to be considered when trying to determine the connections and similarities or dissimilarities between various quality systems. Where the emphasis lies on precision and reproducibility, with consideration of reproducibility not only within one laboratory but between several laboratories, then the testing of methods in ring tests, with regard to their repeatability and robustness, will gain in importance. This is certainly possible for tests where the environmental conditions can be accurately controlled, like in an analysis by any current method. The fate of a pesticide in the environment will depend, however, on a multitude of factors which cannot be influenced, let alone controlled, by the study personnel. Therefore, since such studies cannot be reproduced, it is of absolute and utmost importance that the conduct of the study can be investigated and can be demonstrated to have been beyond reproof.

In relation to this latter point, another difference between the two systems lies in their internal controls. While under the requirements of GLP every study has to be inspected by a study-independent Quality Assurance during its conduct, and every final report has to be audited to confirm its compliance and its reflecting the raw data, ISO and laboratory accreditation do not necessarily include such data inspections for ensuring the quality and reliability of the data, and their quality control units and quality control managers need not necessarily be independent from the studies to be audited (inspected). This fact is furthermore reflected in the recognition by official bodies of a laboratory's status with regard to its compliance with the respective standards and regulations. A laboratory may be accredited for the totality of its activities, but it may also be accredited just for one single test or assay method. To give an extreme example: A laboratory may be accredited for performing melting point, but not boiling point determinations, since it is the purely technical competence which counts in these systems. On the other hand, GLP will decide only, whether a test facility is able to perform studies under the rules of GLP. A laboratory, performing microbial mutagenicity tests under GLP, will at the same time and without question also be able to perform chromosomal aberration tests in mammalian cells and *in vivo* micronucleus assays in mice.

The main similarities of GLP and other laboratory quality systems may be seen in their focus on apparatus and instrument suitability, maintenance and calibration, where the requirements of accreditation systems go beyond what GLP is regulating, since these issues are of the utmost importance for generating accurate, precise and reproducible results. Thus, it has to be possible in every case to trace back the calibrations to the respective national standards of measurement, and the quality control of the measurements has to ensure that trends to deviations form the precision required are detected already early on.

While it may thus be possible that in certain areas, the different quality systems may be similar to one another, it has nevertheless to be recognised that neither the adherence to an ISO or accreditation standard may replace GLP compliance, nor can a GLP compliant test facility claim the same technical competence as a laboratory operating under an accreditation scheme. However, the existing "redundancies" in the different sets of rules can make it possible to implement two such quality systems in one laboratory utilising the common points of the two systems to facilitate the tasks of personnel and quality management. The same can be true for the official compliance monitoring inspections and audits, where audit or inspection results of aspects that are fully covered by one system may be accepted by the other without further investigation.

One aspect that may have bocome rather obvious from the above paragraphs is the fact that the same words may not mean the same things in different quality systems. What "quality" means for GLP in contrast to what it signifies to accreditation has been described above. What an "audit" is for accreditation is an "inspection" for GLP, and the GLP "audit" would be termed a "review" by the accreditation expert. Therefore a dictionary might be needed to bring the various quality systems to common terms. This problem has been already the subject of a number of papers (e.g.: Dybkaer, 1994; Plettenberg, 1994) and discussions, and we need not develop it further here.

In summary it can be stated that, although GLP differs from other quality systems in aspects that are important not only for the traceability of data but especially for the full reconstructability of the study, there are certain overlaps between GLP and other quality systems like accreditation schemes which may allow some "joining of forces", at least to a certain extent.

II. How is Good Laboratory Practice Regulated ?

1. Introduction

Although there are many national guidelines regulating Good Laboratory Practice, which may differ in certain details from each other, the one guideline that is most universally accepted, and in general adopted completely – or at least to a large extent – by the various national guidelines, is the regulation of GLP through the Principles of Good Laboratory Practice of the Organisation of Economic Cooperation and Development (OECD), since these have been discussed by an international panel of experts and have been agreed on at an international level; they also form the basis for the OECD Council Decision/ Recommendation on the Mutual Acceptance of Data in the Assessment of Chemicals (C[81]30 Final, of May 12, 1981), which has to be regarded as one of the cornerstone agreements amongst the OECD member states with regard to trade in chemicals and to the removal of non-tariff barriers to trade.

In this part, therefore, the most important issues in the regulation of GLP will first be discussed, and interpretations of these principles will be presented which should enable the reader not only to learn how to comply with these Principles of Good Laboratory Practice but also to look behind the actual, very generalised, wording of the GLP guidelines and thus to grasp the intention of the respective sentences. The entire text of the Revised OECD GLP Principles will then be reproduced at the end of this part, together with the – similar but slightly deviating and in some respects more detailed GLP regu-lations of the United States Food and Drug Administration (FDA) and the Environmental Protection Agency (EPA).

While interpretation and application of the Principles for a good number of their paragraphs can be regarded as being very straightforward, in other instances the resolution of issues, addressed only by a general statement in these Principles, may become rather controversial. Therefore, already (or more precisely: especially) in the first years of their existence, questions about the applicability of the Principles in special situations, and questions about their practical implementation in such instances arose, which were discussed by the respective monitoring authorities concerned and published as their guidance to the interpretation of these Principles.

Such interpretation aids on a large number of topics were, e.g., published by FDA and EPA in response to specific questions to these Agencies as so-called "Advisories". Other national monitoring authorities answered such questions similarly through Newsletters, and also national associations of Quality Assurance professionals provided analogous services to their members. For various, broader areas such interpretation aids are also available in the form of OECD Consensus Documents (see list of references). These were obtained through discussions amongst experts from national authorities of OECD member states and from industry at Consensus Workshops; when consensus had been reached amongst these experts the documents were subsequently endorsed by the official bodies of the OECD member countries as represented by the OECD Joint Meeting and the OECD Council. While some of these Consensus Documents could be rather clearly and succinctly worded and are unanimously accepted as final guidance for the correct imple-mentation of the GLP Principles, others had to be more cautiously and generally phrased, as their areas are of a more contentious nature owing to their very broad or controversial field of application which makes it difficult to accommodate each and every imaginable situation in a specific manner. They are therefore probably again leaving (too) much open to discussion and interpretation. Furthermore, they are intended to give general guidance on the topics discussed, and they do not intend to provide final and firm guidance to every single, very specific situation arising in practice. There are thus no final answers to be found in these documents for each and every question that may arise in any specific area within the multitude of everyday situations.

However, these Consensus Documents are of value not only in the sense that they are giving somewhat more detailed instruction on how to interpret and apply the general Principles, but they are a very good instrument to deal in a timely and uncomplicated way with changes and developments in the field of "human health and environmental safety studies". This has become apparent especially with the revision of the Principles themselves, whereby the consequent adaptations of the respective Consensus Documents could be easily implemented. These Consensus Documents will subsequently be cited at the relevant places, and their recommendations will then also be scrutinised, not only for their value in a number of often encountered situations, but also for their applicability in more special circumstances.

2. Definitions in GLP

Definitions are an important part of any regulatory document, if not of life itself. They are the key to understanding and to common interpretation of statements and recommendations given in the document. As such they ideally have to be precise enough to avoid the possibility of multiple interpretation, but at the same time to be general enough to encompass all and every conceivable possibility of application. This task may be very difficult to achieve, and indeed it has been proven difficult, if not sometimes impossible, to attain this goal within these Principles of Good Laboratory Practice, as will be seen later in this section.

Some of the definitions given in the OECD Principles will therefore be explored in detail in the following sections with the intent of providing the reader with some more profound and clearer insight into their original meaning, and thus of aiding in their interpretation. There are, on the other hand, definitions which should not need further explanation, and they will therefore not be touched upon in this section.

2.1 Good Laboratory Practice

The definition of the term "Good Laboratory Practice" itself, which identifies GLP as *"a quality system concerned with the organisational process and the conditions under which non-clinical health and environmental safety studies are planned, performed, monitored, recorded, archived and reported."* can be considered as an example of a concise and precise definition; it not only defines GLP as a quality system, but it also delimits it against other quality systems by confining it to the organisational process and the conditions under which the whole process of a non-clinical health and environmental safety study has to be developed, from its inception and design till its final stage of reporting and archiving. In an earlier section (see part I, section 7, page 45), some of the differences as well as similarities between GLP and other quality systems have been briefly described. It will thus suffice to stress here again that GLP *sensu stricto* is not primarily concerned with data precision in a metrological sense, or with data validity in their scientific aspects – although this may be a welcome by-product from the application of GLP – but with the more "administrative", documentary processes and the management aspects under which studies will be performed. In the end, the faithful observation of these guidelines will lead to the full reconstructability of the process of study conduct, and thus to enhanced confidence in the results as they are reported. The judgement on the scientific validity of the study design and the precision of the results is then a matter for consideration by the authorities receiving these data as part of a submission for the registration or for obtaining a marketing permit for the product in question.

2.2 Management

It might seem self-evident that defined (administrative) structures would be necessary for the orderly operation of any human endeavour. This should be especially important in the case of test facilities conducting non-clinical safety studies under the conditions of Good Laboratory Practice, since under these conditions there need to be clearly distinguished levels of responsibility. The responsibility to oversee the general operation of the company and its facilities and establishments lies thus with the management, one specific level of which, namely the test facility management, is defined in the OECD Principles as follows: *"Test facility management means the person(s) who has the authority and formal responsibility for the organisation and functioning of the test facility ..."*.

With the definition of the management the GLP Principles are setting the stage for a successful functioning of a test facility. It stands to reason that there has to be some position of ultimate responsibility for any question or problem that might arise out of the sometimes divergent interests and opinions of the two instances responsible for the GLP-compliant conduct of studies, namely the Study Director and the Quality Assurance.

While the general definition of test facility management may not seem to pose too much problems, there are some practical questions around this term which are dependent on the size of the company or test facility. In large, multinational companies there are several layers of management, with varying levels of responsibilities, and it may, at times, be difficult to clearly ascribe to any single one of these levels the function of management in the sense of the GLP Principles (see figure 1). On the other hand, there are the small to tiny test facilities, which operate with only very few personnel, and there the question may arise, whether management is separated clearly enough from the "operative" part of the test facility (see figure 2). This latter question boils down to the very practical problem of how many persons a test facility has to comprise in order to be accepted as a full GLP facility.

In the former case, the one of the large, multinational company, the most logical solution should be that the level of management immediately supervising the organisation and conduct of the safety-related studies to be run under GLP would be the "GLP-relevant" management. This management level would also be the one which is most aptly termed "test facility management". However, it will be helpful to scrutinise the organisation chart of the whole company and to look at the various functions and duties which the different levels of management have to discharge. In this way, the "top management" of such a company will become able to act as the body discharging one of the first responsibilities of management, namely to *"ensure that a statement exists which identifies the individual(s) within a test facility who fulfil the responsibilities of management ..."*. The necessity for such a policy document will be most obvious for complex situations of managerial structures, where its existence will then clarify the situation with regard to the one part of management responsible for ensuring that "... *these Principles of Good Laboratory Practice are complied with, in its test facility."*

The latter case, the one of the small, probably specialised test facility with a very limited number of personnel, is certainly much more widespread, since there are much more small test facilities than large ones, and therefore the question posed above about the adequate separation of management from study personnel has given rise to many discussions, resulting in quite different answers. With regard to the single part of management, the definition given in the Principles makes it clear that the tasks of the test facility management could certainly be accomplished by one single person. This statement will not answer, however, the question of the minimal size of a test facility; this issue will be discussed later on in the section 2.4 on test facilities and test sites (see page 61).

> In order to attain its purpose, any quality system needs to define very clearly the different levels of responsibility. The definition of the test facility management serves to delineate these responsibility borderlines, and it invests test facility management with the ultimate power for ensuring GLP compliance.

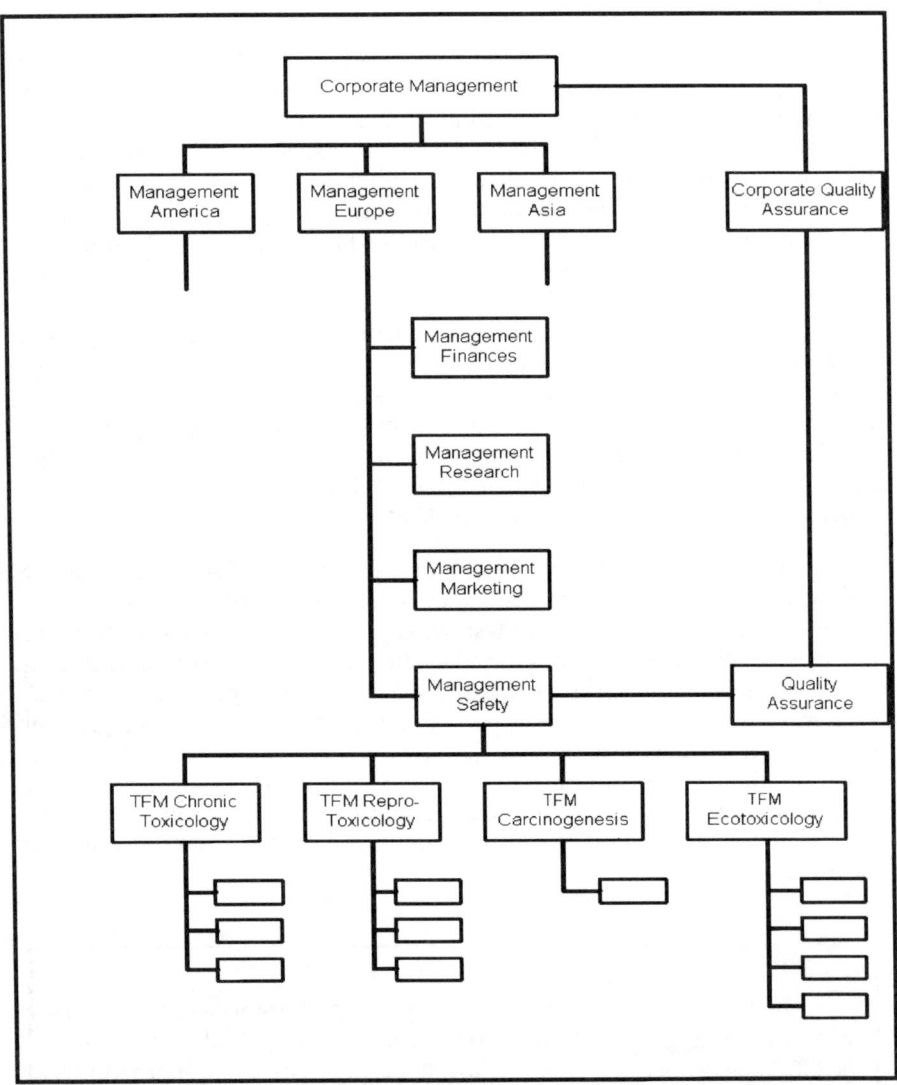

Figure 1: Possible Example for the organisation chart of a large, multinational company with a number of test facilities and several levels of management

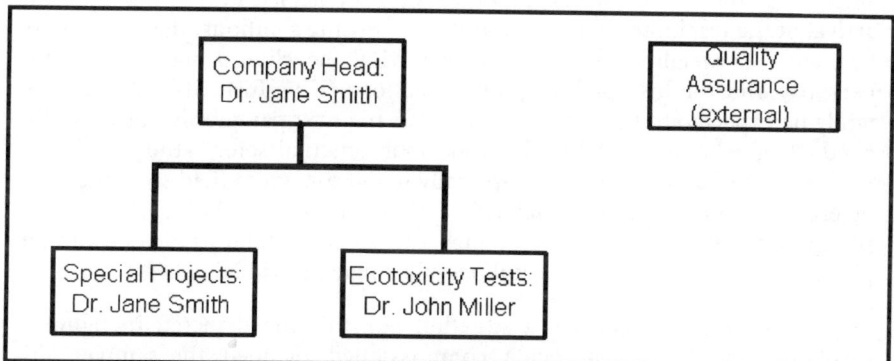

Figure 2: Example for a possible organisation chart of a small test facility with an external Quality Assurance.

2.3 Study Director and Principal Investigator

It can be stated without any exaggeration that the most important individual in the context of GLP compliance is the Study Director. The Study Director, who is defined – in rather innocent looking wording – as "*the individual responsible for the overall conduct of the non-clinical health and environmental safety study*", is the one person who is the central and pivotal individual within, and for, the whole study. The "OECD Consensus Document on the Role and Responsibility of the Study Director" (OECD No. 8, 1999) elaborates on this definition, and it very clearly describes the role of this individual as the "sole point of study control". It is the Study Director who is the only person who bears the ultimate responsibility for adherence to the GLP Principles. In all offices and laboratories around the world from time to time the old joke is displayed on the blackboard about the four people, named Somebody, Nobody, Anybody, and Everybody ("Somebody should do it, Everybody thought that Anybody could do it, and Nobody did it"). The GLP Principles want to avoid such a situation by centring the sole and ultimate responsibility for the GLP compliant conduct of a study onto the shoulders of one single person, the Study Director.

If the study indeed runs completely under the eyes of the Study Director, this may be considered an easy task. In the context of a simple toxicology study, especially a study conducted in a small, easily surveyable test facility, where the Study Director can supervise him- or herself everything from the arrival of the test animals, the preparation of dosing solutions, the daily activities, such as weighing, dosing and observing the clinical signs, up to the necropsy and the histopathological evaluations (if performed), this can certainly be expected to be reality. However, as the next paragraph will show, the conduct of a "non-clinical health and environmental safety study" is by no means always as simple as that. Especially in the context of field studies for the determination of environmental safety, it had already at the beginning been recognised that the supervision of such studies could (and would) exceed the physical capacities and the possibilities of a single individual.

Consider, for instance, a situation like the one depicted in figure 3, involving a study on a pesticide, commissioned, planned, the samples ultimately analysed and the final report written in Sweden, but with the experimental, field part to be conducted in a multitude of different locations as widely apart as California, Brazil, Egypt, Spain, India and Japan. It could not be expected of any single individual to be physically present as Study Director simultaneously at all of these places in order to exercise the required immediate control over the actual conduct of the practical work. The same holds true for the Quality Assurance personnel, who would also be put to extremities, if they should have to inspect simultaneously the critical aspects of study conduct in all these locations. One possible solution to these problems might been to split such studies into a number of "sub-studies", each of which would become a separate, full study from the GLP point-of-view. However, this way may be more or less barred, because of the opinion that such a splitting would create more problems than it would solve (see 2.4, page 61). Only, if these separate studies were to involve the field part in one specific country, e.g. India, coupled with the analysis of the respective samples in Sweden (as in this example), then the concept of a GLP study might still hold. To apportion the whole trial, however, into six singular field studies performed in the various countries, and six separate, even though from the technical standpoint identical, analytical studies conducted in Sweden would stretch the concept of a study somewhat. Other, more recent developments added to the pressure of dealing with such situations. This concerns especially the globalisation of pharmaceutical companies and their restructuring, with subsequent effects also in the context of non-clinical study conduct. The impact of these processes, partly of out-sourcing and partly of internal specialisation, on the possibilities of a Study Director to really and fully control every single aspect of a study under his/her nominal supervision made it necessary also for these areas to come up with a practicable solution to this problem.

2.2 Definitions

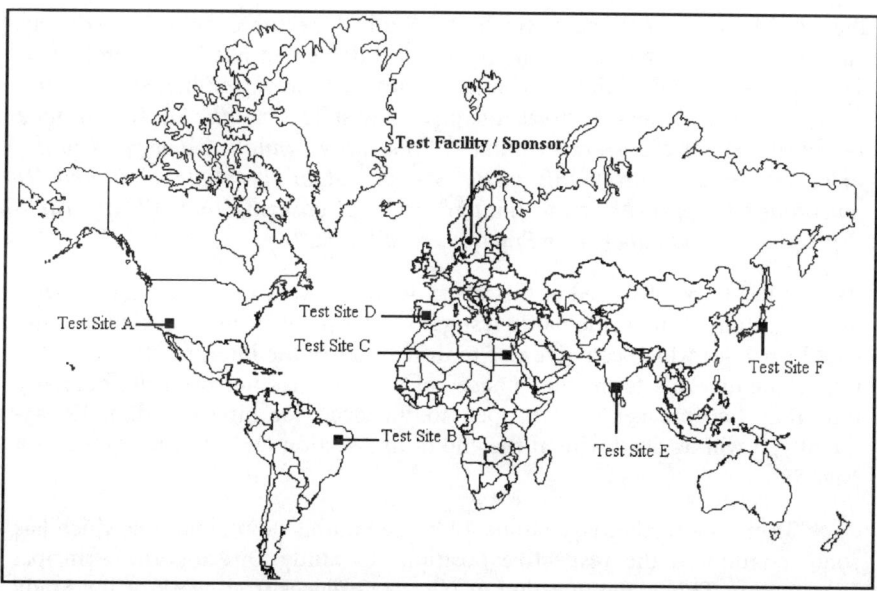

Figure 3: World map showing possible locations for the conduct of the various field parts of a study with one test item and one scientific purpose; the samples from the different areas will be analysed in the sponsor's laboratories.

Therefore, it was first in the context of field studies, where the need was most obvious and pressing, that a new concept was discussed in which some of the Study Director's responsibilities could be transferred to another individual who would then be fully responsible for the GLP compliant conduct of the single, defined tasks within a study deferred to him or her by the Study Director. This position of the "Principal Investigator" was discussed by an expert group of the OECD and formally described in a Consensus Document (OECD No. 6, 1999); the emerging utility of this concept also for other areas of safety testing has then led to its definition and inclusion in the Revised Principles of GLP. Thus, the Principal Investigator *"means an individual who, for a multi-site study, acts on behalf of the Study Director and has defined responsibility for delegated phases of the study."* Thus, the Principal Investigator is an individual, who bears the responsibility for the day-to-day experimental conduct of the one defined study part which the Study Director cannot immediately supervise and for the monitoring of which this person has been nominated. The ultimate responsibility for study conduct, however, still rests with the Study Director, as the Principles unequivocally state, that *"the Study Director's responsibility for the overall conduct of the study cannot be delegated to the Principal Investigator(s); this includes approval of the study plan and its amendments, approval of the final report, and ensuring that all applicable Principles of Good Laboratory Practice are followed."*

It has to be observed at this place, that, in a very unobtrusive way, with this definition of the Principal Investigator a term has been smuggled into the GLP Principles which can give rise to difficulties in the interpretation of other requirements. This term, the "phase" of a study, has not been defined, and thus may become very much open to the need for interpretation. Consequently, it will be dealt with at the end of this section on Definitions (see 2.10, page 88).

There is a further issue to be addressed at this point, and one which has some relation to the respective positions of Study Director and Principal Investigator. This is the question of (short) absences from work of the Study Director and of the consequent lack of immediate control, if there were no "deputy" available to step in for the Study Director (or the Principal Investigator). It is advisable, that management would appoint for any given study, or in any given test facility, not only a Study Director but also deputy for him or her. If such a deputy (or "acting" Study Director) were nominated at the outset of a study, then short absences of a Study Director from the test facility and from the immediate supervisory role would not matter, as long as this deputy could step in and, e.g., sign (or at least acknowledge) any necessary amendments. It would certainly be vital to the correct performance of the Study Director's role that any such occurrences and actions of the deputy would be brought to the immediate attention of the Study Director as soon as he / she would resume the normal functions. Another important point in this issue would certainly be the correct and complete documentation on the appointment and on the actions of this "deputy" or "acting" Study Director. This issue will be taken up also later on, in the part of management's responsibilities (see 3.1, page 94) where the problems and issues around the replacement of a Study Director will be discussed.

> Somebody has to be in complete, one may say absolute, control over a study in order to coherently ensure its GLP compliant conduct. The GLP Principles invest the Study Director with this role which in general must be regarded as indivisible.
>
> There may be special circumstances where the Study Director cannot exercise this immediate control; there, the appointment of a Principal Investigator may be advisable, or necessary, to really accomplish the requirements of GLP.
>
> In any case, however, the GLP Principles intend to have one single, pivotal point of study control, and this position is filled by the Study Director.

2.4 Test Facility and Test Site

Closely related to the question of the relative positions of Study Director and Principal Investigator, is the extension of the definition for the location where the study is being performed.

Studies need one or more physical location(s) where they can be performed. Since GLP is concerned with studies that may be conducted in the laboratory, in the field or in the greenhouse, a term was needed that encompassed all possible locations where studies might become conducted. Thus, GLP is referring to "test facilities", which are taken to mean *"the persons, premises and operational unit(s) that are necessary for conducting the non-clinical health and environmental safety study"*, as well as to "test sites" which are defined as *"the location(s) at which a phase(s) of a study is conducted"*.

It has to be noted that the term "test facility" does not only comprise buildings, rooms and other premises, but that it includes also the people who are working there and are responsible for performing these studies. Although the general interpretation of this term would certainly stress the location aspect, the definition rightly recalls that it is the people who will conduct these safety studies. The premises should only be regarded as the stepping-stone for the GLP compliant conduct of studies.

The possibility of utilising more than one location for the performance especially of field studies, and the developments in company policies resulting in the fragmentation of studies into various phases, which could be conducted separately in different places, necessitated an expansion of the term "test facility", and the definition of the test facility had to be amended accordingly. Thus, there is not only the test **facility**, which has to be considered as the all-embracing location of study conduct, but a part of the study might be run at a test **site**. The definition of a test site acknowledges the fact, most important for Field Studies, that some sites, where a defined part of a study is conducted, will not and indeed cannot fulfil all the requirements laid down for a site to be recognised as a test facility. A field plot, where a crop is sprayed, will most probably have no test facility management, it will be used possibly just once in a while for a field test under GLP, and most importantly no full studies are conducted at this place. It would therefore be difficult to qualify, e.g. a field, an orchard or a vineyard as a full, real test facility. The fact that it will also be in most cases geographically remote from the place where the Study Director is located precludes its inclusion under the term "test facility".

Thus, the definition of the test facility, originally a very simple one, had to be expanded in order to include the possibility of utilising one or more test sites as additional places where parts of a study could be performed, and to separate the requirements for such test sites from the ones needed for a test facility. This new distinction between the "full" test facility and the - in some respect - "restricted" test site might have (regulatory) consequences for the classification of certain enterprises which are offering the conduct of specialised parts of studies (e.g. histopathology evaluation, analytical or clinical chemistry services).

In the wake of this introduction of the possibility of conducting studies at more than one location, another new concept had furthermore to be introduced, namely the one of the "multi-site study". Thus, the definition of the "test facility" now does not only cover the single place where the Study Director is located, but for these *"multi-site studies, those which are conducted at more than one site"* the term "test facility" comprises the site at which the Study Director is located and all individual test sites, where defined study parts are conducted. Again, at the background of this concept is the idea that a study should, if at all possible, be indivisible: Even if distinct parts of a study were to be conducted at different locations, there should be no question about the fact that there is one, single study, albeit a "multi-site" one, and not a series of "sub-studies" which are somehow connected to each other.

One point that has already been raised in the context of the definition of management (see 2.2, page 54) remains to be discussed here, namely the question of how large should, and how small could, a test facility be in terms of number of personnel, in order to qualify as a true GLP test facility. In order to answer this question, consideration has to be given in a first step to the various functions which are necessary for the establishment of, and which indeed make up, a test facility. In a second step, the interdependence of these functions has to be scrutinised, in order to arrive at an estimate of the minimal number of persons a test facility has to comprise. A test facility needs a management, a Study Director, a Quality Assurance Unit, study personnel and a person responsible for the archives. In the case of the smallest imaginable test facility, one may suppose that, where such a facility would conduct only one type of studies, there might be just one Study Director. Since the Quality Assurance personnel has to be completely independent from study conduct, its functions would have to be provided by an external, independently operating company offering Quality Assurance services. The crucial division, as far as responsibilities are concerned, will thus arrive between the management and the Study Director. Since management has to nominate a Study Director for each study, has to approve all SOPs, and has to ensure that the Study Director and the study personnel perform their duties according to the GLP Principles, it has a clear supervisory function which cannot normally be combined with the operational level of study conduct. Management therefore should not participate in study conduct, necessitating theoretically the separation the test facility manager from study personnel. Thus, in the first instance it might be considered that two or more persons would be needed for the establishment of a true GLP test facility. The Principles do call, however, also for the appointment of a person responsible for the archives, whose function might, however, be discharged by management or by the Study Director. Whether a test facility would need some study personnel in addition to the Study Director (who would certainly in the best possible overall control of study conduct, if he or she were to do everything him- or herself) could also be disputed. There are different opinions in this question, ranging from allowing a test facility to consist of one person only, up to the requirement of at least five persons for constituting a real test facility. On the whole a minimal number of three persons (plus an external Quality Assurance) would seem to be a good compromise for the establishment of a GLP-compliant test facility.

A minimal size of three persons for a test facility might of course cause problems, if any one of these three would be temporarily absent for a shorter or longer time: On absence of the manager, new SOPs, or revised versions of existing ones, could not be put into use, however pressing the need for them, since their approval would be on hold, pending the return of the responsible individual. Studies could certainly not be conducted with the Study Director being absent. And, finally, the material from studies terminated during the absence of the archivist would be piling up in some office and could not be archived in an orderly fashion. Thus, the view that, in order to deserve the name "test facility", an establishment should comprise at least five to six individuals could be considered reasonable, while it is self-evident that a test facility comprising one person only would totally avoid these complications by having to stop activities altogether.

> A study can be divided into separate phases which may be conducted at different locations. With the introduction of the terms "test facility", "test site" and "multi-site study", the GLP Principles make it clear that there should only be one, indivisible study, even if the study can be conducted in separate phases at different locations.
>
> The possibility of assigning study parts (or phases) to test sites different from the Study Director's test facility increases the flexibility in study planning and conduct, while at the same time it will be increasing transparency for the purpose of reconstructability. A differentiation has to be made, however, between full test facilities, where whole studies can be conducted from beginning to end, and test sites, where only parts of a study will be performed and where the GLP Principles have only to be complied with as far as its responsibilities in study conduct would require. In this differentiation GLP wants to achieve clarity in terms of the respective requirements for a test facility, where the full GLP Principles will apply, as opposed to those for a test site where not all of the GLP Principles might necessarily be applied.

2.5 The Study

The definition of a "*non-clinical health and environmental safety study*" takes up again the scope of Good Laboratory Practice in reiterating that it "*means an experiment or set of experiments in which a test item is examined under laboratory conditions or in the environment to obtain data on its properties and/or its safety, intended for submission to appropriate regulatory authorities*". While it should be clear in general, what a "study" is, there may be some need for additional clarifications regarding what may constitute a "study under GLP".

A GLP study involves clearly an experimental activity, but not every single experimental activity may also be regarded as a study from the viewpoint of GLP. There may be instances where one single experiment may be considered sufficient to constitute a study in the GLP sense: The determination of the melting point of a given substance can be conducted in one, single experiment, and without the necessity of having to perform multiple repeats of this determination in order to collect sufficient data. There may be other instances, where the reverse is true, where only repetition of an experiment may yield sufficient, or sufficiently reliable, data: The determination of genotoxicity in a bacterial mutagenicity test involves not only the testing of the test item in a number of different bacterial strains, and the testing in the presence or absence of an exogenous mammalian metabolic activation system, but it involves also a preliminary toxicity test for the determination of the usable concentration range as well as one (or even more) repetition(s) of the mutagenicity test itself. Every one of these single parts of the bacterial mutagenicity study is an independent experiment, but not each of these experiments can be regarded as a full study, since only the combined experiments will allow the Study Director to express any statement regarding the genotoxicological safety of the test item.

On another level, there may be experiments, in which there would be no test item to be administered to the test system, but in which baseline data for other sets of experiments will be gathered. These baseline data will not be intended for submission to Regulatory Authorities in their own right, but they will be used to support the interpretation of "regulatory" GLP studies. As an example, the determination of the microbiological status of test animals may be considered. This status has to be investigated on a regular basis, since the GLP Principles require that biological test systems (i.e. in this case the respective animals) "*should be free of any disease or condition that might interfere with the purpose or conduct of the study*"; the same would hold for viral or mycoplasma contamination of mammalian cell culture systems. It is not strictly required that such tests should mandatorily be performed under the rules of the GLP Principles, since they would more resemble quality control in the manufacturing environment, but since these data might be used later on to aid in the interpretation of findings from a GLP study, they should nevertheless be collected under circumstances as near to GLP as possible.

Thus, it is the objective of the study that determines whether one experiment, or a set of experiments has to be regarded as a study. On the other hand, this opinion, that a study is defined and held together by its objective, may be questioned in special circumstances. Indeed, there have been practical problems which need to be addressed in the context of study definition, and which are discussed in some detail below.

In studies involving a number of separate activities (scientifically as well as locally), as e.g. in residue and environmental fate studies, the question of the admissibility of the splitting of such studies into a number of "sub-studies" could be pondered. Such studies may involve very diverse tasks (spraying the fields and harvesting the crops or sampling soil, as opposed to residue analytical chemistry), constituting scientifically and experimentally so divergent activities, that the field part on the one hand and the analytical laboratory part on the other could really be considered as separate studies. Furthermore, especially in field studies, there might even be a number of different endpoints to be addressed which would have no apparent connection between them, short of demonstrating the effects of one and the same test item. For instance, a test item might be applied to a field with some crop growing on it; residues on this crop could then be determined, as well as residues in soil and the metabolic fate of the test item in the soil and on the crop. Thus, there would be a number of analytical determinations to be made which would differ from each other by the methodology to be applied as well as by the purpose for which they are conducted. It might therefore be reasoned that a breaking-up of studies with such divergent parts (technically as well as scientifically) could be an advantage from the point of view of GLP, since then a truly competent Study Director could be appointed for each "sub-study".

However, the treatment of study parts as complete and separate studies may be problematic and the applicability of such a splitting has to be judged carefully. If such a breaking-up of a study into various component studies, each of them being treated as a full, independent study with its own study plan, Study Director and study report, were to be contemplated only because of its apparently greater ease of control, the difficulties arising from this opinion should not be underestimated.

There may first be technical difficulties arising from the breaking of studies into components and their being identified separately as "studies". For example, in any study, where study material (samples or specimens) is transferred from one facility or site to another, there is the problem of control over these transfers; if such a transfer would not only involve different test sites or facilities, but also the transfer from one study to another, i.e. from the area of responsibility of one Study Director to the responsibility of the next Study Director, difficulties could arise in the assessment of who has had the responsibility for the respective samples or specimens during these critical phases of material transfer. If the whole endeavour is forming one indivisible study, then it is only one Study Director who should be in absolute control also over all of these transfers, and who then will be ultimately responsible for the GLP compliance.

Furthermore, the Study Directors of each part of such a subdivided study would, individually, have the authority to make protocol changes by amendments. This could be considered to be expedient, but with the probable lack of co-ordination between the various Study Directors with regard to such changes that could be foreseen in such cases, there would be a loss of assurance that the amendments and changes made on an individual basis would conform to the overall purpose of the study. Furthermore, if these subparts of a study were to be considered as separate, stand-alone studies, it would then not be required of the Study Director of any one of these single studies to inform the respective Study Directors of the other parts about changes in the protocol of his part. This lack of information might have truly disastrous consequences, if such a change were of relevance for the assessment of results from subsequent investigations. When, e.g., the time intervals for the different bleedings required for the toxicokinetic part of a toxicology study were changed by the in-life Study Director without notification of the respective person in the toxicokinetics area, then the calculation of the kinetic parameters would yield completely wrong results. The same holds for a field test, if the time of harvesting the crops, foreseen for a number of fixed time points, with the goal of defining the degradability of the test item and thus of arriving at an estimate for the latest spraying time before harvest, would be changed without notice to the analytical laboratory. The analytical "study", proceeding from the originally fixed time points, would again yield results that could only be distorted by the unknown changes made in the field part.

There would also be the question of the test item, which term might have a different meaning from study to study: Take, e.g. a simple toxicology study, which could be easily broken down into three separate studies, namely the in-life part, the toxicokinetics part, and the histopathology part. First of all, there would have to be a study plan for each part, delineating not only the way of conducting the "study", but foremost to identify the test item and to define its purpose. In the first part, the test item would certainly be the substance, the safety properties of which are to be tested. In the toxicokinetic part, it could be debated whether the test item would still be the chemical substance, the safety of which should be investigated. Alternatively, the plasma samples from the test animals of the in-life part could be regarded as the test item, since the substance applied to the animals in the in-life phase, analytically speaking, could be considered to play the role of a reference item in this second study. The purpose would be less clear, however, since, if the plasma samples were to be defined as the test item, it would not be their properties with regard to safety that were to be investigated. Finally, in the third "study", the slides prepared from the animal tissues, normally regarded as specimens in the context of a single study, would turn out to be the test item, since these slides which have to be read by the pathologist would be subject to investigation.

The same enumeration of different "test items" could be made for a field study, which could be split into a study involving the preparation and the testing for stability and homogeneity of the test item in the application vehicle, another one involving the application of an analogously prepared tank mixture to the respective crop, a third one involving the analytical determination of residues on this crop and in the environment, and finally a fourth one dealing with the investigation of the breakdown-products of this test item in crop and soil. Again, in every one of these "studies", while being part of one and the same design and investigation, the test item could be defined differently. Furthermore, the "descriptive title" of the respective study reports, required by the GLP Principles, would then have to mention different test items, which might lead to difficulties for the Regulatory Authorities to identify with confidence those parts that belong to each other.

But the question of the test item would not remain the only one. In such situations as described above the different "studies" would need different study plans, and would receive different study numbers and other means of identification as required by the GLP Principles ("*A unique identification should be given to each study*"). Consequently, the study parts would eventually lose their coherence, and it would later on become more and more difficult to overview the whole of such a study. And finally, the question could be asked, whether such a study part, in which only the starting material is prepared for a subsequent study, as it would be in the case of a field study with the crop spraying and the preparation of samples for analytical purposes, could - in the light of the study definition - indeed be regarded as a full study, since no "*data on (the test item's) properties and/or its safety*" will be obtained directly from these experiments.

In general, the disadvantages, also from the point of view of overall study control inherent in such a splitting of studies, would seem to outweigh any advantages that could be imagined. Only in such instances, where the results of the split-off-part of a study would not be expected to significantly impact on the interpretation of the "overall study", such a breaking-up could be considered to have its merits. Solutions to the problem of studies with such vastly different parts have thus to be looked for, and have been found, in other means of coping with these situations, and which are briefly touched below.

In the first place, the concept of the Principal Investigator has to be mentioned, which has already been described in a preceding paragraph (see 2.3, page 57). In this case, even though the study remains a single entity, some study parts can be conducted in relative independence from the rest of the study. The coherence of the whole study is guaranteed here through the still pivotal role of the Study Director, who has to be kept informed about the progress of those parts that are running under the responsibility of the PI, and who has to acknowledge all changes to the study plan asked for by the PI. In the second instance, for studies where this is feasible, the concept of the "Senior" or "Contributing Scientist" could be applied. In this concept, the full responsibility for study conduct lies with the Study Director, who may rely, however, for the scientific aspects of single study parts on the expert judgement and knowledge of these specialists.

> The "human health and environmental safety study" forms the basis of all the assessments on which decisions about the safe use of the tested items, and the products derived from them, will be made by the relevant authorities, and these studies are therefore to be regarded as the building blocks for the general safety assessment. They have thus to be conducted in a meaningful way, according to a pre-defined procedure, and for a pre-defined purpose.
>
> Therefore, a "study" can be either one single experiment, or, in other instances, it can consist of a series of experiments: One experiment may yield either a complete, safety-relevant answer, or only one single data point, where then only the data from several experiments together (or from a "set of experiments") may finally address a safety endpoint in a relevant way.

> With the definition of the study as "an experiment or set of experiments" the Principles of GLP clearly want to express this notion that a study has to have a defined objective with regard to safety evaluation. Depending on the declared purpose of the study, a single experiment may or may not be sufficient to fulfil this requirement. This is a major reason why it should be advisable not to break down a study into a series of sub-studies, since none of these could then claim to pursue such an objective in the strict sense. Only in the coherent assembly of all parts, of the whole "set of experiments", can a study become recognised as a full "human health and environmental safety study".

2.6 Short-Term Studies

Some of the most difficult problems in defining expressions used in the GLP Principles can be seen as originating from the wide diversity of study types that are to be covered by these Principles. One of the most hotly debated issues has been, and still is, the problem of "short-term studies".

The expression "Short-Term Studies" encompasses in itself such a wide variety of study types that it is indeed very difficult, if not impossible, to arrive at a comprehensive list of short-term studies, let alone a meaningful, all-embracing and clear-cut, but nevertheless concise definition. The revised OECD Principles could thus do no better than to define Short-Term Studies as "studies of short duration with widely used, routine methods", a definition which still leaves the expression "short duration" open to interpretation.

This definition, therefore, does not seem to clarify the situation very much, since in its first half it states the obvious, while the second half does not seem to add any really helpful information. There are, however, great difficulties in identifying short-term studies in a concise manner. Labelling a study as "short-term" only because of its limited duration would leave the question unanswered "duration of what?" (in this respect also the issues around start and end dates may be looked at, see section 2.7, page 76). Owing to the wide diversity of such studies and study types, it has not been possible to link the expression "short" to any definite time of study duration which would define exactly and comprehensively a short-term study: What might be considered "short" in the context of biological studies may not be regarded as "short" in a physical-chemical study setting.

2.2 Definitions

As an aside: This latter point even should make it advisable to treat biological studies differently from physical-chemical ones with regard to the application of the provisions for short-term studies. Indeed the Consensus Document on short-term studies (OECD No. 7, 1999) started to differentiate these two types without, however, carrying this principle fully through.

One may consider, for instance, the determination of a melting point as a truly short-term study, because the whole experimental activity is terminated within a few minutes, and nobody would object to calling such a study "short-term". But what about an environmental fate study in a lysimeter? There the test item is applied to the test system (an activity taking a few minutes), then the whole system is left undisturbed for two years, after which the distribution of the test item in the column of soil is determined, again an activity of a few days at most. Is this a "2-year, long-term" study, or has it to be labelled "short-term", because for 99.9 % of the time the test system is just left alone with no experimental activity whatsoever occurring?

For this reason it is certainly more useful not to become fixed exclusively on the issue of defining "short" in precise "time of duration" terms, or on editing a comprehensive listing of "Short-Term Studies", but rather to concentrate on two pertinent questions:

- Firstly, why should, in these studies, certain adaptations in, and exceptions from, the general rules of the OECD Principles of GLP be necessary?
- Secondly, under which conditions should such facilitations, which have been introduced into the revised OECD GLP Principles, be admissible?

Generally speaking, the Principles of GLP require that certain formal procedures be respected on different levels of test facility organisation and study conduct in order for a study to become GLP compliant. A number of these, however, if fully observed, can be considered as out of proportion with regard to their labour and time requirements in comparison to the respective requirements for the actual, experimental conduct of studies of the "short-term" type. Thus, the definition of a short-term study becomes important for several aspects of GLP compliance.

First, there is the requirement of retaining a sample of the test item for all studies. A second aspect is the requirement of having to write a study plan, to have it checked by Quality Assurance, to have it approved by the test facility management and / or the sponsor (if required by national legislation) and to date and sign it, before the study can be started experimentally. After the termination of the experimental phase of the study, there is the obligation to write a full study report, which again has to be checked by Quality Assurance and to be dated and signed by the Study Director. And thirdly there is the requirement for the Quality Assurance to inspect each study in order to ensure that it is conducted in compliance with the GLP Principles.

These three issues can become very cumbersome in the context of conducting short-term studies, because they place administrative burdens on the test facility management, on the Study Director and on the Quality Assurance personnel which may indeed be out of proportion with the actual experimental work, e.g. for the determination of a melting point, or the investigation of the mutagenic activity in a bacterial test system. This means that a balance has to be achieved between

a) the administrative and the experimental "time- and labour-expenditure" necessary for the GLP compliant conduct of any specific study and

b) the "returns" in terms of the assurance of quality and reconstructability of the study on the one hand and its mutual acceptance as a valid study on the other.

Foremost in all these considerations for balancing "expenditures and returns" must, however, be the consideration of the ultimate goal of GLP. Assurance has to be provided that neither quality nor reconstructability of such studies could be jeopardised by the application of the special provisions for Short-Term Studies which the GLP Principles are presenting as possibilities for a simplification of control mechanisms. As for this balance, it is furthermore recognised that common sense has to be exercised with regard, on the one hand, to the feasibility of the full application of the GLP Principles in special situations, but that the same common sense has to be applied when judging the advisability of taking advantage of the provisions for short-term studies. In consideration of the three areas cited above, where such facilitations for short-term studies are provided, it can be observed that the question of which type of study might be looked at as "short-term" is of real importance for the amount of "expenditure" needed.

If a test facility is performing very many studies on a great number of compounds, as e.g. in screening-type investigations, the requirement of retaining a sample of the test item will become burdensome because of the sheer volume of storage space needed, and because of the storage and retrieval logistics necessary for keeping track of these samples. This is on the one hand one of the reasons why the GLP Principles have exempted short-term studies from this requirement. On the other hand it may be envisaged that a short-term study, because it can be conducted and completed with not too much effort and in a short time, might be repeated with not too much effort, should any questions arise later on about, e.g., the purity of the test item used. Therefore it may not be considered necessary to retain an analytical sample of the test item for such studies. Thus, for a test facility conducting studies with very many test items, it becomes an important question, whether their studies could indeed be labelled as "short-term".

Secondly, the Principles allow for such short-term studies the use of general, standardised study plans and study reports which have just to be amended by the study specific details. Thus, it is not necessary for the Study Director to write, date and sign a study plan for every single melting point determination, of which dozens could be performed within a single working day; instead, he may write and sign a standardised study plan, containing all the general points of the respective study and which can then be used together with an amendment, giving the details of the specific test item that is to be investigated, in the place of an ordinary study plan. In the example cited above, the standardised study plan might for instance contain the information about the apparatus used for the determination of the melting point, the way of preparing the samples for the determination, the necessary calibrations to be performed, as well as other general information required by the GLP Principles such as the address of the test facility, name and address of the Study Director, and type and storage location of raw data and other study records to be retained. Also the management of the test facility is somewhat relieved in its responsibilities, in that it does not have to nominate a Study Director for each individual study, as it is the case for e.g. long-term, chronic toxicity studies, but can do so in a general way by its signature to the general, standard study plan.

And finally, for Quality Assurance, there is the possibility of inspecting such studies in a "process-based" way, which means that not every single such study has to be inspected, but that inspection of every x^{th} study can be considered sufficient for ensuring that all of them are indeed conducted in a way compliant with the requirements of GLP. However, there is an additional consideration, namely the one about the frequencies with which such studies are performed at a specific test facility, an issue which is addressed by the phrase *"with widely used, **routine** methods"*. Concomitant with the frequency of study conduct, also aspects like study complexity and personnel routine will have to be taken into account. Thus, if a particular test (e.g. a single dose, acute toxicity test) is conducted by the score at one test facility, this test facility would certainly be justified in applying (some or all of) the "simplification" measures to this test. If, however, this same test were to be conducted, at another test facility, every other leap year only, this second test facility would certainly not qualify for any of the special provisions with regard to the application of the GLP requirements for this test. In this respect, any study of "short duration" may or may not qualify for the special treatment foreseen for such studies "with widely used, routine methods", and thus be (or not be) eligible for the various measures facilitating the GLP-compliant conduct of it. On the other hand this "facilitation process" should certainly not lead to losing sight of the ultimate goal of GLP, the complete reconstructability of any particular study.

Therefore, the question of which study types may qualify as short-term studies is a matter of some debate.

There may be a number of study types, for which this qualification is undisputed, but in other cases the situation may be far less clear. It is difficult to draw a definite and exact line between the short-term studies and the other ones. There are, on the one hand, certainly some study types, which can be generally agreed on to be truly "short-term". The Consensus Document of the OECD on Short-Term Studies (OECD No. 7, 1999) provides in its Introduction a few such examples. There are, on the other hand, a number of ways for any single test facility to define its own package of short-term studies, or for a Compliance Monitoring Authority to recognise a list of study types. Therefore, in cases where there is no general agreement on the status of a certain study type, it may be possible for test facility management, in co-operation with Quality Assurance, to formulate a general policy on how to handle this problem. This policy should be documented as such and should be applied to the study types concerned in a documented way. The decision on which study types to recognise as "short-term" should be based on the test facility's own statistics on duration, frequency and complexity of the studies generally conducted there. The reasons for such decisions would, however, have to be clearly stated and documented.

Analogous considerations would have to be made by the Quality Assurance for the planning and conduct of their inspections: Infrequently conducted studies of a type of great complexity would certainly deserve a closer look, and a relatively greater number of the respective studies should therefore be inspected. This assessment may result in the decision that every single one of these studies is to be inspected. On the other hand, very simple, straightforward and truly routine studies, conducted at great frequencies might be inspected, e.g., only once every four months. This aspect of the short-term

2.2 Definitions

study will, however, be dealt with fully later on (see section 4.2 Quality Assurance Inspections, page 128).

In summary, while a study has to fulfil certain requirements, such as a study plan agreed to by dated signature before the start of the study, and it has to undergo some control by inspections of the Quality Assurance there may be instances where certain of these requirements would become neigh impossible to be fulfilled. Where a study is, however, of so short experimental duration, that it would be neigh impossible to schedule the respective inspections, and where the administrative workload would become much higher than the workload for the actual conduct of the study itself, then, GLP would by right be considered a burden only. Therefore the requirements of GLP are adapted by the Principles to short-term studies in a way that allows for a certain alleviation for Study Directors, study personnel, and Quality Assurance.

If any study type can be considered to allow for the application of a highly standardised study plan, if this type is of short duration and so frequently conducted at a test facility, that the study personnel therefore can be expected to possess a high degree of routine in the conduct of this study type, then the GLP procedures may be alleviated to some degree, as it is provided in the Principles.

In this respect, it becomes no more important to define an exact length of time, where "short-term" borders to "longer"-term, but common sense has to tell, which study type can, for the purposes of GLP compliance, be called "short-term". It will depend not only on the study type itself, but also on the test facility, since, among other factors, it is the frequency with which such studies are actually conducted that will have to be considered. A study type that might profit from these special requirements at one place might, because of the rareness of its performance, not be able to do so at another. In any case, and for each individual test facility, there should be a policy document, stating which studies are to be regarded as "short-term" and the reasons for including the chosen study types in this list. The same holds of course for the Quality Assurance Programme which will have to define in its SOPs those study types that will be regarded as "short-term" and thus will be treated accordingly.

> GLP aims at a full reconstructability of any study conducted under its rules. Certain studies may be too limited in duration as to warrant the application of the Principles to the fullest extent. Feasibility considera-

> tions may lead to the view that certain pragmatic facilitations should be advisable. Nevertheless the spirit of the GLP Principles needs to be preserved, even when certain "short-cuts" may be allowed. Thus, the GLP Principles strive to lessen the burden of their (administrative) requirements for short-term studies to a certain degree without, however, jeopardising quality, integrity and reliability of these studies.

2.7 Initiation, Starting and Completion Dates

Everything has a beginning and an end, and so have studies. This obvious truth does not seem to be the source of any problem, but when it comes to defining exactly these points, then difficulties arise. Again, these difficulties stem from the wide variety of study types performed under GLP, and the concomitant variety of time points that could define start or end of a study; the importance for an exact definition of these time points is connected to the practical issue of when the GLP Principles have to be fully applied, since documentation before the start of a study may be less extensive than the one required once the study has begun.

Four time points have been defined in the OECD Principles for start and end of a study. The so-called "study initiation" and "study completion" dates do not contain any ambiguities that could give rise to interpretation difficulties. The dates when the Study Director signs study plan and final report are unambiguous calendar dates, and they clearly mark the two time points between which a study has been conducted. However, more difficulties are encountered with the experimental starting and completion dates for themselves and for their connection with initiation and termination dates.

While there are definitions given for these four different study dates, their definitions lack the logical connection between them. On the one hand, it is obvious that, since "a written (study) plan should exist prior to the initiation of the study", and since study initiation is defined as the date on which the Study Director signs this study plan, therefore, logically, this study plan has to be in physical existence at this time. On the other hand there is no obvious, formal connection between study initiation and experimental starting date such as a paragraph indicating that this latter date should follow the former one, or that experiments (data collection) should be performed only after the initiation date. The intention, however, was certainly that a study plan should

2.2 Definitions

exist prior to the start of study activities in order to conduct the study in an orderly and planned fashion. Common sense would also dictate such a procedure, although this logical connection can only be inferred from the fact that the study plan should contain, among other information, the proposed experimental starting date.

While this is more of a semantic exercise, there are more important questions that can be raised in the context of these definitions.

Consider the statements that an experiment starts at *"the date on which the first study specific data are collected."* and is completed on *"the last date on which data are collected from the study."* Could these dates not be defined in a more straightforward way? Why, for instance, should the experimental starting date not be the date on which the test item is first applied to the test system, as it is stated in the Draft GLP Regulations of the Environmental Protection Agency (40 CFR 806, see Appendix II.III, page 307)? And, conversely, why should the experimental completion date not be the date on which the test item is last applied to the test system? The answer to these questions is, however, not as simple as it would seem at the first glimpse. It may be true that in the majority of cases, test item application can best define the experimental phase, i.e. its beginning and end. However, consider two situations, one of which has been already used to elucidate another point.

Consider the 2-year lysimeter study which has been described above, and in which there is just one application of the test item. Could this be the experimental starting as well as already the experimental completion date? For the experimental starting date, this may well be correct; but is it equally true for the completion date? Certainly not, because the main experimental effort in such a study is the analytical determination of the test item concentration in the effluents and its distribution within the core of the lysimeter. Therefore, it is completely out of the question to set the experimental completion date as the date on which the test item has last been applied to the test system.

On the other hand, consider an embryotoxicity/teratogenicity study as it is normally conducted in rats. There, the test item is applied to the pregnant animal from day 6 to day 16 of gestation. Should the determination of the day 0 of gestation, which is either done by visual inspection of the copulation plug in the early morning (timing of this activity is very important, since mating occurs during the night, and the copulation plug falls off after a few hours, thus observation just before noon would certainly be too late), or by microscopical examination of a vaginal swab for the presence of sperm, not be conducted under the strict prescriptions of GLP, in order to achieve a sufficient level of confidence? Not only does the start of treatment depend on the exact determination of the mating date, but also the scientific result depends completely on this. Some sensitive periods of organogenesis will last only a few hours or days, and an error in the determination of the copulation day may severely jeopardise the results of the study and the value of the conclusions to be drawn therefrom. But in this case it is not only the experimental starting date that is not coinciding with the application of the test item, but also the experimental completion date. After the last application on day 16 of the pregnancy the animals will be kept until they are either killed on day 20, when the uterine contents are examined for live and dead foetuses as well as for signs of early implantation losses (embryotoxicity), and the live foetuses are examined for external and internal malformations (teratogenicity). Other pregnant females will even be left alive and be allowed to litter, and their pups will then be examined for survival as well as for developmental landmarks, experimental activities which will continue up to weaning (28 days after littering). These activities have certainly also to be ranged under the heading "experimental", and therefore, the experimental completion date will not be described in a sufficient way (and certainly not in any way relevant to the study goal) by the date, on which the test item had last been applied.

It is certainly to be recognised that this definition, by using the terms "(when) ... *study-specific data are collected*" and "(when) ... *data are collected from the study*" instead of setting the limit by a very clear-cut activity, like the first or last application of the test item, can and will be open to discussion: Which data should be regarded as "study-specific" in order to fix the experimental start of the study? Do health checks after the arrival of animals or the weightings used for stratification and randomisation belong to this category? Does the acclimation period generate study-specific data, or does only the first application of the test item so?

The same practical questions arise with the "experimental completion date". One of the main questions, where opinions clash, has been in the context of toxicology studies, namely, whether the necropsy is the last occasion on which study-specific data are collected, or whether activities after the necropsy of the animals, i.e. the preparation of histological slides and their reading by the histopathologist, constitute "collection of data from the study".

From one possible interpretation of the Principles the latter opinion would seem to dominate, as the reading of slides also could be regarded to constitute "original observation", while from another, possibly more pragmatic point of view, the necropsy, as the end of the in-life phase of a toxicology study, could be regarded as the end of experiments (with a possible extension to the cutting, embedding, sectioning and staining operations, till the final slide preparation), since reading the slides may not be viewed as "experimenting" anymore, but could be regarded rather as "interpreting the data" (i. e. tissue sections) by the pathologist. This activity may be compared to the Study Director's task, who, on writing the final study report, is interpreting the data output from haematology, clinical chemistry and urinalysis determinations. This activity of the Study Director cannot, at this stage, be considered an experimental activity anymore.

Another point to be made is connected with activities associated to the "main" study performance. The toxicokinetic part of a chronic toxicity or carcinogenicity study may well be delayed with respect to the experimental end of the in-life phase of the main study, especially when blood samples from the last treatment day are to be analysed. Should the study plan assign the end of the analytical determinations as the experimental end date, or should still the necropsy determine this date? In analogy to the field studies, where analysis of the samples will constitute, scientifically speaking, the main part of the study, and where therefore the completion of the analyses will mark the experimental end of the study, this standpoint could also be taken in a toxicology study.

However, what is true for a carcinogenicity study may be wrong for an analytical chemistry study; what can be applied to an *in vitro* genotoxicity study could be completely out of question for a field study. Therefore, it would seem to be important to interpret these definitions flexibly and with well considered regard to the study type and the "experimental" activities connected with it. Thus, it would seem that in order to arrive at a clear situation, the Standard Operating Procedures for the conduct of the various study types, or else the study plan, might address these issues and define the dates in a concrete way.

> Studies have a beginning and an end. This self-evident fact may, however, give rise to uncertainties, if it comes to the definition of the exact dates of starting and finishing the experimental work on a study. Neither start nor end dates of the experimental activities within a study can be described in general, and at the same time accurate, terms, because of the variety of study types, where different considerations may be applicable for defining these dates in terms of actual activities. To

> define the experimental phase of a study is of high importance to GLP, however, since it is in this period, where the study relevant data are collected. The definitions leave therefore some latitude for the exact determination of the respective (calendar) dates, maintaining at the same time the necessity for fixing them in the study plan so as to leave no doubt about the beginning and the end of the interesting and important study period.

2.8 Study Plan, Amendments and Deviations

For a study to be started, there has to be a written study plan, dated and signed by the Study Director. There cannot be much discussion about the definition of the study plan, which is given in the Principles as meaning "*a document which defines the objectives and experimental design for the conduct of the study*". The study plan thus has to delineate the conduct of the whole study in an as detailed way as possible. It is a common truth, however, that nobody can tell the future with absolute confidence, not even weather forecasts are infallible. As the study proceeds, there may be instances or conditions which should make it imperative that the study plan be momentarily adapted or permanently changed. While the study plan itself will not be dealt with here, but will be discussed in more detail in section 11.1 (see page 240), the definitions for the two possibilities of introducing changes into it may be in need of some interpretation and are to be considered at this point.

It is well known, that even with the best intentions of faithful adherence to a pre-conceived plan, there are instances, where a deviation from the written procedure would not only be a necessary act, but one based on a rational decision. It depends, however, on the exact circumstances how such changes are to be treated in a GLP compliant way. The definitions in the OECD Principles distinguish two different cases, namely the Amendment and the Deviation.

While the deviation is defined as an "*unintended departure from the study plan after the study initiation date*", the amendment is an "*intended change to the study plan after the study initiation date.*" In both cases, the change occurs after the study initiation date, which is of course logical, since any change to the (draft) study plan before the initiation date could be incorporated into the final study plan and would thus no more constitute a change. But there is more to the difference between deviation and amendment than just the two words "unintended" *versus* "intended". An amendment is required to be kept with the study plan ("*Study plan ... includes any amendments*"), and it also has to follow a defined pathway of approval in an analogous way to the one of the study plan itself. A deviation, however, will not be approved, but just acknowledged, and it will furthermore only be kept in the raw data, and will not become appended to the study plan. Another difference between the two expressions may be seen in the fact that a deviation is mainly a "once-only" event. A computer breakdown for instance, which necessitates a deviation to be noted, will be fixed within a short time, and the activities for which the computer system is indispensable, can then continue as before. On the other hand, an amendment describes in many cases something of a more permanent nature. If the spraying apparatus that the Study Director had been named in the study plan had been taken out of operation a few weeks before the start of the crop treatment, then the replacement, as a planned change, will have to be indicated in an amendment, and the change will, furthermore, be permanent. The same holds for any changes to personnel or test facilities / test sites which might occur during the course of the study: The Principal Investigator who has to be named in the study plan may be changed, the histopathologist may be overburdened with work at the time when the slides are ready for evaluation and might have to be replaced, or the analytical work could be contracted to another laboratory than the one named originally in the study plan. All such changes would thus call for an amendment to be issued.

There are, however, a number of other problems which lie not only with the interpretation of the expressions "intended" and "unintended", but also with the concrete application of the tools "amendment" and "deviation" in some special situations.

An example for a clear-cut unintended change may be the following: The study plan may call for rats of a certain weight range, which has been determined and fixed in writing through experience with the rats normally delivered by the breeder. However, this time the breeder is, by chance or bad luck, unable to deliver animals in the necessary numbers within the usual weight range, and he sends therefore rats which are a little bit younger as usual and therefore also lighter. The Study Director certainly did not intend to start with animals of this weight, but he has to cope with what he could get, and thus, he acknowledges, by dated signature, this deviation from his study plan.

An as clear-cut case for an amendment would be the following situation: In a chronic toxicity study in rats, the test item is more toxic than foreseen, and the high dose animals do very badly; morbidity and mortality is increased, and the study faces an abrupt end, if nothing is done to curb the impending dying off of all the high dose animals. In this situation, and to save as much of the study as possible, the Study Director decides that the daily dose of the test item in the highest dose group should be reduced for the remaining duration of the study. While the toxicity with the accompanying morbidity and mortality is certainly an unintended event, it is as patently clear that the decision by the Study Director to decrease the daily dose is an intended change, and that therefore this new dose is to be fixed by an amendment to the study plan.

These are the more or less straightforward cases, but there can be more contentious ones. In a field test, the exact date of the application of the test item can in most cases not be predicted with sufficient accuracy, since the time point of application depends on a good number of variables that cannot be influenced by the Study Director or the Principal Investigator, such as weather conditions and growth stage of the crop to be treated. It is therefore customary for field studies not to specify an exact date for the spraying of the crop, but to define the time point in terms of crop development. There are numerous ways to do so, depending on the crop or the nature of the test item, from the "two-leaf stage" to the "tassel formation" of a cereal, or from the "swelling bud stage" to the "fully opened blossom" of an apple tree. If the stage intended for treatment is described in the study plan in such a way, and if the treatment can indeed be delivered to the crop at the intended stage, there is only the exact calendar date to be noted in the raw data, but there is no need for neither a deviation, nor an amendment. If, however, weather conditions do not permit the treatment to be applied at this exact stage at all, there will be a change to the study plan, in that only the next developmental stage of this crop could be treated. Since this is not an intended change of the study plan, it will definitely become a deviation, which, however, will have to be duly acknowledged by the dated signature of the Principal Investigator and the Study Director.

Other possibilities of more or less unintended changes to the study plan may occur, when additional interventions may become necessary, such as pesticide treatments of crops in field tests to combat some disease that is not the target of the study, or veterinary interventions in toxicity tests to treat some lacerations in a dog.

Another, sometimes really necessary, sometimes but only tempting, possibility in specifying certain issues in the study plan, especially with regard to fixing dates of intended activities, is to give them not as exact calendar dates in terms of the specific day (e.g. "September 25 and 26") foreseen for this activity, but to provide an interval, say "week 38", for the conduct of the respective activity, which is mainly done to retain some measure of flexibility. There are a number of intricacies connected with this use of intervals instead of fixed dates, which we shall not go into in any detail. Suffice it to say that it requires sharp reasoning to distinguish those instances where just a note to the raw data will suffice to satisfy the GLP Principles from those where a full-blown amendment will become necessary. In an aside, it has to be mentioned here that this manner of providing dates as time intervals in study protocols has become more and more frequent; it indeed is a pragmatic approach to combine the rigidity of GLP requirements with the flexibility needed in today's business environment. There is nothing in the GLP Principles that would forbid the use of such time intervals for the "relative" fixing of dates; neither is there any indication or answer to the question of which length of time it would be admissible to provide as such an interval. In the extreme, all calendar dates (with the probable exception of the date for sacrifice and necropsy of animals, or any other defined final study endpoint) could be given as "sometime during the course of the study", which would be absolutely contrary to the intentions of GLP, which are to guarantee an orderly planning and conduct of any non-clinical and environmental safety study.

Incidentally, there should really be a third category of change documents: The Compliance Failures. While for a deviation, being an unplanned change to the protocol, it is not uncommon and indeed even understandable, that the document specifying this deviation may be written and signed by the Study Director some time (maybe a long time!) after the event, an amendment, being a planned change to the protocol, should, by necessity, be treated like the study plan itself, i.e. the change could not lawfully be introduced until the Study Director has given his consent by dated signature under the amendment. The OECD Principles clearly recognise this difference; in the case of a deviation, it is only required that they *"should be described, explained, acknowledged and dated in a timely fashion by the Study Director and/or Principal Investigator(s) and maintained with the study raw data"*, while for the amendment, the requirement is worded in a more stringent way: *"Amendments to the study plan should be justified and approved by dated signature of the Study Director and maintained with the study plan."* Approval by dated signature means nothing else than that an amendment is to be treated exactly like the study plan itself, so that the change cannot be introduced into the conduct of the study before the Study Director has signed the amendment as part of the study plan.

This may certainly prove to be difficult at times, especially in the context of a study of relatively short duration. In such a case, one can imagine that the Study Director would possibly have had to decide on the spot about some necessary (but nevertheless planned) change. To dictate the amendment and to have it typed by the secretary, or to go back to the office and to type it himself, to date and sign it, and to distribute the necessary copies to all the persons concerned, including the Quality Assurance, would in such a case take too much time. Unless the change could be immediately implemented (and the amendment written later), the study would in all probability have to be aborted. Thus, a certain latitude in the interpretation of the resemblance of an amendment with the study plan can certainly be exercised. When, however, in longer term toxicity or field studies amendments are uncovered that are concerned with some start parameters of the study, but which are written (or at least dated and signed) well after the experimental start date, or maybe even after the experimental study completion date, then the correct term for such a document is certainly not "Amendment" anymore, because in this instance it has failed the intentions of the GLP Principles, namely that the amendment would serve to "amend", i.e. change or complement the study plan. Like the study plan, the amendment is a document that has to be available at the work place and in a 100% complete form to study personnel for the correct and GLP compliant study conduct.

In summary, while the deviation serves to list unintended alterations or to record necessary changes due to circumstances beyond the control of the Study Director, the amendment really serves to correct the study plan in instances where intended and permanent changes to the original design have to be made. The only question a Study Director would have to ask himself, if placed before the (nearly Hamletian) problem "to amend or to deviate", is thus: "Would I have written the study plan differently, had I known what I do now?"

> Good planning is the greater half of success. Only with a clear purpose in mind and with a well thought-out and delineated testing procedure is it possible to obtain an evaluable outcome of a study. Therefore, GLP places a high degree of reliance upon creating and following a predefined study plan. Unforeseen events may, however, necessitate changes in even the best prepared and best designed study. To deal with such situations, the GLP Principles have introduced the instruments of the

> Deviation and the Amendment. The distinction between these two instruments lies in their propensity for affecting study conduct which makes a different handling by the Study Director possible.
>
> While the (temporary) deviation from the study plan (or from the procedures prescribed in an SOP) needs only an acknowledgement by the Study Director, the amendment has a more fundamental function in terms of the future reconstructability of a study. In being part of the study plan it has not only to document the actual conduct of the study, but it provides also evidence that the Study Director had indeed been in control of the study at all times during its conduct, and it provides for the GLP compliant documentation of important decision points as they occur during the conduct of the study.

2.9 Raw Data

Formerly, raw data were defined as *"all original laboratory records and documentation, or verified copies thereof, which are the result of the original observations and activities in a study."* While this general definition is still valid, with the technological advances in recording and documenting it has become more and more difficult to agree on which type of records or documentation would constitute real raw data. In order to evade these questions and the problems of raw data storage and retrieval, it has been customary for data recorded on-line, e.g. for animal weights, to define the paper print-out of these data as "raw data", even though they would not be the truly "original records" of the activity. The same procedures were employed for analytical records like chromatograms, recorded in digitised form by the computerised analytical instrument, where also the paper print-out (hard-copy) of the resulting chromatogram had been customarily defined as raw data.

Reasons for adopting such procedures were manifold. First of all, there was the problem of storage in terms of stability of the medium, i.e. of the disk or tape where the information was stored in digital form. Tightly connected with this question was the problem of decipherability of the electronic data in the long run, mainly because software updates and changes would create a new "environment" in which the data might become unintelligible to the system. But not only the change of software versions but the very simple reason of hardware changes might make data unreadable. Just remember the transition from the 5½" floppy disk to the 3¼" diskette and to the CD-ROMs. No PC has nowadays a slot for 5½" floppies, and any data saved on such media not transferred to the newer ones may irretrievably be lost. Thus, any of these developments, from the software to the purely instrumental ones, may contribute to data loss through loss of decipherability. Furthermore, also the ageing of the storage media themselves, whether they be disks, tapes or other magnetic or optical storage media, with the concomitant loss of single bits of information, might corrupt the information content of these media.

Technological advances have superseded these concerns to at least a certain, if not even a large extent. Already the great drop in the price of computer memory, with harddisks even in PCs of the lower price class attaining storage capacities of several gigabytes, has allowed for the secure storage and retrieval of the complete data base of a whole test facility on the computer system itself, without the necessity of downloading data on tapes to make room for new data. In this way, software updates are no longer a big problem, and do not necessitate the retrieval of data from the tapes, their re-introduction into the system and the re-saving on the archiving tape. There are some additional points to consider, when computerised systems are used in the generation of data; these will be dealt with in a later section (see section 7.3, page 180).

There may be some special questions and issues that have to be considered when dealing with the application in certain situations of the definition of raw data. All these, however, can be easily resolved when keeping in mind the logics behind the idea of GLP and the definition of raw data as given in the GLP Principles. Thus, for instance, if data, which had been collected manually on laboratory data sheets, are transferred to a computer system, then neither the computer file nor the print-out of this file can be considered raw data. Such a situation may easily happen, if during a certain activity, which is normally computer-supported, the electronic system goes on strike, and therefore e.g. the weights of animals have to be recorded manually. In such a case, it is clear that these manual records are the original observations, even if the weights will have to be entered into the data base of the computer system at a later time in order to keep the respective file complete and up-to-date. A print-out of the respective computer entries may, however, be retained along with the "real" raw data for examination by the Quality Assurance and for the demonstration of an accurate transfer of these data into the system.

Another issue arises with such data that may not be specifically connected to any particular study. Instrument calibrations, for instance, should be performed regularly in order to guarantee the correct functioning of the respective apparatus. Animal rooms, but also other rooms, like the computer rooms or refrigerated storage rooms, have to be monitored continuously with regard to the continuous and continued adequacy of their environmental conditions. Although not directly part of any specific study, such environmental records are original observations which may have a bearing on the integrity of any study, and they have to be considered therefore as raw data.

On the other hand, examples can certainly be found of information which is not considered to be raw data. One such example may be animal cage cards: Cage cards bearing just the usual information to enable study personnel to perform their duties correctly, like animal number, study number, study dates, and cage number, are not raw data, since this information is not the result of original observations and is not necessary for study reconstruction (although it may serve as further proof of the correct cage occupancy). However, if a technician uses the cage card for recording an original observation, then all cards must be saved as raw data.

It may also happen that it becomes necessary to file the same raw data in more than one place separately. Let us consider, for instance, the case of an investigation with a number of test items being run in parallel, and thus using the same controls; there will be one set of "original observations" for the control values, which will be valid for all of the separate studies conducted in parallel. Since these studies will have to be reported and archived as separate entities, each of them should have its own record of the control values. The single record of the control data, constituting the "real" raw data, will therefore have to be copied (possibly a number of times) in order to accommodate the different studies.

In the same way are those instances to be regarded where all original observations are entered in a bound laboratory notebook which ultimately will then contain raw data from several studies. The original raw data will certainly be contained in this notebook, which consequently will have to be archived in a proper way as soon as it is full. At the time of completion of any such study the ultimate location of the notebook will not be known, and therefore, "verified copies thereof" will have to be made and archived with the study raw data. These records should, by the way, also bear the identification of the notebook itself, so that later on a comparison between the various types of "raw data" may be possible.

Finally, such copying may become necessary also in other instances, e.g., where the recording has taken place on a medium that will deteriorate rather rapidly, as is the case for certain light-sensitive paper records. However, these copies should represent truly accurate copies of the original, with no corners or edges cut off; thus they have to be verified, normally by dated signature of the person who did the copying.

The GLP Principles recognise this need for copying certain raw data in that they allow *"verified copies thereof"* to be treated as raw data. For the purpose of facilitating Quality Assurance control or the work of compliance monitoring inspectors, it would be advisable also to mention the location of the "true original" on all copies made and filed with other studies, or the designation of the notebook, in order to allow an easier comparison between copy and original.

> The "human health and environmental safety study" derives its conclusions from an evaluation and assessment of the observations made and the parameter values obtained during its conduct. The importance of these primary records is underlined with the definition of the term "raw data" and the examples given thereto. In principle only the "really original" records representing the original and unadulterated observations can be regarded as the primary data from which the reconstruction of a study will be possible. The permission to use "verified copies" of original raw data is a pragmatic approach to situations, where the same data have to be available for more than one study, or in other such instances.

2.10 The Phases of a Study

The term "phase(s) of a study" is used at various places in the OECD Principles; however, this term is not defined, neither in the Principles themselves nor elsewhere, and its use has given rise to questions about its meaning, its applicability, and indeed its value in the context of GLP. Even though a definition for this phrase is lacking, it has to be dealt with in this section on definitions, since a clarification of its meaning should be of some importance for the discussion of various other issues in the application of GLP.

2.2 Definitions

In a colloquial sense it may be well understood what the term "phases of a study" should signify, namely the clearly and logically delimited parts or portions into which a study may be subdivided. Such divisions can be performed in various ways, either by temporal associations, or by defined activities. The difficulties with this term therefore do not stem so much from any problems with subdivisions of a study, but from the (not defined, thus unexplained) dual use of the word in the text of the Principles as well as in some of the Consensus Documents.

There is for instance the possible division of a study into a preparatory phase, followed by the experimental phase; finally, the reporting phase concludes the process. The experimental phase may then be subdivided further, according to the nature of the study itself. In a field study, for instance, a logical subdivision could be the preparation of the tank mixes and the analytical work connected with this activity, the field phase of spraying, the harvesting and sampling of the material to be analysed, and finally the laboratory work involving the various analytical procedures. Also in a toxicology study, the various phases can be described as acclimation of the animals, in-life phase with test item application, recovery (if applicable), necropsy, and histopathology, while a number of additional activities such as haematology and clinical chemistry investigations might constitute sub-phases to the in-life phase of the study.

When the definitions of Test Site and Principal Investigator mention this term in the context of the *"location(s) at which a phase(s) of a study is conducted"* and the *"defined responsibility for delegated phases of the study"*, it should be clear that this with the term "delegated phase" one of the gross subdivision of a study, as exemplified above, is addressed.

Even the requirement for Quality Assurance to inspect studies and subsequently to *"prepare and sign a statement, ... which specifies ... the phase(s) of the study inspected"*, may still be interpreted with these divisions in mind. The situation with respect to the term "phase(s) of the study" becomes less clear, however, if it is read in conjunction with the respective Consensus Document (OECD No. 4, 1999). There, Quality Assurance is required to *"identify the critical phases of the study"* and consequently to conduct inspections which should cover those activities that are most critical for the assessment of GLP compliance and for the quality and integrity of data and study. Study-based inspections should thus be *"scheduled according to the chronology of a given study, usually by first identifying the critical phases of the study"*. There is a difference between the "phase" that is subject to supervision by a PI, which is to be conducted at a test site and which probably encompasses a whole and in itself closed part of the study, and the "phase" which has to be inspected by Quality Assurance, being further qualified in the Consensus Document as "critical". In this latter case the term "phase" is utilised more in the sense of a single activity that has to be closely monitored through inspections, and the "critical phase" may thus involve any activity being regarded as of special and pivotal importance by the Quality Assurance. This can, in a toxicology study, be any activity on which data reliability may hinge critically, from the dose preparation, the weighing and dosing of the animals, their daily observation, up to special activities like sampling of blood or other biological samples, and ending with the necropsy procedures. Likewise, in a field study, these "critical phases" may involve the weighing of the test item and the preparation of the spraying solutions, their application on the crop, as well as the sampling of the respective crop, and subsequently any one of the various analytical procedures.

There is a further critical point connected with the term "phase of the study", which will be dealt with later in this book (see section 4.5, page 150), but since the respective phrase of the Principles has already been cited above, this point shall be mentioned here for completeness' sake, too. The Principles require the Quality Assurance to *"prepare and sign a statement, ... which specifies ... the phase(s) of the study inspected"*. The problem with this requirement lies again in the interpretation of the term "phase of the study": Should Quality Assurance limit the enumeration of its inspectional activities to the relatively gross subdivisions, such as protocol check, in-life phase, necropsy and report audit, or should it specify all of the single activities observed during the inspection, which would, especially in the in-life "phase" of a toxicology study result in a large list of a huge number of small items and activities inspected. Here, too, the disputed phrase can be interpreted in one way or another. In this case, it is in the responsibility and in the interest of Quality Assurance to provide for a clear-cut definition in order to remove any possible ambiguities from its statements.

> For the purpose of following the Principles of Good Laboratory Practice it must be recognised that the term "phase of a study" may be applied to two different levels of study conduct. It may be connected with the gross divisions of a study which follow the time course of study development, but it may also be used to denote single activities within the

> experimental conduct of a study. They are furthermore directed towards two different areas of study conduct and study control.
>
> In delegating parts or "phases" of a study, it has to be clarified, through the terms of appointment for the Contributing Scientist or the Principal Investigator, what the exact area of responsibility is for this individual, what exactly is to be done at those test sites where such "phases" are conducted. In this way, the respective "phase" will become identified. The main GLP requirement here would seem to be a clear delimitation of one single, closed in itself, part of a study from any other study activities, in order to clearly define the limits and borders of responsibilities and to avoid the appearance of "grey zones" at the boundaries between neighbouring activities.
>
> On another level, the "critical phases" which Quality Assurance is required to inspect, are much more restricted parts of a study, down to single, but highly important activities, on which the quality of a study is "critically" dependent. Their definition in terms of specific activities will enable the Quality Assurance to cover these "critical" phases in an adequate way and to ensure that the really pertinent study activities will obtain the attention they deserve.

2.11 The Master Schedule

One of the responsibilities of management is the maintenance of a master schedule on which studies that are planned to be, or actually being, conducted at the respective test facilities are to be entered. This may be regarded foremost as an organisational, managerial tool, since it would allow management to keep control over the activities at the test facility. But the definition of this tool, namely that the "*Master schedule means a compilation of information to assist in the assessment of workload and for the tracking of studies at a test facility*", goes further than that. Indeed it is not only management to which master schedule is of importance, but this tool is also eminently valuable to Quality Assurance.

The master schedule, according to its definition, has in the first instance to give information about the workload at the test facility. For it to become a true instrument to gauge the actual workload, the master schedule has to contain information on all studies performed or planned at this test facility, and not on the GLP studies only. It would certainly be advisable to mark or distinguish GLP conforming studies from others, since these will be those most relevant for the workload of the Quality Assurance, whereas those studies that are not to be conducted under the provisions of the GLP regulations will not need to be inspected nor will their final reports need to be audited. If the master schedule is kept in this way, then it allows the test facility management to judge whether the acceptance to conduct or to plan still another study in addition to those already scheduled could indeed be considered, or whether such an additional study would have to be postponed because of a work overload at the respective test facility. Management has the obligation to provide a sufficient number of qualified personnel for the GLP compliant conduct of studies (see section 3.1, page 94). The master schedule will allow management therefore to judge the total workload in terms of concurrently conducted studies against the sufficiency of the means for carrying them out. Consequently it will further allow management to draw conclusions about the possibility to decrease, or the necessity to increase, the number of study personnel in its test facility. Furthermore, it will also allow a judgement on the availability of suitable test facilities, i.e. rooms or areas that are necessary for the proper, GLP-compliant conduct of the studies. If a test facility has only one animal room large enough to house a full carcinogenicity study, then it should certainly be considered a folly for management to enter contracts for three different two-year rat carcinogenicity studies to be started within one year. In the same manner, the master schedule will allow management sufficient time to contract early enough for an adequate number of additional fields or other agricultural areas on which to conduct field tests, if the normally used plots would turn out to be of insufficient size for the planned studies.

But let us look at the issue of the master schedule also from the other side: Its existence will allow the Compliance Monitoring Authority to judge whether the test facility management did fulfil its obligations, having provided adequate facilities and technical resourses as well as a sufficient number of personnel for the performance of the studies that have been conducted.

The other aspect addressed in the second half of the definition, the tracking of studies, may be regarded also as a managerial tool, since it allows management to judge the time-points on which information on certain safety aspects of a test item would become available, and thus to provide the sponsor with exact dates on which to expect the final information, or to set the respective deadlines for decisions on whether and how to proceed with the development of the respective test item. But this is not the only advantage of the master schedule in the tracking of studies. The information presented on it, if detailed enough, will enable the Quality Assurance personnel to better plan their activities, i.e. to time the inspections of critical study phases in a much more exact and prospective way. While it is the responsibility of management to see that a master schedule is maintained, the physical task of keeping the master schedule may be delegated to any suitable person or department in the test facility. Although the GLP Principles are requiring only that Quality Assurance should "*have access to an up-to-date copy of the master schedule*", the Quality Assurance would have some assets in store for performing the actual book-keeping: Quality Assurance is not only the best equipped defined unit in terms of expertise with all kinds of studies conducted at the respective test facility, but the placing of the master schedule into the hands and care of the Quality Assurance will simplify communication ways and kill two birds with one stone:

- Quality Assurance has anyway to be notified by Study Directors of all GLP studies planned through submit of their draft study plans; and
- Quality Assurance has anyway to report to management at regular intervals.

At the same time, it will enable Quality Assurance to better plan their respective activities in relation to the conduct of GLP studies. In laboratories, where only rarely "true" GLP studies are performed, the existence of a full and up-to-date master schedule at the Quality Assurance office will allow for the planning of inspections at such times, when non-GLP studies of a similar type are conducted, thus allowing the Quality Assurance inspector to better judge the GLP compliance of the everyday work at this test facility.

> The master schedule is an organisational instrument which allows management, as well as Quality Assurance, to fulfil their obligations towards GLP adequately and in the required way. Management has to ensure that adequate resources are allocated to every GLP study, which can only be guaranteed if the appropriate information on the availability of such resources is present and up-to-date. Quality Assurance on the other hand has to plan its activities, and to allocate its resources in such a way as to ensure the adequate coverage, through inspections and audits,

> of all studies for their GLP compliant conduct, which again is critically dependent on the exact and complete information available on the studies and their progress.

3. Responsibilities in Good Laboratory Practice

The responsibilities of the various partners within the GLP system have been defined in the Principles in order to distribute and assign the various tasks in a clear-cut way. These descriptions and clear delineations of the respective responsibilities form a very important part of the whole system of GLP. There are some mutually exclusive tasks where responsibilities have to be unequivocally fixed in order to create a real quality system. While it may be taken for granted that a Principal Investigator should know his responsibilities as well as the person who is tending the archives, and that the technician who is performing an analysis, or the field worker who is spraying an orchard should know theirs, there are also limitations to these responsibilities to be observed. Thus, neither should management interfere with the work of the Study Director, nor should the Study Director be able to influence the decisions of the Quality Assurance. It is this clear separation of tasks which should guarantee that all partners involved in a GLP study could perform their duties in an unhindered and correct way.

3.1 Management

In a way, it is certainly common sense that, on each level of a test facility, responsibilities are well defined, and this common sense practice ought to start with the head of a test facility. Under normal circumstances, an organisation chart clearly delineating the structure of the test facility would be sufficient to ensure this point. However, this chart should not only consist of a graphic representation of the test facility structure, but it may also be important that all relevant individuals, at each level of competence and responsibility, should be named on it. This entails of course the obligation that any such document (organisation chart) has to be updated as soon as a change occurs.

The responsibilities of management derive in general from the commitment, imposed by the GLP Principles, to provide an optimal environment for the GLP compliance of the test facility. An obvious endeavour of any manager will be to ensure the quality of work performed in his or her test facilities. In pursuance of this goal he or she will strive to have well educated, trained and experienced personnel. Although management will most probably have delegated the task of hiring personnel to a specialised personnel department, it is the cues from management which will be important for the hiring policy of this department. Especially in times of financial stress, the "Department of Human Resources" may become induced or tempted by the perceived stand of management in this issue to hire less well qualified and thus less expensive personnel without regard to the probability that this will tend, in the long run, to jeopardise the quality of the work performed at the test facility. As important as the qualification and expertise of the personnel is of course their sufficient number and the appropriateness of facilities, equipment and materials required for the performance of the activities at the test facility. Therefore, it remains the ultimate responsibility of management to ensure that *"a sufficient number of qualified personnel, appropriate facilities, equipment, and materials are available for the timely and proper conduct of the study"*. Although it is generally recognised and acknowledged that it is the management who is responsible in the end for any decisions taken with regard to the functioning of a test facility, this part of the GLP Principles just reiterates this fact, in order to make it unequivocally clear that in no case it will be possible for management to blame somebody else for inappropriate numbers of personnel, for insufficient laboratory space or for outdated, unsuitable equipment.

The qualification and experience of a technician or scientist is subject to change: Once useful skills may not be needed anymore, when the work or the equipment change, and new skills have to be developed and new experience has to be acquired. These changes have to be documented, and management has the obligation to ensure the maintenance of such records by appointing some person (or office) to collect this information and document it, to keep the records up-to-date and to archive the records of persons who have left the test facility. The obligation of management goes, however, further than to simply and passively maintain documentation on the skills and working experiences of the test facility personnel. management has to pursue an active policy of continuing education and training, as it has to ensure that *"personnel clearly understand the functions they are to perform and, where necessary, pro-*

vide training for these functions". Although not explicitly stated in this requirement for providing education and training opportunities for the test facility personnel, this obligation of management entails as an important part the training in the application of the Principles of Good Laboratory Practice. Management has to ascertain that any changes in GLP-relevant areas, e.g. when new or revised SOPs are issued, are clearly communicated to the personnel, and that the personnel thus will maintain, or even improve, its standard of GLP compliance.

One of the main and most important responsibilities of management with regard to study conduct is the appointment of the Study Director for every single study. While the respective phrase in the section on management responsibilities might be read as if a formal decision would be needed every single time a study is planned (*"ensure that for each study an individual with the appropriate qualifications, training, and experience is designated by the management as the Study Director before the study is initiated"*), this responsibility can in practice be fulfilled by the approval to the study plan. Since the study plan has to be approved in the first instance by the Study Director through dated signature, approval by the test facility management will entail the appointment of the Study Director. It will also clarify the situation if those individuals who may be able to act as Study Directors, were to be simply identified on the organisation chart. All this holds of course in an analogous way for the appointment of Principal Investigators.

However, the responsibility of management does not end with the appointment of the Study Director; it has also to make allowance for the possibility that, once appointed, a Study Director could become temporarily or permanently unavailable for the continuous or continuing supervision of the study. In such cases, the replacement of the Study Director, whether temporary or permanent, will have to be considered by the test facility management, taking into account the prospective duration of absence. If the absence of a Study Director could be foreseen to last for a longer period of time (up to a permanent absence), the need for a permanent replacement would certainly become obvious. This might be of importance only in longer-term studies, however, since in the case of a study of only short duration, the continuing presence of the designated Study Director can certainly be assumed. Otherwise it might be regarded as throwing an unfavourable light on the planning practices of the test facility. While the circumstances under which a Study Director would be replaced are not defined in the GLP Principles, there are indeed only two possibilities in which the replacement of the Study Director would have to be considered.

When a Study Director is temporarily absent because of holidays, a scientific meeting, illness or an accident, his or her replacement will depend on the foreseen duration of absence. An absence of short duration might not necessitate the formal replacement of the Study Director, if it is possible to communicate with him or her if problems or emergencies arise. It would certainly be a good idea in such a case to at least also nominate a deputy, i.e. to delegate temporarily some or all responsibilities to competent staff. If the study is planned so that critical study phases would fall into the period of absence, these activities may then either be moved, if feasible, to a more suitable time (with study plan amendment, if necessary), or a temporary replacement of the Study Director should then be considered. Should the unavailability of the Study Director be of longer duration, the solution with a replacement should be preferred rather than just to delegate the tasks to competent staff.

The second circumstance, where a replacement must not only be considered, but is an absolute requirement, needs no discussion: In the event of termination of employment of a Study Director, the need for replacing this key person is obvious.

The responsibility of the management for the replacement procedures consists again in drawing up the policies and principles to be followed in such a case, since the GLP Principles require that "*replacement of a Study Director should be done according to established procedures*". Management will of course be responsible for the final decision for replacement or temporary delegation of the Study Director's tasks, and this decision, as well as the reasons for it, has to be fully documented in writing.

The very same responsibilities are applicable for the case, where the management has to appoint, in addition to the Study Director, also one (or more) Principal Investigator(s). Of course, here is a distinction to be made: The management which appoints the Study Director can only appoint Principal Investigators in its own test facilities or test sites. When out-sourcing study parts to an independent CRO, the management of this CRO becomes then responsible for the nomination of the respective people. The management of the test facility where the Study Director is located is then, however, responsible for ascertaining that there are clear lines of communication between the Study Director and the Principal Investigator(s) on the one hand, and between the Study Director and the respective Quality Assurances on the other. This may not be a very easy task, since it may involve also the necessity of communication between the different managements at the various levels and sites of study conduct.

In connection with the appointment (and replacement, if necessary) of the Study Director, management has some other responsibilities as well. It has to ensure that the study plan is approved through dated signature by the Study Director, and it has to ensure that the Study Director has made the approved study plan available to the Quality Assurance personnel. In the former case, the test facility management might require that every study plan also be signed by a (nominated) member of the management itself. This possibility can even be a national requirement, and the GLP Principles have formally recognised it in section 8.1.1 on the study plan, where this possibility is expressly mentioned. Through the latter requirement, on the other hand, management will be enabled to assure that the Quality Assurance can properly fulfil its role in controlling the conduct of the study. It would certainly not be expected that any continuous, supervisory activities from the part of the management should be needed; all that is necessary is that management should issue a clear directive with regard to this area. In practice, this responsibility may be exercised in a number of ways, of which the most direct one would be to have a management representative receive all approved study plans; a paragraph to that effect in the SOP on study plans, their generation, approval and distribution, would serve this purpose. In this way, management can easily control whether the study plan has been distributed prior to the experimental start of the study, and whether the distribution list covers all the necessary personnel, including Quality Assurance personnel.

It might be added here, that it is also management who has to ensure that an individual is identified as responsible for the archive. This is an organisational matter and as such lies without any question in the realm of management; the respective paragraph just serves to remind management that one of the major elements of GLP is the archiving of all pertinent documentation and other study-related items, which are necessary for the full reconstructability of any study, and that therefore the person responsible for the archives holds another pivotal position in the GLP system.

Another very important responsibility of management is to "*ensure that there is a Quality Assurance Programme with designated personnel*" and furthermore to "*assure that the Quality Assurance responsibility is being performed in accordance with these Principles of Good Laboratory Practice*". As will be seen later on in the section on Quality Assurance (see section 4, page 121), this part of the GLP system has to act independently from study conduct and has to report all findings to management. Thus it is customary that Quality Assurance is organisationally placed directly under the wings of management. Although management has to assure that Quality Assurance can, and does, perform its tasks according to the Principles of GLP, management's responsibilities with regard to Quality Assurance activities do not end there: Management's primary responsibility is to ascertain that *"these Principles of Good Laboratory Practice are complied with, in its test facility"*. Therefore, any reports by Quality Assurance on deviations from the GLP Principles, or deficiencies in observing them, have to be followed up by management, which has to ensure that the necessary corrective measures are taken and are fully implemented by the respective study or test facility personnel. While this requirement that management should act on such reports from the Quality Assurance is not expressly mentioned in the OECD Principles, it can be found in other regulations, e.g. in EPA's GLP Guidelines as follows: *"Assure that any deviations from these regulations reported by the Quality Assurance unit are communicated to the Study Director and corrective actions are taken and documented."* It will also have to assume an arbitration role in cases, where the opinions of Quality Assurance and study personnel with regard to the correct interpretation of a GLP requirement might clash. As has been described at the beginning of this section, it is not only the concrete action taken by management which will determine the extent and depth of GLP compliance, but also the cues relating to the inner conviction of management with regard to the value ascribed to GLP will play an important role. Even a nominal support of a Quality Assurance complaint through a management memo will lack its convincing power, if it is not followed by a continued show of interest over the progress of the corrective measures.

There are a number of other managerial responsibilities mentioned in the respective section of the GLP Principles, the concrete execution of which, in practice, will be delegated to some specialised functions or personnel in the test facility. Thus, the maintenance of the required historical file of all Standard Operating Procedures and the maintenance of a master schedule may be given to a secretarial function within test facility management, but it may also be delegated to become the responsibility of the Quality Assurance. The appropriate characterisation of test and reference items will of course be in the responsibility of the analytical chemists; while management should have to watch over the implementation of this requirement, the responsibility for ascertaining that test and reference items are indeed properly characterised to make them fit for use in a GLP study is in practice mainly delegated to the Study Director. The same holds also for the task of ensuring that test facility supplies meet specifications and requirements appropriate to their use in a study.

A function of management is also to ensure that Standard Operating Procedures are established, to ensure that they are being followed, as well as to approve any new or revised SOPs. The role of management in the process of the establishment of SOPs may be regarded as a difficult exercise. The Principles call for the establishment of *"appropriate and technically valid"* SOPs, and it might be questionable, whether management could have the technical knowledge to ascertain this property of an SOP. However, this point in the enumeration of responsibilities again can be seen as one, where management will just delegate the technical responsibilities for the proper, scientific and technical content of an SOP to the respective specialist, while retaining the (managerial) responsibility to declare, with its approval, any such document as a standard procedure to be used and followed by the whole test facility.

These supervisory functions of management are extended by two further responsibilities, in that management has to *"ensure that the Study Director has made the approved study plan available to the Quality Assurance personnel,"* and that it has also to *"ensure for a multi-site study that clear lines of communication exist between the Study Director, Principal Investigator(s), the Quality Assurance Programme(s) and study personnel"*. The former responsibility has already been dealt with in connection with the relations of management with the Study Director. The setting-up and maintaining clear lines of communication between all the different individuals involved in a multi-site study is a task that needs the involvement of upper hierarchical levels, especially when the different test sites do not belong to the same organisational unit, or form part of another company. Management has to have first of all a formulated, documented policy for such cases; it would then have the direct responsibility to discuss, negotiate and fix these lines of communication with the other managements involved, and finally to bring the results of these discussions to the attention of the personnel concerned, e.g. by distributing a memorandum explaining the required flow of information and detailing the communication pathways to be followed. While this management responsibility concerns the "official" lines of communication, through which the GLP compliant flow of information would be expected to run, in all probability there would, and there indeed even should, exist "unofficial" lines of communication between the various individuals in a multi-site study. The existence of the "official" line of communication between a Study Director and the various Principal Investigators would by no means relieve the two parties from their direct and personal responsibilities to procure or to provide information in any other suitable way.

A responsibility which has gained more and more importance in the past few years, and which therefore has rightly become a special responsibility for management is the task to *"establish procedures to ensure that computerised systems are suitable for their intended purpose, and are validated, operated and maintained in accordance with these Principles of Good Laboratory Practice"*. While apparatus generally are required to be *"of appropriate design and adequate capacity"*, which would include the requirement that they be also suitable for their intended purpose, it is specifically for computerised systems that management is held directly responsible for their design, validation and operation. The allocation of responsibility to the management level can certainly be regarded as the obvious way for large (network) systems connecting and serving a whole company, or at least a whole test facility. In these cases, it is anyway - especially with regard to the costs involved - the decision of the management whether or not to go into computerisation and buy such a system, or to replace the existing one by a more recent version. The responsibility of management extends further than that, however. The heart of the matter and the most crucial point in GLP compliance of any computerised system, starting from the simple electronic balance or the computer-controlled HPLC system up to the most complex information technology network or the test facility's LIMS (Laboratory Information Management System) is the question of the extent and depth of the validation of such systems. Owing to this wide diversity of instruments, apparatus and systems utilising electronic controlling and data processing, it is most important that there should be a general policy and general guidelines, establishing common procedures for dealing with all the problems in connection with any computerised system, irrespective of how small or big this system would be. Therefore, while of course the technical responsibility for writing the respective (technical) SOPs lies with the computer specialists, it is at the level of management, where these policy decisions have to be made, and where thus the ultimate responsibility for the establishment of the necessary general procedures has to be situated.

> Managing a company entails a number of responsibilities, from determining the business area(s) in which the company will be active, to procuring the necessary financial means for running it and nowadays for buying up other enterprises. To ensure the right implementation of GLP it is highly important that management should be aware of the responsibilities it has to shoulder in this respect. In enumerating these responsibilities, the GLP Principles do address the crucial points which a

> test facility management has to adhere to and where it cannot shun an active role. The listing of those issues in which test facility management has to assume responsibility serves also to address the boundaries between those of management on the one and those of the other players in GLP on the other hand, thus ensuring a clear distribution of tasks.

3.2 Study Director and Principal Investigator

The responsibilities of the Study Director stem from the axiomatic approach of the GLP Principles that the Study Director represents *"the single point of study control"* and that he has *"the responsibility for the overall conduct of the study and for its final report"*. This concentration of the study control in the hands of one single person originates from the experience, which, in more than only this specific area of human activities, has demonstrated that, unless responsibility is assigned to one single person, there is a potential for conflicting views and instructions, as it is illustrated in the famous military dictum "order - counterorder - disorder". In the case of a safety study, such conflicts might result in various deficiencies, e.g. in poor implementation of the study plan. Therefore, the GLP Principles have firmly maintained that there can be only one Study Director for one study at any one time, and irrespective of study phases or parts separated by location or types of work performed. Although in present times there is the widespread custom of occupying any positions in job-sharing, this is absolutely no possibility in GLP as far as the position of the Study Director is concerned. There is certainly the possibility of delegating some of the tasks of a Study Director to other responsible persons, as will be detailed further down, but the ultimate responsibility of the Study Director as the single, central point of control cannot be delegated. In this regard, the powerful position of the Study Director serves to assure that the scientific, administrative and regulatory aspects of the study are controlled to the fullest extent possible.

First of all, the Study Director has the scientific responsibility for study plan design and approval. It is his dated signature on the study plan which approves the study and sets it in motion. Since the study plan becomes the official working document for a study, it is important that this document should clearly define the objectives and the whole conduct of the study, which calls for the scientific input and judgement of the Study Director, while not forgetting to address all the practical points of study conduct. In the special case of a multi-site study, this involves also the requirement that the study plan should *"identify and define the role of any Principal investigator(s) and any test facilities and test sites involved in the conduct of the study"*. For the Study Director, this may require a good deal of additional planning and interaction with the respective managements of the test facility as well as of the proposed test sites in order to become able to exercise this responsibility.

At the end of the study stands the final report. The dated signature of the Study Director under this document indicates *"acceptance of responsibility for the validity of the data"* and furthermore also indicates *"the extent to which the study complies with the(se) Principles of Good Laboratory Practice"*. In order to be able to accept the validity of the data, the Study Director, as the lead scientist, has not only to co-ordinate the activities of other study scientists, he has to keep himself informed of their findings during the study. The respective individual reports received from other study scientists have then to be evaluated for inclusion in the final study report. Only if the Study Director has assumed the required pivotal role in the conduct of the study he will be able to sign with confidence the study report. This confidence in the quality and integrity of the data reported should finally be reflected in a positively worded "Statement of GLP Compliance": It is not unusual to see in such statements "disclaimer-like" wording such as "To the best of my knowledge and belief...", wording which, however, should be avoided. There is nothing in the GLP Principles that may be interpreted as relieving the Study Director from this ultimate responsibility for the overall integrity and quality of the data gathered in the study and reported in the final study report. Instead of distancing himself in this general way from his responsibility the Study Director should, in the final study report, address all circumstances which might have affected the study and the quality of the data (for some examples see figures 30 and 31, pages 252/253). In this way, the assessor at the Regulatory Authority will be in a much better situation for judging data quality and study reliability, than if the Study Director just asserts these properties "to the best of his belief". Such "disclaimers", though, may have a legal background, in that the Study Director might fear to be held legally liable with respect to the accuracy, completeness and compliance of the data reported, if he would unconditionally sign such a compliance statement. However, although the Study Director, with his signature, assumes responsibility for the performance of the study in compliance with GLP Principles and for the accurate representation of the raw data in the final report, his legal liability is established by national legislation and legal processes, and not by the OECD Principles of GLP. In this sense, from a GLP point of view, there is no reason for shying back from the assumption of full responsibility for data integrity and quality, and for the accurate representation of data in the final report, if indeed the Study Director has interpreted and exercised this role as intended by the GLP Principles.

It might be added here, that these responsibilities will certainly be transferred to their full extent to an eventual replacement Study Director. Since a permanent replacement will entail the privilege – or the burden, whichever describes the situation better – of accepting, with the signature under the statement of GLP compliance of the final report, the full responsibility for the quality, integrity and reliability of the data and the report, it will be one of the first activities of this person to assure him- or herself, as soon as practicable after taking over the new position, and preferably with the assistance of Quality Assurance personnel, of the GLP compliance in the study as conducted to date. If this were to be done by means of an interim review or data audit, the results of such a GLP review should be fully documented, especially in such cases where deficiencies or deviations have been found.

Besides these scientific aspects of the Study Director's responsibilities, there are a number of more managerial responsibilities, which are, however, as important for the proper conduct of a study as are the scientific principles involved. Already before any work on the study is undertaken, the Study Director should ascertain that management have committed adequate resources to perform the study, and that adequate test materials and test systems are available. In other words, a Study Director should never take on the assignment for the conduct of a study, for which there are no adequate resources. It would therefore certainly lie in the Study Director's responsibility to alert the test facility management about any such deficiencies and to insist on their remediation, as it would, *vice versa*, become the responsibility of test facility management to react to such an alert with proper measures.

To conduct a GLP compliant study means that there are requirements for documentation and recording, activities for which the personnel has to expend time and efforts, and any inadequacy in resource allocation would therefore negatively affect the proper and GLP compliant conduct of a study. In the same way the supplying with copies of the study plan of all key personnel which are involved in the study, including Quality Assurance staff, has to be considered a necessary prerequisite of a duly conducted study. With regard to the Quality Assurance, as well as to any Principal Investigator involved in study conduct, it is of great importance that clear lines of communication not only are instituted, but are effectively utilised. Thus, it is the responsibility of the Study Director to inform Quality Assurance in a "timely manner" about any changes in the study plan and in the timing of activities connected with the study conduct, as well as to supply Quality Assurance with any amendments to the study plan. The same holds for the interaction between Principal Investigator and Study Director, where the Study Director has to be kept informed about any changes – whether foreseen or unexpected – that may happen at the Principal Investigator's test site.

Within this area of responsibility the Study Director has to ensure that the experimental procedures laid down in the study plan are followed, that the study activities are performed at the proper times and that all observations, test item applications and samplings are conducted in an orderly and study plan compliant fashion. This includes the constant overviewing of the study procedures and data to ensure that there is compliance with the relevant Standard Operating Procedures and the study plan procedures. As all decisions that may affect the integrity of the study have ultimately to be approved by the Study Director, it is important that he remains aware of the progress of the study, which is another important cause for maintaining effective communication with all the scientific, technical and administrative personnel involved. Again, the proper establishment and functioning of the lines of communication should ensure that, e.g., changes in the study plan can be rapidly transmitted, and that any issues arising in the course of study conduct are fully documented. As this responsibility for the overall conduct of a study according to the GLP Principles, theoretically, can only be fulfilled if the Study Director is present all the time during the whole study, the problem of regulating absences of the Study Director from the study becomes also an obvious one. It is certainly not always feasible in practice that the Study Director could be physically present at all study-related activities during the whole time of study conduct. It will be unavoidable that there will be periods of absence from the study. It is clear that, on returning from such short-term absences, the Study Director must inform himself about the progress of the study. Furthermore, he has to ascertain as soon as practicable whether or not deviations from GLP Principles have occurred, which then should be documented and acknowledged in a timely manner.

In connection with the actual, experimental conduct of the study, it is the responsibility of the Study Director to "*ensure that all raw data generated are fully documented and recorded*" in compliance with the GLP Principles. For data recorded manually this entails ensuring that the data have been recorded "*promptly and accurately and in compliance with these Principles of Good Laboratory Practice*"; if data are recorded electronically through the utilisation of a computerised system, the Study Director should "*ensure that (the) com-*

puterised systems used in the study have been validated", and are fit for use in the study.

Compliance with regulations is also the responsibility of the Study Director. In this role the Study Director is responsible for ensuring that the study is carried out in accordance with the Principles of GLP, which require the Study Director's signature on the final study report to confirm compliance with the GLP Principles. And, finally, after the end of the study, the study Director is still responsible for ensuring "*that after completion (including termination) of the study, the study plan, the final report, raw data and supporting material are archived*".

It has already been mentioned that, although the overall and ultimate responsibility of the Study Director for the study cannot be delegated or in any way be split among different persons, there may be instances, where the Study Director could delegate some of his responsibilities to (a) person(s) who may be in a better position to immediately supervise some part(s) of a study. This is certainly true for any tasks for which the Study Director lacks the necessary education, experience and training. A toxicologist as Study Director may not be able to follow and supervise the procedures in an analytical laboratory, where by means of an HPLC-MS method the concentrations of the test item in a biological matrix are determined. An analytical chemist may be at a loss when he should have to overview the spraying of a field or the sampling of soil or crops, and a histopathologist might not be able to judge the correct conduct of the assessment of the results from a sensibilisation assay by Draize grading. An agricultural scientist, entomologist, plant physiologist or mycologist may lack the technical expertise to perform the kinetic calculations on the analytical residue data for defining the degradation curve of a pesticide, while being the experts to determine its biological effects. Thus, in any study, besides the Study Director, there will be other scientists involved who could be held responsible for certain parts of the study. The Study Director, as already mentioned above, would have to maintain a close communication with these scientists in order to be able to overview the general proceeding of the study and to be kept informed about any findings of these specialists. If these speciality scientists were situated at the test facility of the Study Director, the maintenance of communication would not be too difficult. However, in the case where the respective activities were to take place at some geographically remote test facilities or test sites, or in test facilities or test sites not belonging to the same organisation and thus not amenable to his/her direct control, a practicable solution to the problem of the required "immediate control" over the study becomes urgently needed.

As already described in the section on Definitions (see section 2.3, page 57), the GLP Principles have addressed this problem and attempted to solve it with the creation of the function of the Principal Investigator, a kind of "Secondary Level Study Director", responsible for a well defined, restricted part of the study. Thus, the Principal Investigator *"will ensure that the delegated phases of the study are conducted in accordance with the applicable Principles of Good Laboratory Practice"*. The Principal Investigator will receive the assignment for the study part or phase, which he/she will be expected to perform, from the Study Director, who delegates the respective study part or phase to the Principal Investigator. The role of the Principal Investigator at a test site is thus to direct the work on the delegated phase of the study and to ensure that this phase is conducted in compliance with GLP Principles, and in accordance with the study plan and with all relevant Standard Operation Procedures (SOPs). The Principal Investigator, in these respects, bears the responsibility of a Study Director and should ensure that all raw data generated are fully documented and recorded, and that all raw data, records and specimens are adequately maintained to assure their integrity. Furthermore, he has to ascertain that they are transferred in a timely manner to the Study Director, or to any other person or location, as directed in the study plan. The Principal Investigator is not, however, allowed to issue and sign any amendments to the study plan; any circumstances necessitating an amendment would have to be reported to the Study Director who then would decide about the necessity of any action and would issue and sign the respective amendment. At the completion of the study part(s) entrusted to him, the Principal Investigator may write, sign and date a report of the delegated phase(s) of the study. Irrespective, however, of whether he provides such a full report, or just delivers the raw data collected, and results obtained, to the Study Director, the Principal Investigator should write and sign a statement indicating acceptance of responsibility for the validity of the data and for the extent of compliance with GLP.

There is some debate as to whether the concept of the Principal Investigator is really necessary or even desirable in areas of safety testing outside of the area of field testing of pesticides. While for such field studies (see figure 3 on page 59) this concept certainly brought about a very welcome alleviation of the Study Director's burden, by distributing some of the responsibilities onto other, better situated, shoulders, the need for this concept may not be obvious in the same way for the conduct of "classical" toxicology studies: There, the complete study is conducted at the test facility where the Study Director is physically located, and so he would normally be able to fully control the whole process (see figure 4). Even when, in order to speed up the development of regulatory studies, these toxicology studies could be subdivided into various

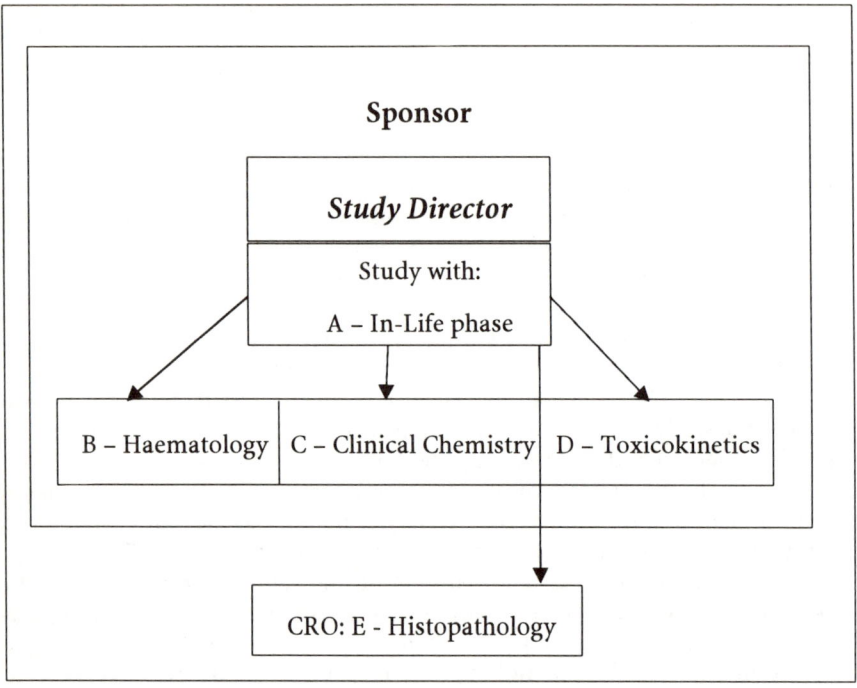

Figure 4: Schematic representation of the organisation of a toxicology study, where the Study Director may be assumed to have immediate control over most parts of the study; since the histopathology part is conducted at a CRO, the sponsor Quality Assurance may also have some control over this part (Figures 4 and 5 adapted from Beernaert et al., 2000).

phases (e.g. in-life, toxicokinetics, haematology and clinical chemistry, histopathology, etc.) which then were to be conducted at different places, and which thus may be seen to employ something like a multi-site approach, the involvement of collaborating investigators would normally cause no insurmountable problems to the Study Director. When the Quality Assurance unit responsible for controlling the work of the Study Director would also

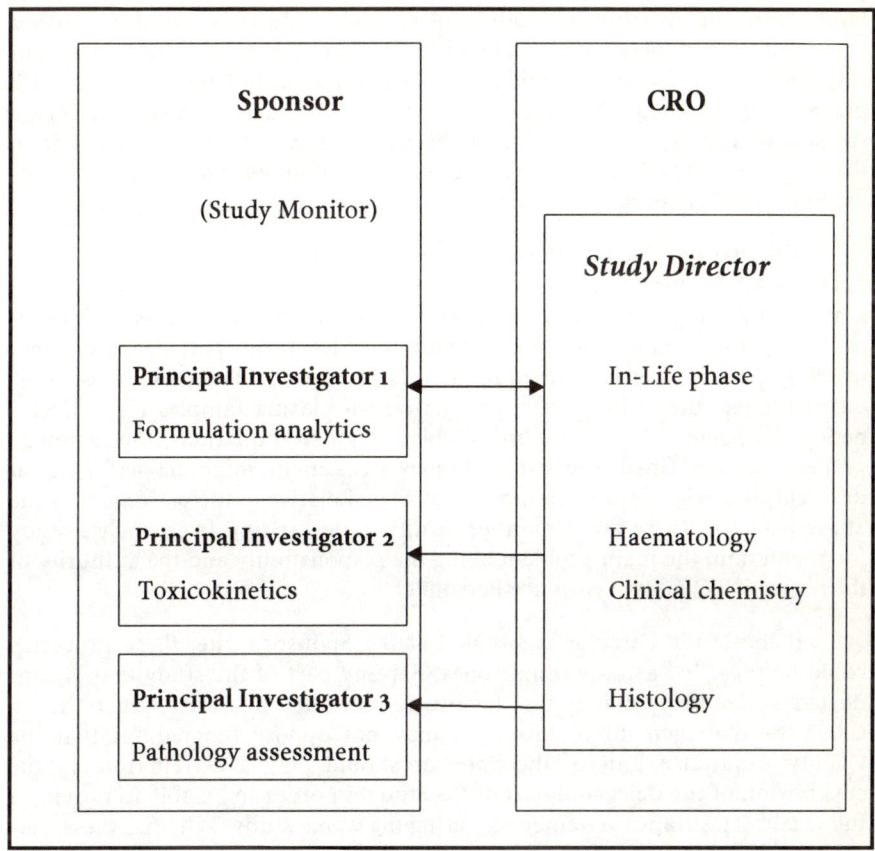

Figure 5: Schematic representation of a contracted study, where the sponsor conducts certain parts of the study in his own test facilities; for the Quality Assurance of the Study Director, it may not be possible to perform its functions with regard to these parts.

be able to conduct the appropriate inspections at these other test sites, it could normally be assumed that the Study Director could exert full control over all parts of the study. In such a case, the appointment of a Principal Investigator may certainly not be considered necessary. On the contrary, it might be seen as a nuisance: Such an appointment would create an additional hierarchical level within the structure of study control, and might thus introduce further communication problems into it. On the other hand, there are certainly instances, where this possibility should not be dismissed out of hand, where it could prove to be advantageous even for the conduct of such "simple" toxicology studies. Where there are practical problems of study control and supervision, they will need attention in order to judge the possibilities and pitfalls in the use of either "Participating" or "Senior Scientists" or "Principal Investigators".

The main problem with regard to the GLP compliant planning and conduct of a safety study lies in the possibility of its "fragmentation". Any not too limited study might be split into various fragments or study parts. A toxicology study, for instance, may be broken into at least four parts, each of them needing special expertise and equipment (see figure 4). Therefore, these various parts, e.g. the in-life phase, the analysis of plasma samples for toxicokinetics, the haematology and clinical chemistry determinations, the histology preparations and finally the histopathology assessment, might be performed at different, specialised test facilities or test sites. Relative to the placement of the Study Director, there are a number of difficulties arising from such a study fragmentation, the main problem being the responsibility and the authority of the respective Quality Assurance Personnel.

If the Study Director is situated at the Sponsor's site, these problems could be regarded as only minor ones. For any part of the study that is conducted at an independent test facility (a Contract Research Organisation, CRO) the management of this CRO may not oppose the request that the Quality Assurance unit of the Sponsor should play a certain role in the supervision of the delegated part of the study, in order to be able to provide a full Quality Assurance statement covering the whole study. Whether these outsourced activities would then be supervised by a Principal Investigator or simply by a participating investigator as "Principal Scientist", the matter of the study control could thus probably be resolved without further difficulties. Only in the case where, for confidentiality reasons, the CRO management would be reluctant to let the sponsor's Quality Assurance perform the necessary inspections directly, the ensuing problems would have to be resolved in a GLP-compliant way.

On the other hand, the situation may become more difficult when a whole study is contracted-out to a CRO, with some parts or phases of it to be subcontracted to another CRO or even back-subcontracted to the sponsor.

Obviously, in this case the Study Director has to be located at the "primary" test facility, i.e. at the CRO, while a Study Monitor (a term or position which is not defined in GLP and which has no relevant place there) will exercise some additional control on behalf of the Sponsor. If, subsequently, some phases other than the in-life part of this study, e.g., analytical chemistry, toxicokinetics or histopathology, are back-sub-contracted to Test Sites at the Sponsor's facilities (see figure 5), the problem of both the Study Director's and the CRO's Quality Assurance unit's supervisory possibilities over the participating investigators at the Sponsor's facilities will arise. The Sponsor may, e.g., not be willing to accept being inspected by the Quality Assurance unit of the CRO. The Study Director could then certainly not assume the full responsibility for these parts which he had not been able to supervise, and for which his Quality Assurance unit had not had the opportunity to perform the necessary inspections. Actually, such a study part would, at least theoretically, have to be formally excluded from the Quality Assurance statement issued by the Study Director's Quality Assurance. In such an instance, the situation might possibly be improved through the appointment of a Principal Investigator. By signing and dating his report, accompanied by the respective GLP statement, and by including a Quality Assurance statement of his Quality Assurance unit, the Principal Investigator would assume the responsibility for the GLP compliant conduct of the delegated phase or part of the study. The Study Director could then rely on these statements and in turn assume the responsibility for the whole study without any problems. If, in such a case there had only a "Participating Scientist" been nominated, without any clearly defined responsibility in terms of GLP compliance (though certainly with responsibility as to the scientific side of study conduct), problems with the GLP status of such a study would certainly arise. Still, the Study Director at the CRO might be well advised to take out an "insurance policy" in the form of a written and signed declaration on the part of the Principal Investigator regarding the adherence to the GLP Principles (see such an example of a Principal Investigator sheet in figure 6).

There are further questions involved in these instances of splitting studies into various, practically independent parts. The question of the various problems of Quality Assurance involvement with regard to the Quality Assurance statement in such "fragmented" studies will be again investigated in more detail in section 4.5 (see page 150). A further problem, which cannot be resolved in the context of GLP, is the question of confidentiality and of various shades of competitor relationships. These have to be dealt with in the context of the contractual provisions and need not be considered further in the context of this book.

CompliantLab, Inc.

Principal Investigator Data Sheet

Study title	
Project no.	
Name and address of Principal Investigator (PI)	
Facility of PI	
Objective of subcontracted study part	
Key dates of experimental phase	Start: End:
Methods	
Subcontracted study part under GLP?	GLP Compliance Statement for PI facility available? yes no Are SOPs available? yes no Has the PI facility a QA Unit? yes no
Archiving	Raw Data of study part archived at the PI facility? yes no Raw Data of study part archived at the Study Directors facility yes no
Reporting	Will the PI report include a Statement of Compliance with GLP signed by the PI? yes no Will the report be audited by the PI QA and will a QA Statement be included? yes no

....

Place and Date   Signature of PI

Figure 6: Possible "insurance policy" for a Study Director, with the PI having to acknowledge the points of relevance to GLP compliance (Courtesy Dr. G. Menne, RCC, Itingen, Switzerland)

In summary, the GLP Principles are maintaining that the Study Director is the pivotal point of study control. Nobody else than the Study Director can be charged with the ultimate control of the whole study; there is no way for a Study Director to entrench him- or herself behind explanations like "This other guy did it, and since I have no special or expert knowledge in this field, I had to assume this to be correct". This statement may apply to the science behind the study, but the Study Director is the one individual who has to assume full responsibility of the GLP compliant conduct of the entire study. All information has to be passed to him, all decisions have to be made or at least to be acknowledged by him. Only in special circumstances, where the Study Director cannot exercise his immediate control, the responsibilities of a Study Director may be extended to other individuals. Such individuals may be specialised scientists, whose knowledge of the special matter may make it advisable to let them assume the role of the supervisor of this specific part of the study. Such individuals may be temporary replacements, when the Study Director is absent for a shorter or longer period of time. There may be circumstances, however, where the appointment of a Principal Investigator may be advisable or necessary to really accomplish the goals of GLP. However, even with this appointment of a Principal Investigator, who is acting on behalf of the Study Director, the ultimate study control will remain in the hands of the Study Director. The Study Director will, on the other hand, necessarily have to rely, for the GLP compliant conduct of a delegated phase of the study, on this individual as a secondary point of control. Only in this way can it be ensured with high probability that the whole study has indeed followed the rules of GLP and has been recorded and reported faithfully, that the quality of the study is such that its data and results are reliable, and that its conclusions reflect the data obtained in a truthful way.

> A study can only be fully controlled in all aspects, if there is one single point of study control. In GLP, this position is assumed by the Study Director, who has to bear the full responsibility for a GLP compliant conduct of the study and for the quality and integrity of the data reported. The Study Director has to be aware of all circumstances, facts and occurrences that might affect the quality and integrity of a study. Thus, the Study Director has to ascertain that clear lines of communication exist between himself and all scientists (and other personnel) involved in study conduct, in order to be kept at the forefront of de-

> velopments in a study, and to be able to act, as deemed proper, on unforeseen developments.
>
> In those instances, where the Study Director is physically unable to exert this immediate control over the actual conduct of a study, the concept of the "Principal Investigator" can be put to use. The Principal Investigator, as a "Secondary Level Study Director" bears the same responsibilities as the Study Director himself for the experimental, GLP compliant conduct, although only for a defined part, of a study. The Principal Investigator is not responsible for the study plan, nor can he approve any amendments to it. The Principal Investigator has, however, to assume responsibility for the defined, delegated part of the study by signing his own compliance statement.
>
> The ultimate responsibility of the Study Director for the overall quality and integrity of the study cannot be shared with any other individual involved in the study. The GLP Compliance Statement signed by the Study Director in the final study report is the declaration that gives the Regulatory Authority the reassurance for a properly conducted, valid study, the results and conclusions of which can be trusted to reflect the real data obtained in the study.

3.3 Study Personnel

The responsibilities of study personnel, as defined in the GLP Principles, may be regarded to some extent as truisms. It seems to be clear that personnel involved in a GLP study should be *"knowledgeable in those parts of the Principles of Good Laboratory Practice which are applicable to their involvement in the study"*; it is also self-evident that it is *"their responsibility to comply with the instructions given in (the study plan and appropriate Standard Operating Procedures)"* to which they have to have access. As they are required to comply faithfully and exactly with these instructions in the conduct of the study, it follows logically that it will also become their responsibility to document any deviation from these instructions and to communicate such deviations directly to the responsible Study Director (or Principal Investigator, if appropriate). It has to be stated clearly here that this responsibility is absolutely confined to the documentation of deviations but does not entail any evaluation of the deviation in terms of its scientific significance or its relevance to the integrity of the study. This evaluation task lies solely in the responsibility of the Study Director him- or herself. This means that any deviation irrespective of its apparent importance or unimportance has to be documented and communicated to the Study Director.

While it is the Study Director who is ultimately responsible for the quality of the study as a whole, he / she cannot be held responsible for the quality of each and every data point and each and every single record. This is clearly the domain of responsibility of the study personnel, who are "*are responsible for recording raw data promptly and accurately and in compliance with these Principles of Good Laboratory Practice, and are responsible for the quality of their data*". Certainly, these raw data will be scrutinised by the Study Director, who will use them for the preparation of the final study report; certainly these raw data will be checked by Quality Assurance for their compliance with the provisions of the GLP Principles. However, data quality cannot be obtained retrospectively by control measures, data have to be recorded in such a way that quality is an intrinsic characteristic of these records. Thus, this responsibility of study personnel for the quality of their data cannot be emphasised strongly enough.

A last, only seemingly minor aspect of study personnel responsibilities is concerned with the influence of their health on the study. The integrity of a study should not be jeopardised by any external disturbance, and the health of the person(s) working with the test system might be one factor which could affect a study. On the other hand, working with a certain test system might lead to conditions which could possibly endanger the health of the personnel. Thus, the GLP Principles mandate that "*study personnel should exercise health precautions to minimise risk to themselves and to ensure the integrity of the study*". This requirement addresses not only the possibility that a diseased person could bring in some viral infections to the test system, or that an infection of the test system could spread to the personnel involved with it, it also applies as much to work in field tests, where the "normal" precautions against bad weather or the protection against contamination with spray from the application of the test item are considered. Although the precautionary principle is addressed in the first place, it remains as a possibility that conditions could develop under which an undue influence on the study could not be excluded. Therefore, study personnel is furthermore responsible for communicating "*to the appropriate person any relevant known health or medical condition in order that they can be excluded from operations that may affect the study.*"

> All the important positions of management, Study Director and Quality Assurance notwithstanding, it is the test facility personnel who are actively working with the test systems, who are making the observations, who are recording data, and who are finally providing the Study Director with the raw material for writing the study report. In consequence, the personnel has an important responsibility for the GLP compliant conduct of the study which is not to be underestimated. The influence of the personnel, through their daily, constant contact, on the test systems with which they are working, and on the study in general, however, goes beyond the requirements of faithful recording of "original observations", and the integrity of a study is contingent on additional factors, which the GLP Principles are addressing at this point.

3.4 The Sponsor

Although the sponsor had been given a definition in the original OECD GLP Principles, its role within or around a GLP study remained relatively obscure. There were no indications as to which, if any, responsibilities were to be borne by the sponsor, a situation which has been leading to a number of difficulties. One of these will be dealt with in detail later on (see section 9, page 202); for the present it may suffice to state that in numerous cases of studies conducted at CROs the question of identity, purity and stability of the test item could only insufficiently be addressed, or even remained unresolved, because the sponsor – as the supplier of the test item – did not disclose these data to the CRO and the Study Director. On the other hand sponsors did not look very favourably at study reports which excepted these data from the GLP Compliance Statement, as they had to under the strict application of the GLP Principles.

It was therefore obvious that the sponsor should become more involved in the issues of GLP compliance, although as the *"entity which commissions, supports and/or submits a non-clinical health and environmental safety study"* the sponsor itself is not directly involved in the conduct of a study under GLP. Thus, it is questionable whether the GLP Principles could have imposed any responsibilities directly to the sponsor, although several explicit references to the sponsor have been introduced into the revised GLP Principles. However,

2.4 The QA Programme

certain aspects on the role and responsibilities of the sponsor have been compiled in an Advisory Document (OECD, No. 11).

The responsibilities of the sponsor with regard to the GLP-compliant conduct of studies may first of all be seen as dictated by self-interest. Since the sponsor may be the one entity who finally submits the whole package of nonclinical health and environmental safety studies in support of a product registration to a Regulatory Authority, who is demanding the conduct of these studies under GLP, the sponsor should be aware of the exigencies of GLP. It is, however, not sufficient that the sponsor should simply ask for a GLP-compliant study. but the sponsor should as well assume an active role in facilitating the GLP compliant conduct of such studies. This role is not confined to one direction, as it may be interpreted from the respective paragraph in the FDA regulations which state that "*When a sponsor conducting a nonclinical laboratory study intended to be submitted to (a regulatory authority) utilises the services of a consulting laboratory, contractor, or grantee to perform an analysis or other service, it shall notify the consulting laboratory, contractor, or grantee that the service is part of a nonclinical laboratory study that must be conducted in compliance with (GLP)*". This sentence implies that the sponsor is itself a test facility working under GLP, and thus interested in not jeopardising its own study by parts that are not GLP compliant. There may be situations, however, where a sponsor commissions a whole study, and where the responsible person may not quite know what GLP compliance at the contractor means in terms of required actions by the sponsor. Therefore, three points have to be observed by the sponsor in order to play the expected part in the GLP game.

First of all, the sponsor should be knowledgeable in the requirements of the GLP Principles. Since the connections between sponsor and study will be mainly through the test facility management and the Study Director, and/or the Principal Investigator, the sponsor should be aware of their responsibilities towards the study, and especially of the fact that the full responsibility for the whole study remains with the Study Director, including the validity of the raw data and the report, notwithstanding any elements of the study which may or may not have been disclosed to the Study Director, as they have been obtained at the sponsor's own test facilities.

The second responsibility may again be seen as dictated by self-interest. Since the sponsor should commission a GLP-compliant study, it lays squarely in his responsibility to ensure that the test facility to be chosen for the conduct of the study should indeed be able to perform the study under the conditions of the GLP Principles. Furthermore the requirement that the study should be conducted under GLP should certainly be fixed in the contract between sponsor and the test facility, i.e. the CRO. However, for assessing the ability of a test facility to conduct a study in compliance with GLP, the sponsor would be ill advised to rely solely on the assurances of the test facilities contacted for the possibility to perform such studies. The sponsor should either monitor the selected CRO prior to the initiation of the study, or the respective National Compliance Monitoring Authority may be contacted to determine the current GLP compliance status of the test facility.

The third area of responsibility centres around the submission of the full dossier to a Regulatory Authority. Such a submission will consist of the totality of all studies necessary for the assessment by the authority, which are presented to the authority in a single package of final reports. These final reports may originate from a variety of test facilities, and they certainly may not be altered except through report amendments by the respective Study Director. Since the Study Director may not be anymore in control of his/her report which has been delivered to the sponsor, it becomes self-evident that the responsibility for the integrity of the assembled package has to lie completely with the sponsor. There is an exception to this rule, which is also mentioned in the GLP Principles: The re-formatting of a report, or other formal modifications to it, which may be necessary to fulfil certain specific, formal requirements of a Regulatory Authority. The addition of a further title page by the sponsor will not alter the content of the report, and such alterations are therefore admissible under the provisions of the GLP Principles without the need of adding an amendment (which would again require the signature of the Study Director). Of course, such re-formatting will have to be in the hands of the sponsor who is submitting the assembled package of studies. Insofar as such sponsor signed items included in the final report do not constitute intrinsic alterations related to the performance of the study there is no reason to demand that the Study Director should acknowledge and sign them, but such contents should be clearly identified as non-data items, and the signature should be clearly identified as the sponsor signature.

Additionally, the overall responsibility for the integrity of the submission dossier is to be distinguished from the Study Director's responsibility for the scientific validity of the single study, for which the sponsor cannot be held responsible. On the other hand, the sponsor has to make the decision, based on the outcome of the studies with the respective test item, whether or not to submit a product for registration to a Regulatory Authority.

A number of obligations of the sponsor are also explicitly mentioned in the GLP Principles. They centre around the study plan and the study report on the one hand, and the test item on the other. Due to legal considerations related to the responsibility for validity of test data, some countries require that the sponsor should sign the study plan along with the Study Director in order to acknowledge agreement with the planned study and its methodology and conduct. Thus, the Principles require that *"the study plan should also be approved by ... the sponsor if required by national regulation or legislation in the country where the study is being performed"*. Even when the signature of the sponsor is not legally required, the sponsor cannot shy away from his responsibility by remaining anonymous, since the GLP Principles clearly require that the name and address of the sponsor to be included in the study plan as well as in the final report. On the other hand, the responsibility of the sponsor for the study itself ends with signing, if required, the study plan. The sponsor, e.g., need not approve the choice of the Study Director, since this is the responsibility of the respective test facility management. The sponsor will also not receive inspection or audit reports of the test facility's Quality Assurance and will thus have no direct role in the assertion of GLP compliance within the study.

The responsibility of the sponsor for the characterisation of the test item supplied to the test facility has sometimes been a sore point, as has been already mentioned. In order to ensure that there would at least be no mix-up of test items, the Principles require that *"In cases where the test item is supplied by the sponsor, there should be a mechanism, developed in co-operation between the sponsor and the test facility, to verify the identity of the test item subject to the study."* In this respect, the sponsor is directly involved in the GLP-compliant way of study conduct, whereas in other areas related to test item requirements, the sponsor is only indirectly addressed. The GLP Principles call for careful identification of the test item and adequate description of its characteristics. Such characterisation may be carried out by the contracted test facility, but it will mainly be the sponsor who is in possession of these data. Since the Study Director is responsible for the GLP-compliant conduct of the whole study, the characteristics of the test item should be known to him; if, however, the characterisation of the test item has been conducted by the sponsor, this fact should be explicitly mentioned in the final report, because the Study Director would in this case have had no direct control over the GLP compliance of this part of the study. Furthermore, if characterisation data are not disclosed by the sponsor to the Study Director, this fact should also be explicitly mentioned in the final report. Naturally, sponsors should on the other hand be aware that failure to conduct characterisation of the test item, possibly even failure to do so under the conditions of GLP, might lead to the rejection of a study by a Regulatory Authority.

Implicitly contained in this obligation of the sponsor to provide data on test item characteristics to the test facility and the Study Director is the requirement that every available information on any known potential risks of the test item to human health or the environment should be transmitted to the contractor; this obligation would include of course also information on any protective measures which should be taken by the test facility staff having to handle this test item.

There is a final, explicit, but not immediately applicable, responsibility of the sponsor mentioned in the GLP Principles. When a sponsor commissions a study at a CRO, he will receive a copy of the final report, but the entire study documentation will probably remain with the CRO and be archived there. This may have practical reasons and may be considered as advantageous in terms of GLP, because the CRO as the test facility will then not only be in possession of the single study's raw data but will also dispose of all additional and circumstantial documentation like the respective SOPs and the environmental data of the facility itself. However, a contract research laboratory may go out of business, and its archive could then be destroyed, if nobody would claim proprietary rights to it. Therefore the GLP Principles have charged the sponsor in a certain sense with the task of guarding over the raw data of their studies, in requiring that *"if a test facility ... goes out of business and has no legal successor, the archive should be transferred to the archives of the sponsor(s) of the study(s)."* Thus, in such a case, the sponsor is expected to arrange for sufficient and adequate archiving space for the appropriate storage and retrieval of study plans, raw data, specimens, samples of test and reference items and final reports of all studies conducted at this CRO on behalf of the sponsor. For a more detailed discussion of this topic see section 12.4, page 269.

> In a way, the sponsor may be regarded as some kind of a bracket around the study: At the start there is the commissioning of the study by the sponsor, and at the end, there is the submission of the study by the sponsor to the Regulatory Authority. GLP is not involved in either of the two points. If the sponsor is not directly involved in the experimental study conduct, this entity may therefore be seen as independent from, and unconnected to, any responsibility in the field of GLP. On the other hand, the sponsor may greatly help or hinder the GLP-compliant conduct

> of the studies commissioned by him, and it should be in his well understood self-interest to do the former rather than the latter. The GLP Principles, and the respective Advisory Document of the OECD, do address certain aspects in the relationship of the sponsor with the test facility, in order that sponsors should fully understand and correctly interpret their role in the area of GLP compliance.

4. The Quality Assurance Programme

"Trust is Good, Control is Better" says an old proverb. However, the Quality Assurance in GLP is not intended to act as a Quality Control entity, its responsibilities lie more in the direction of helping and guiding the test facility in the ways of GLP. Even the final audit of a study, that has to be performed by the Quality Assurance unit in order to check whether the report truly reflects the raw data of the study, cannot be used to introduce qualities into the study which had not been there before. A maxim of the Quality Assurance in GLP is therefore that one cannot "control quality into a study". This is in contrast to the quality control as exercised in a manufacturing environment, and which has also already been discussed in another context (see part I, section 7, page 45). A company may manufacture screws to a certain specification with a defined tolerance range being applied. Quality control, in this case, will control the observance of the specification and reject those individual screws falling outside the tolerance range. Thus, depending on the quality of the work, a more or less large proportion of manufactured screws will be rejected. In this way, quality will have been controlled into the output of the manufacturer, since all screws leaving the facility will now conform to the specifications. The problem is different, however, in the area of safety testing, where the studies do not yield results that can be checked against some pre-determined specifications. In this area, the term "quality" has to be differently interpreted as an intrinsic property of these studies, allowing an estimate of the reliability of the results to be made. All the different responsibilities and activities of Quality Assurance have to be viewed, therefore, in this light.

Quality Assurance work consists of conducting inspections at the test facility to ascertain that all activities conform to the GLP standards, of auditing reports of studies to ascertain that the report reflects the raw data of the study, and of checking pertinent documents for their compliance to the GLP Principles. In this latter area may fall tasks like the checking of draft SOPs and of draft study plans. While it is management that is held responsible for ensuring that Standard Operating Procedures (SOPs) are produced, issued, distributed, revised and the originals archived, it is advisable that Quality Assurance personnel, who are not normally involved in the writing of SOPs, will review the new or revised versions of SOPs before they are signed, issued and distributed, in order to assess their clarity and compliance with GLP Principles. In the same way, although the Study Director is responsible for the writing and the approval of the study plan, Quality Assurance personnel should be able to check the format and content of all study plans for their compliance with the GLP Principles.

One obvious presupposition for the competent and efficient work of Quality Assurance is the expertise and experience of Quality Assurance personnel that is necessary to fulfil their responsibilities. Like any other, this expertise cannot be got out of the blue, but it has to be acquired through training and experience. Thus, it is the duty of management to ensure that there is a documented training programme for the Quality Assurance personnel, encompassing all aspects of Quality Assurance work. Ideally, the training programme for novices to this profession should include on-the-job experience under the supervision of competent and well trained staff.

However, the whole of this training and experience should not simply cover the Quality Assurance work only, but the Quality Assurance personnel should also be, or become, familiar with the test procedures, standards and systems operated at the test facility. This familiarity with test systems and test procedures should be extensive enough to allow the individual inspector, who is monitoring some part of a study to choose the most suitable approach to the inspection, to ask the most pertinent questions and to judge the importance of any deviations from study plan or from SOPs. On the other hand, it is certainly not necessary that Quality Assurance personnel should possess highly expert knowledge in the conduct of the tests and the test systems they have to monitor, since the Quality Assurance inspector is neither expected to perform a scientific evaluation of a study nor to judge the validity of the scientific procedures that are used.

Since the role of Quality Assurance personnel involves to a great extent personal interactions with study personnel through the controlling activities which may be connected with teaching and may finally lead to admonishing study personnel and Study Directors, the individuals selected for Quality Assurance work should have very good communication skills. In this respect, and in view of the highly sensitive nature of their work, training in communication techniques and conflict handling would certainly also be advisable.

4.1 General Considerations

There are a number of activities that may be seen as being not, or only indirectly, related to study conduct, or they may at least not be ascribable to any individual study. On the other hand, many of the activities and responsibilities of Quality Assurance are directly related to the supervision and control of specific studies. It is this latter interconnection of the work of Quality Assurance with the work of the Study Director and the study personnel that necessitates two important provisions for Quality Assurance personnel: The independence from the study conduct, and the familiarity with the study procedures.

The first of these two provisions may be seen to originate from the common experience that nobody is perfect in recognising his or her own faults and errors. From the biblical word of "noticing the splinter in thy neighbour's eye, while not noticing the beam in thy own" to the experience that one has better let somebody else proof-read one's manuscripts for typographical errors, it is recognised that for control functions an independent observer is much more useful than a person deeply involved in the activity or process to be controlled. Therefore, the Principles are very clear about the necessity of independence of the individual who is to assure the GLP compliance of a specific study from an actual, personal involvement in the conduct of this study ("*This individual(s) should not be involved in the conduct of the study being assured*"). In the majority of situations, this requirement poses no special problems. At the "upper end" there are the test facilities big enough to dispose of their own full-time, professional Quality Assurance units, even if such units may be composed of just one individual. At the "lower end" there are those test facilities which are too small to employ a full time Quality Assurance inspector of their own, and which therefore have to resort to an external Quality Assurance company. Of course, there may also be mid-sized test facilities which may choose this possibility out of other reasons. This latter solution of hiring an external Quality Assurance can be a good, cost-effective solution and is perfectly admissible, if the necessary effectiveness, required to comply with the GLP Principles, can be ensured. Problems may arise in such "intermediate" situations, where the size of the test facility would make Quality Assurance activities only a part-time job, but where, for one reason or another, the use of an external Quality Assurance unit would be ruled out. In this instance, management has to give permanent, even if part-time, responsibility for carrying out the Quality Assurance functions to at least one individual within the test facility. For the sake of accumulation of expertise and in order to ensure consistent interpretation of the GLP Principles with regard to the test facility's activities, continuity in the Quality Assurance staff is certainly desirable. In such a case, the individual (or individuals) charged with the duties of Quality Assurance may, for the remaining part of the job, also be involved in some of the test facilities' other study activities. From the viewpoint of GLP this can be tolerated, if this person is not involved directly in the study which he or she is going to assure. There is a pitfall in this statement, however: It is acceptable for an individual involved in GLP studies to perform the Quality Assurance function for other GLP studies conducted in the test facility, but these supervised studies need to be performed in another department within the test facility. Thus, the specialist for HPLC analysis may not act as Quality Assurance person for a GC-MS analysis within the same analytical test facility, even if he or she would never perform such an analysis. On the other hand, an analytical technician may be employed to perform the Quality Assurance function for ecotoxicology studies. It is not allowable, however, that an individual who is involved in one part of a nonclinical laboratory study would perform Quality Assurance functions for another part of the study in which the individual is not involved; thus, this analytical technician should not be allowed to inspect phases of the one ecotoxicology study, samples from which would be analysed by him.

These various situations are graphically represented in figure 7. It lies in the responsibility of the test facility management to investigate the possible interrelations between the various test facility units and their common involvement in studies to select the most appropriate individuals for performing the respective Quality Assurance functions.

The second general requirement for Quality Assurance personnel is the necessity of their having, in addition to a thorough understanding of the Principles of GLP, a knowledge and understanding of the basic concepts underlying the activities being monitored. Not only can such knowledge help to focus on the really crucial activities within a study when inspecting it, but it will

2.4 The QA Programme

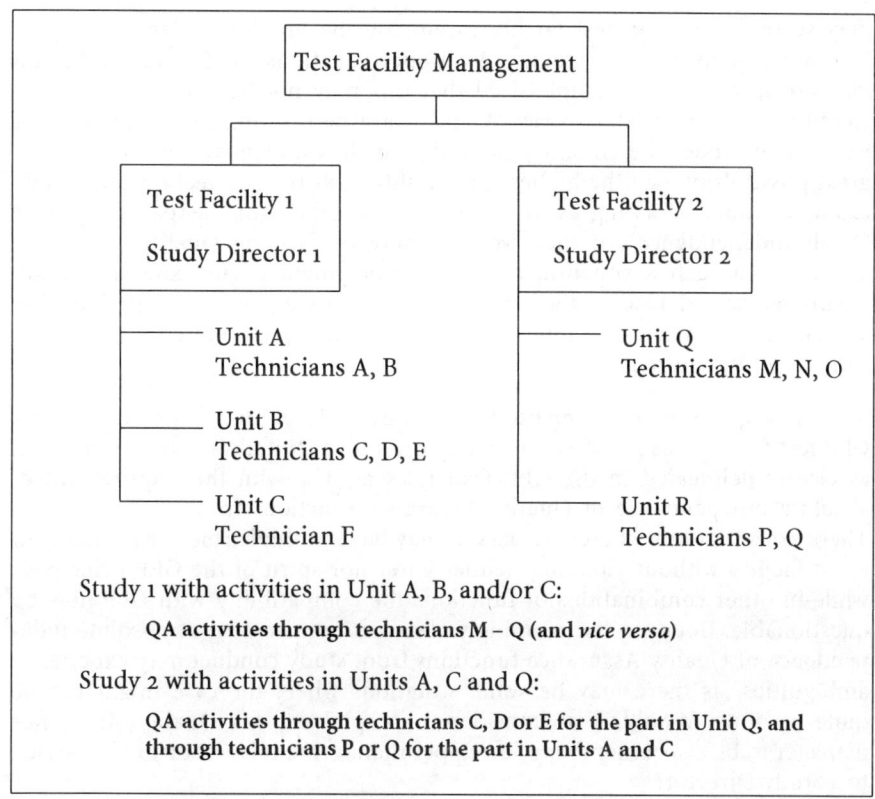

Figure 7: Schematic representation of possibilities for the reciprocal performance of Quality Assurance functions in a company with two different test facilities.

foremost also help to determine the critical phases of a study (see section 2.10, page 88), and it will furthermore help to ascertain the coverage of the pivotal activities that are to be described in SOPs which the Quality Assurance should check for compliance with the GLP Principles (see below).

This requirement that the individuals responsible for the Quality Assurance functions should be knowledgeable in the activities to be monitored by them may, especially in situations encountered in smaller test facilities, clash with the first requirement for Quality Assurance personnel, the one that calls for independence from the studies to be monitored. However advantageous it may seem for a small test facility to employ the one technician working in laboratory A to perform Quality Assurance functions for laboratory B, and *vice versa*, it has to be emphasised that this may not be considered a valid solution. If the two laboratories A and B are performing the same types of work, then, under the circumstances of a small test facility, there is the very great possibil-ity, nay the highest probability, that the two technicians in this example would be acting as deputies for each other, thus jeopardising their "study-independence", if they were actually performing Quality Assurance activities. In such a situation only the employment of an external Quality Assurance would resolve the issue, and it points again to the problematic aspects of too small test facilities with regard to their possibilities of complying with the GLP requirements.

This question of "compatibility" of various levels and functions within a GLP test facility has given rise to many questions. Not always is the situation as clearly delineated in the GLP Principles as it is with the requirement of absolute independence of Quality Assurance functions from study conduct. There are situations where one person may have multiple roles or positions in a test facility without violating neither letter nor spirit of the GLP Principles, while in other combinations of functions the compatibility with GLP may be questionable. But even the seemingly clear-cut requirement of absolute independence of Quality Assurance functions from study conduct may experience ambiguities, as there may be some situations where the case might not be quite obvious: Would it be considered acceptable for a Quality Assurance manager to be also a member of an upper management level? or to be married to a Study Director?

Looking at the various responsibilities on the one hand, and the requirement for independence from study conduct for the Quality Assurance personnel, it will become clear that the spirit, if not the wording of the GLP Principles prohibits the personnel assigned to the Quality Assurance unit from performing any other function at the testing facility which may impact a study. Therefore, members of the Quality Assurance unit cannot work on any aspect of a study, even in an advisory capacity. A Quality Assurance manager most certainly cannot be a Study Director. Given the supervisory role of management over the activities of both the study personnel and the Quality Assurance, a Quality Assurance manager can neither be part of the test facility management.

The interesting question concerning the compliance status of a test facility at which a Quality Assurance employee would be married to a Study

Director, or maybe even to the president of the company, could be considered a tricky one. There is indeed reason to suspect a possible violation of GLP requirements. The Principles, however, only require that Quality Assurance should maintain independence from the conduct of any given study. Steps may be taken for instance by the test facility management to assure compliance with the Quality Assurance requirement of independence by stating in an SOP that Quality Assurance functions cannot be performed by the husband for any given study conducted by his wife (or *vice versa*). In general, personal interconnections may be hard to avoid in such situations where the human element plays a role in the relationships between the various partners connected by the GLP Principles. But then, questions of a similar or related nature might be asked for a number of other combinations of tasks, functions or hierarchical levels, too: Can a Study Director be part of test facility management? Can a member of management, a Study Director, or a laboratory technician perform the functions of the person responsible for the archives? Can a Principal Investigator, situated at the sponsor's test site, act as the supervisory "study monitor"? Some of these questions will be dealt with in later parts, since they are not connected to the Quality Assurance functions which are the theme of the present section. In a general sense, however, they can already be answered here: Any such situation should not negatively affect the integrity of studies and should not infringe the ultimate purpose of GLP.

> The quality which is intended to be achieved in GLP is not a quality which can be controlled by easy, numerical or other means, but it is the control over the intrinsic quality of a test facility and its studies. Therefore, it is of utmost importance that the entity which is responsible for ensuring this type of quality should be able to express its opinion in an unbiased way, which means that it should be absolutely independent of study conduct. Only through this independence a reliable assurance of the studies' inherent quality can be achieved.
>
> In this sense, Quality Assurance might be compared to the independent auditors of a company's financial status, who should also, unbiased by any involvement in the company's well-being, deliver their verdict in no uncertain terms.

4.2 Quality Assurance Inspections

To assure the GLP compliance of testing performed in a test facility, Quality Assurance has two main instruments: The inspection, and the audit. By watching people perform their daily duties and their assigned tasks within studies, and by asking questions about these activities, the requirement that Quality Assurance should be able to assure that "*all studies are conducted in accordance with the(se) Principles of Good Laboratory Practice*" can be fulfilled. An inspection, however, will provide only for a snapshot impression of the GLP-compliant manner of working at the test facility, and a study audit will give insight into the continuous, day-to-day adherence to the GLP Principles for a single study. Since both of these instruments of Quality Assurance for the monitoring of GLP compliance serve to observe different sides of one and the same picture, they have to be considered as being of equal importance, and they have to be formally described in the SOPs of the Quality Assurance Programme.

With regard to Quality Assurance inspections, it can be observed that, in principle, there are two kinds of them: Either an inspection is directly concerned with the conduct of a specific study or study part, or the inspection is taking stock of the general state of the test facility in a manner not related to any study. Both of these will have to be described in detail in the respective Quality Assurance Standard Operating Procedures.

In the former case the inspections are scheduled according to the chronology of a given study, usually by first identifying the critical phases of the study. The identification of the phases of the study to be inspected may on the one hand be based on the experience of the Quality Assurance inspector with the specific study type. On the other hand this task could be done by the Study Director and participating scientists working in concert with the Quality Assurance; for this latter case, coverage by an SOP of the procedures to be used would be advisable. In the course of such an inspection the inspector may watch the preparation of the test item, control the maintenance records and the calibration of the balance used in this task, interview the technician about the procedures to be followed if the calibration would show the balance to be out of range, control the labelling of the test item and so on, till the final application of the test item to the test system. One of the prerequisites for conducting such an inspection in a meaningful way, covering as many of the "critical phases" as possible or as required by the Quality Assurance Programme, is the correct planning of the full array of inspections for any given study. This in turn requires that management has fulfilled its obligation to ensure that the Study Director has made available the study plans to Quality Assurance before the start of any study. Only this "timely availability" will allow Quality Assurance first to monitor compliance of the study plan with GLP and to assess the clarity and consistency of the study plan. It will then allow to identify the critical phases of the study and their chronology. In this way Quality Assurance will finally be able to plan a monitoring programme in relation to the study. The same requirement for "timely availability" holds for any amendments to the study plan; only by providing such information to Quality Assurance in an efficient way will facilitate effective study monitoring.

The second type of inspections covers the general facilities and activities within a laboratory (installations, support services, computer system, training, environmental monitoring, etc) without being based upon any specific studies. Thus, the Quality Assurance inspector might verify, again by interviewing test facility personnel and by checking the relevant documentation, that animal rooms are regularly cleaned and sanitised, that the environmental control instruments in test areas are maintained, that the respective records are archived and proper measures taken upon the transgression of the normal limits, that apparatus are maintained and validated, that Standard Operating Procedures are immediately available to the personnel, and that the versions encountered at the work places actually correspond to the current version, that measures are provided to deal with power failures, and so on.

There is a general point to be made with respect to the conduct of inspections. While the OECD Principles do not specifically address this point, it is nevertheless logical that inspections should be performed at intervals that are narrow enough to allow Quality Assurance to become reassured that GLP is indeed followed in the inspected test facility. If inspection intervals are too widely spaced, adherence to the GLP rules might relax due to several reasons. First of all, any single instance of non-compliant behaviour would be excused as a "singular lapse" that never happened before and would never occur again. Also, test facility personnel could easily guess the advent of the next inspection, especially if relatively regular intervals were maintained. Then test facility personnel could start to take a little bit more care in their daily working patterns, with the result that the inspection would show compliance, though possibly a mediocre one. Therefore, inspections have to be performed frequently in order to enable Quality Assurance to really judge the GLP compliance in a particular test facility. On the other hand, inspections may become a nuisance, if conducted too frequently, and personnel would become immune to the too frequent remonstrances of the Quality Assurance personnel. There is a fine balance to be achieved between annoying people with constant bickering and letting things run as they will, and it is probably one of the most exacting tasks of the Quality Assurance manager to strike this very balance. Whatever the balance, however, the inspection intervals have to be defined in the Quality Assurance Programme's SOPs, and they have to be followed.

As an aside: Inspections by national GLP Compliance Monitoring Authorities will probably, in the majority of instances, belong to the category of "expected" inspections, which will result in all the accompanying side-effects: Test facility personnel will be doubly on the alert for doing everything correctly and according to the relevant SOPs, folders with study plans and all the raw data collected up to the time point of the inspection will be available at a wink, and all SOPs will have been revised miraculously just a fortnight ago. Malicious as this may sound, it is not all that bad, since, at least, SOPs will have experienced the long overdue revision. But let's turn back to Quality Assurance inspections.

Already the Consensus Document on Quality Assurance (OECD No. 4, 1999) had recognised that these two possibilities would not sufficiently describe the "real-life" inspection activities of a Quality Assurance unit, and the revised GLP Principles have acknowledged this extended view of inspection types. Inspections dealing with the conduct of, and with the activities within, a study are now further subdivided into two different types, so that there are finally three types of inspections, which are just named in the OECD Principles, but which have to be further specified by Quality Assurance Programme Standard Operating Procedures. These three types are:

- **Facility-based inspections,**
- **Study-based inspections, and**
- **Process-based inspections.**

The first two types have already been dealt with above and they do not need any further comments. It is the term "process-based" inspection that remains to be explained in some detail.

Process-based inspections are conducted to monitor procedures or processes of a repetitive nature. They are utilised when a process is undertaken very frequently within a laboratory, and when it is therefore considered inefficient or impractical to undertake inspections of the respective activity for each and every individual study. They are therefore primarily performed independently of specific studies. It is clearly to be recognised that this performance of process-based inspections will result in some studies not being inspected on an individual basis during their experimental phases.

It has to be recognised on the other hand that even in study-based inspections there is some element of "process-based" inspecting, since activities of a repetitive nature, like the application of the test item to the test animals in a chronic toxicity study, will only be inspected on a random basis once or a few times during the conduct of the study. While it is thus customary to assume that, for the purpose of covering phases which occur with a very high frequency within a single study, a limited number of inspections will be sufficient to assure that this activity would always be performed in a similar, GLP-compliant way throughout the whole study(if the single inspection finds so) within this study, this assumption might be contested for the extension to multiple studies being monitored on a random basis only. Therefore, Quality Assurance has a very delicate task at hand, namely to balance the needs for ensuring GLP compliance in as many individual studies as possible against the pragmatic approach of inspecting in a process-based manner. One of the problems might be bias. In order to evade such bias, the process-based inspections should be performed on a random basis with regard to the actual studies chosen for the inspection of the respective process. This holds especially for short-term studies for which this instrument of the process-based inspection is very handy in that it allows the Quality Assurance to monitor them on a random basis only.

There is another general consideration that has to be taken into account when planning for process-based inspections. The acceptability of such inspections is contingent on assuring that the facilities, personnel, methods, and any other items which are inspected are representative of those used in the studies. Process-based inspections would, for instance, no longer provide a valid picture of the general GLP compliant conduct of a certain study type, if the centrally important SOP for this study type had changed. In the same sense, a change of personnel might have a subtle influence on the way, this type of studies is conducted. Therefore, it is not only necessary for Quality Assurance to re-inspect facilities periodically to account for changes in personnel and equipment, but Quality Assurance has to remain alert and to be constantly aware of changes in methodology, which would necessitate immediate and repetitive inspections in order to ascertain the GLP compliant conduct of the studies under these new parameters and new standards.

A further consideration - which follows from the one mentioned above - in terms of the validity of process-based inspections for the ascertainment of

GLP compliance in the conduct of any single study not specifically inspected, is the problem of temporal relationships between inspections and the actual study activities. Thus, the more remote the last inspection has been from the study in question, the less reliability might be placed upon the inspection results in terms of their ascertaining the GLP compliance of study conduct. This issue has therefore to be given due consideration in the Quality Assurance program as well as in the Quality Assurance SOPs.

While thus the term and the use of process-based inspections have been defined, the applicability of this inspection type with regard to study types may seem less clear. Therefore it is of utmost importance to define clearly and unequivocally those study types which would qualify for this facilitation of the Quality Assurance function. To this end, the Quality Assurance has to develop SOPs which should primarily define the circumstances under which process-based inspections may be performed and which should also present a final list of the respective study types. It probably needs not to be specially mentioned that it would certainly be advisable to develop these SOPs in collaboration with the respective Study Directors and on the basis of historical data regarding study frequencies.

The main considerations in the qualification of study types, apart from their duration, for the application of process-based inspecting would certainly be the respective frequencies of studies performed and their complexity with regard to critical phases. Thus, studies with a rather complex design and with activities needing advanced knowledge and skills should be inspected more frequently than studies of a very routine nature. Furthermore, the frequency of inspections has to be based on the frequency of study conduct within the various study types. In these cases limits would have to be given in both absolute and relative terms in order to ascertain that a minimum frequency of inspections is maintained. This means that the Quality Assurance SOPs should clearly spell out, for each of these specified study types, the percentage as well as the minimal number of single studies to be inspected per year, in order to ascertain that the GLP-compliant conduct of these studies will be controlled with sufficient frequency. The Table on the next page shows an example of an ecotoxicological test facility, where a system of determining inspection frequencies based on the rolling statistics of the last two years has been introduced.

It can be seen in this example, that not all of the different tests are inspected with equal frequency, and that for some single tests only one or two inspections have been performed within the relevant two years period. It can

2.4 The QA Programme

Table 1: Determination of inspection frequency of short-term tests by study type, average study frequency ("rolling statistics" over two years) and study complexity (Simplified table by courtesy of Dr. R. Vogel, Novartis CP, Basel, Switzerland)

Test System	Study Duration	Studies (per 2 years)	Inspections (per 2 years)	(%)
Aquatic Toxicity				
Algae	3 - 4 days	67	13	19
Daphnia	2 days	34	8	23
Fish (96 hr)	4 days	44	8	18
Total		145	29	20 (min. 10)
Physico-chemical Properties				
Boiling Point	< 1 day	18	4	22
Melting Point	< 1 day	23	2	10
Solubility water	< 1 day	21	7	33
Vapour Pressure	7 - 14 days	19	4	19
Total		91	17	19 (min. 10)
Acute Toxicity				
Dermal	14 days	82	13	15
Ocular	14 days	32	9	28
Oral	14 days	61	9	14
Total		175	31	17 (min. 10)
Biol./chem. Degradation				
Oxygen Demand	1 day	12	1	8
Respirometry	28 days	28	4	14
Biodecomposition	29 days	42	7	16
IC$_{50}$	1 day	21	2	9
MITI (modif.)	1 day	46	6	13
Total		149	20	13 (min. 10)

also be seen, however, that studies of higher complexity or of higher "criticality" have been inspected more frequently. In the area of acute toxicity studies, ocular toxicity studies have been twice as frequently inspected than studies with either oral or dermal application, which may reflect the greater difficulty and higher demand on technical skills of the application mode in ocular toxicity studies.

In a way, this possibility to have process-based inspections for certain study types instead of study-based ones being performed by Quality Assurance may have seemingly been contradicted by the GLP Regulations of the US FDA and EPA (21 CFR 58; 40 CFR 806, draft), which are requiring the inspection of each and every study. This point of view had even been confirmed in one of the EPA "advisories", where the answer to a respective question was that *"the GLPs state ... that the Quality Assurance unit "inspect each study at intervals adequate to ensure the integrity of the study." While this does not specify the number or intervals we* (i.e. in this case EPA) *believe that in any case where a study is not inspected, i.e. at least once, there is a clear GLP violation."* However, in another of these advisories, the position was extenuated in a way that only enough coverage to all aspects of testing should be provided for each test facility given. Although this coverage should have to include *"facilities, equipment, protocols, personnel, methods, practices, records, and controls"*, the statement concluded that *"it is not necessary to separately address all of these aspects for each study, as long as overall coverage, that is, the sum of all inspections, is balanced to include all aspects"*. Therefore, for any given single study, it would be sufficient to address these issues in process-based inspections, and the requirement that each study should be inspected at least once could be fulfilled by the – in any case performed and also for short-term studies necessary – final report audit. Whether it is thus indeed the well-understood intention of GLP that each and every study, however short its duration, and however repetitive and routine its nature, would have to be inspected "at least once" in order to achieve full compliance, should, with regard to the wordings in the OECD Principles and the respective Consensus Document, not be a topic for debate anymore.

In the context of inspections, Quality Assurance may face a number of additional problems when dealing with field studies. One of these is the number of staff needed for the correct execution of the Quality Assurance functions. While in a small test facility performing only a limited number of toxicity studies, and all of them in its own laboratories, one single person may be considered sufficient for the performance of Quality Assurance functions, a single individual will, on the other hand, probably not be able to perform all the necessary Quality Assurance functions for field studies. Thus, Quality Assurance will need to consist of a number of persons in order to conduct its work properly. There are several reasons for this need:

Firstly, field studies may involve similar activities at separate locations but at similar times. The exact timing of inspectional activities may furthermore depend upon local weather or other conditions, which would necessitate flexibility in the Quality Assurance procedures. Especially when such a study would involve not only multiple sites, but narrow crop windows and field sites without Quality Assurance units, this would severely impact upon the possibilities of a Quality Assurance unit.

Secondly, the geographical spread of test sites may mean that Quality Assurance personnel will also need to manage language differences in order to communicate with local study personnel, the Study Director, Principal Investigators and test site management (see also section 11, page 240). The same problem may be seen to arise with other studies performed in a multi-site setting, where it might be considered impossible for a single Quality Assurance unit to conduct inspections at all individual test sites in the context of one single study.

Lastly the requirement that inspections should be conducted in sufficient frequency, in order that they might be considered adequate to ensure the quality and integrity of any one study, could seemingly multiply the problems. If all parameters would have to be verified to an adequate degree for each site at the same time, this task might well exceed the manpower possibilities of any one single Quality Assurance unit. Here, too, a number of aspects in the monitoring of GLP compliance for each of the individual test sites may be considered to be covered through "process-based" inspections performed during the conduct of similar parts of other studies, or as "facility-based" inspections.

Information flow and communication lines may also be a greater problem in field studies and other multi-site studies than elsewhere. There is, on the one hand, the necessity to assure effective communications between Study Director and/or Principal Investigators, and the Quality Assurance personnel, e.g. for notification of critical activities. On the other hand, many more persons or entities at the same organisational levels (i.e. different test site managements, different Principal Investigators, different Quality Assurances) may be involved in the conduct of a field study, and therefore the flow of information from the different Quality Assurance persons has to connect these different parties: The responsible test site management, the responsible Principal

Investigator(s), the Study Director, the Study Director's management, and the latter's Quality Assurance. It is not without good reason that the GLP Principles require that management should ensure that "*for a multi-site study ... clear lines of communication exist between the Study Director, Principal Investigator(s), the Quality Assurance Programme(s) and study personnel*". Although defined as a management responsibility, the day-to-day realisation of the maintenance of information flow and the actual utilisation of these communication lines rests in the hands of the various interconnected individuals, of whom the Quality Assurance personnel would certainly be considered to hold a key position.

The key to the successful operation of Quality Assurance functions lies, as everywhere, in a good planning. Logically, prospective planning should concern not only Quality Assurance inspections, but it has to be applied to all of the various Quality Assurance activities. However, due to the special nature of study- and process-based inspections, and to the time restrictions connected with them, the need for proper instruments and procedures is very much obvious in this special area. Certain critical study activities may be performed in only a few minutes; if the Quality Assurance inspector would wish to inspect exactly this specific phase of the study, he or she had better be present at the precise time, when this activity would be going on. Since this is a prerequisite for the required precision in the attendance at specific activities or study phases, a well developed communication network between Quality Assurance and study personnel can be seen as an obvious necessity.

If organised according to the GLP Principles, the situation is relatively simple. In the first instance, Quality Assurance should be informed on all studies which are planned in the test facility. This necessity for information is formulated again as a primary management responsibility, in that management has to "*ensure that the Study Director has made the approved study plan available to the Quality Assurance personnel*". It is reiterated in the description of the Study Director's responsibilities, who has to "*ensure that the Quality Assurance personnel have a copy of the study plan and any amendments in a timely manner*". While it may be debatable what time frame is considered to be defined by the expression "timely manner", logic would dictate that at least the study plan should be available to Quality Assurance before the commencement of the respective study. Since the study plan contains all the information about the planned study and the prospective dates of all the critical activities, the submission of the approved study plan will allow Quality Assurance to arrange its inspection activities in a meaningful way, by entering all the important dates into its master schedule.

2.4 The QA Programme

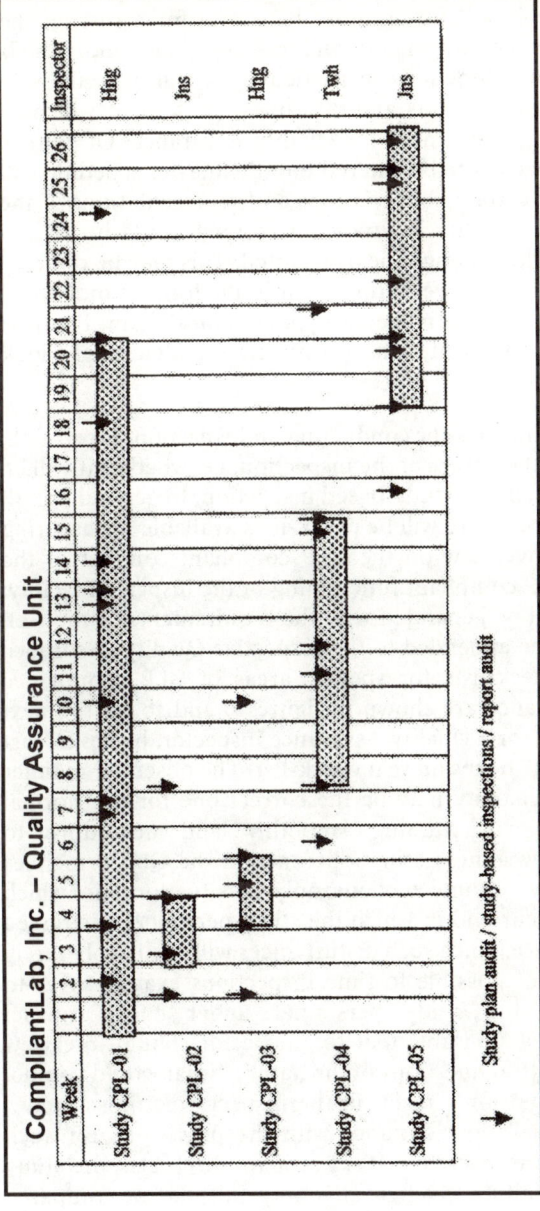

Figure 8: Graphic representation of a Quality Assurance master schedule with activity dates / points and initials of the responsible inspectors, providing for the possibility of a rapid overview of functions and activities to be performed, and of the workload in the respective time period.

The resulting master schedule will thus allow for successful planning of Quality Assurance activities; it will furthermore be useful for assessing the workload of Quality Assurance in general and of the single individuals chargedwith Quality Assurance functions in particular. It will also allow to keep track of Quality Assurance reports and deadlines, in short, it will be a useful and important working instrument for Quality Assurance. Of course, the volume of information that has to be entered into the master schedule will necessitate the production of certain abbreviations and representations of the master schedule in order that Quality Assurance personnel would be able to check at a glance how the workload might be distributed. This may be done by extracting the important points and presenting them in the form of individual daily or weekly activity sheets, or in an agenda-type graphical form. Figure 8 shows how such a graphical representation of a master schedule might look like.

There are several prerequisites for conducting an inspection. First of all, the inspector has to consider the nature of the inspection, i.e. whether it will be a facility-based, a process-based or a study-based inspection. In general, for all of the different inspection types there will be check-lists available for assuring an as complete as possible overview on the GLP compliant conduct of the inspected activities or the GLP compliant functioning of the inspected facility. Such check-lists generally will be generated with the requirements of the GLP Principles in mind, and will be appended to the respective Quality Assurance SOPs. Examples for such check-lists for specific areas of GLP compliance monitoring by Quality Assurance are shown in figures 9 and 10 on the next pages. It will not be sufficient for a Quality Assurance inspector, however, just to tick off the respective "ok ?" boxes on that check-list. The observed manner of performing the inspected activity may be the correct one for the normal situation, where everything is running smoothly and according to expectations. Failures of, or weaknesses in, adherence to the GLP Principles will, however, most probably become obvious only in situations, in which deviations from normality occur and in which the study personnel is stressed by some unexpected occurrence. Since such disturbances will occur only rarely and at random times, it is not possible to time inspections exactly so as to observe the behaviour of the study personnel under such unusual circumstances. Therefore, it is advisable that the inspector should interview study personnel with hypothetical questions of the nature "what would you do, if ...". Such questioning will not only result in the inspector getting a better picture of the study personnel's acquaintance with the provisions for such emergencies contained in the respective SOPs, but it will keep also the study personnel alert to such possibilities. In a figurative way, this may be compared

2.4 The QA Programme

	CompliantLab, Inc. – Quality Assurance Unit		
	Item	Ok ?	Remarks
1	Sufficient number of test system rooms or areas ?		
2	Separation of test species and/or studies possible ?		
3	Are separate rooms for test system quarantine available ?		
4	Are separate rooms for treatment of test system individuals available ?		
5	Are test system rooms climatised, and are surveillance records available ?		
6	Are storage rooms for feed and bedding available ?		
7	Adequate separation of storage from test system rooms ?		
8	Security of storage rooms from infestation ?		
9	Are storage conditions adequate (cold rooms available for sensitive goods) ?		

Figure 9: Example of a (partial) Quality Assurance check-list for the inspection of facilities

with the situation of two hotel guests in the case of a fire; the one who has familiarised himself with the position of fire extinguishers, and the emergency exits, as well as with the precautions to be taken in case of such an incident, will have a much greater chance to come through unharmed, than the one who never spent a single thought on such a possibility. While the former one might,

CompliantLab, Inc. – Quality Assurance Unit			
No.	Item	Ok ?	Remarks
1	Is the study plan available ?		
2	Are the relevant SOPs available at the workplace ?		
3	Does the application mode correspond to the study plan ?		
4	Does the application volume correspond to the study plan ?		
5	Is the amount and time of application recorded directly ?		
6	Are the controls, measurements and observations performed as required by the study plan ?		
7	Are the records legible, dated and initialled, and are corrections made in a GLP compliant way ?		

Figure 10: Example of a (partial) Quality Assurance check-list for the inspection of the dosing procedure and other relevant activities in a toxicology study.

upon the sound of the fire alarm, still keep his nerves, dress calmly and take the most precious belongings with him to the outside, the latter would probably jump out of the bed in panic and run down to the street barefoot and in his pyjamas only. Inspections therefore have not only a controlling function, but an educational one as well. Probably the latter one is even more important than the former, since it is easy to follow rules when everything is running according to schedule; it is important, however, that everything is still

2.4 The QA Programme

running according to GLP, when things go astray, tension is rising, and nervousness is at its height.

As a final point to this section on inspections, it should be kept in mind that - as it is the case for any other operative procedures covered by the GLP Principles - the Quality Assurance Programme of inspections (and audits) should be subject to management verification. Management has to *"assure that the Quality Assurance responsibility is being performed in accordance with these Principles of Good Laboratory Practice"* which means that the Quality Assurance has to provide management with periodic reports on its activities, not only for single studies as already detailed above, but for the totality of its activities. This involves furthermore the ability of the Quality Assurance inspectors to justify the methods chosen for the performance of their tasks, and to completely and confidently back the Quality Assurance Programme.

> One of the instruments of Quality Assurance for ascertaining and ensuring the continued adherence to the rules of GLP in a test facility as such, and within the studies performed, is the inspection of facilities and experimental activities. The GLP Principles do not require a constant supervision, since it is recognised that randomly conducted inspections will be sufficient to ensure compliance with the GLP rules. These inspections have, however, to meet an important point: They have to cover those parts of a study which are of special importance for the validity of the data and the conclusions to be drawn therefrom, or where deviations from the rules of GLP would most heavily impact on the integrity of the study. This holds for studies which undergo multiple inspections during their conduct as well as for those which may be inspected only on a process basis. Quality Assurance thus has to strike a fine balance in their inspectional activities, taking into consideration such aspects as inspection frequency, study type and "critical phases", in order to achieve a well supported view of the GLP compliance at the test facility and within the studies conducted.

4.3 Quality Assurance Inspection Reports

Quality Assurance personnel is required by the GLP Principles to *"promptly report any inspection results in writing to management and to the Study Director, ..."*. Since it is also required that *"records of such inspections should be retained"*, the work of the Quality Assurance personnel exhibits in many aspects the resemblance of a study: There is a plan (exacted by the master schedule), there are raw data (the inspection results, e.g. the filled-in check-lists, see as an example figure 11), and there will be a final report on the findings which has to be submitted to the relevant authorities, in this case the Study Director and the responsible management. However, since the Quality Assurance reports serve an important function, the attention of the addressees of these reports should not be blunted through continuous inundation by reports full of irrelevant details. Therefore, in the case of the Quality Assurance inspection report, this document will describe only the deficiencies noted, while all the other observations noted on the various check-lists which indicated full compliance with (or at least no obvious divergences from) the GLP Principles would remain in the Quality Assurance raw data only and would not be reported. Furthermore, the Quality Assurance program may exempt reports with the conclusion of "no findings" from being sent to the Study Director and to test facility management. This possibility is illustrated below with an example of a Quality Assurance statement, where a number of inspections resulted in "no reports" being submitted (see figure 12). It is then the obligation of the recipients of Quality Assurance reports to take the necessary corrective actions, if any deficiencies are noted and reported; therefore it is important that in these reports such deficiencies are clearly and unambiguously described, and that they are presented in such a way as to really catch the attention of the responsible persons. This is expressed in a more clear-cut and detailed way in the FDA and EPA GLP Regulations, where the requirement to report is limited to *"any significant problems which are likely to affect study integrity found during the course of an inspection"*. Of course, if small problems can be dealt with immediately at the workplace, e.g. the admonition of a technician who is observed to do data corrections in a non-compliant way, they need only to be recorded for the internal use of Quality Assurance (*"maintain written and properly signed records of each periodic inspection showing the date of the inspection, the study inspected, the phase or segment of the study inspected, the person performing the inspection, findings and problems, action recommended and taken to resolve existing problems"*, as it is formulated in the draft EPA Regulation, 40 CFR 806), but such deviations may not need to be reported to the Study Director and the management, or only, as provided for also in these US regulations, as

2.4 The QA Programme

I. Protocol Compliance				
Item	Yes	No	NA	Comment
A. Test System				
1. The test system is that specified by the protocol (e.g. species, sex, age, and/or weight range).	X			
2. The test system is identified in the manner specified by the protocol (e.g., tattoo, ear-tag, ear punch).	X			
3. The diet for the test system is that specified by the protocol.		X		Prot to be amended
C. Equipment				
1. Equipment used for the generation, measurement, or assessment of data is adequately tested, calibrated and/or standardized and written records are maintained of all inspection, maintenance, testing, calibration and/or standardization procedures.		X		S.O.P.s not complete, but in process. To be corrected.
E. Test and Control Articles				
1. The identity, strength, purity, and composition or other characteristics which will appropriately define the test or control article is determined for each batch and is documented before the initiation of the study.	X			By the sponsor
2. Methods of synthesis, fabrication, or derivation of the test and control articles are documented by the testing facility.	X			Reconstitution on site
3. The stability of each test or control article is determined by the testing facility before initiation of a non-clinical laboratory study.	X			By the sponsor. Reconstituted drug used immediately.

Figure 11: Example of some excerpts from a Quality Assurance inspection report, with additional remarks by the Quality Assurance inspector.

"written status reports on each study, noting any problems and the corrective actions taken." These status reports may serve the management as an indicator for the proper functioning of Quality Assurance.

Quality Assurance Inspections

Dates of Inspections	Phase Inspected	Dates of Reports to Study Director	Dates of Reports to Management
11/08/91	Protocol Review for Compliance with GLP Regulations	NA	NA
11/13/91	Test Material Preparation	11/08/91	11/08/91
11/13/91	Prepared Test Material Sample Collection	11/14/91	11/14/91
11/13/91	Test Material Administration	NA	NA
11/20/91	Body Weight Measurement	NA	NA
11/20/91	Food Consumption Measurement	NA	NA
12/02/91	Detailed Animal Observations	NA	NA
2/04/92	Ophthalmoscopic Examinations	NA	NA
2/12/92	Blood Sample Collection	NA	NA
2/12/92	Necropsy	NA	NA
2/12/92	Clinical Pathology Operations – Serono Baker Operation	NA	NA
2/12/92	Preparation for Blood Sample Collection	NA	NA
2/26/92	Histology – Tissue Trimming	NA	NA
2/27/92	Histology – Embedding	NA	NA
2/28/92	Histology – Microtoming	NA	NA
8/14/92	Data Review	8/14/92	8/14/92
8/14/92	Report Review	8/14/92	8/14/82

NA – Not applicable, no reportable findings

Figure 12: Detailed Quality Assurance Statement from a study report, where a number of inspections did not result in reportable findings

In the most simple case, where the Quality Assurance is located at the same place as the Study Director and the test facility management, there should be no problem with reporting and the subsequent responses from the two addressees. There are, however, more complex situations where the

Quality Assurance has no direct connections with the Study Director and his or her management. One might consider for instance the situation in a field test, where parts of the study are performed at test sites remote from the Study Director's test facility. Even if these test sites were to belong to the same company as the Study Director, the Quality Assurance unit serving these test sites might be a different one from the Quality Assurance unit which is directly connected with the Study Director's test facility, i.e. it could well be a local one as contrasted to the central one at the Study Director's test facility. Such a situation might therefore put an additional strain on the different Quality Assurance units concerned. Therefore, it has to be ascertained that, irrespective of where the test sites are located, the written reports of Quality Assurance personnel must reach both management and the Study Director. In such cases, where there is no direct connection between the different parties concerned, additional care has to be taken by the Quality Assurance unit, where the respective report has been written, that it was actually received by management and the Study Director, and this should be documented in the raw data.

While it lies in the responsibility of the Study Director to react to such reports and to take corrective action if necessary, and while it lies in the responsibility of management to ensure that the Study Directors actually will implement corrections required by these Quality Assurance reports, it is nevertheless in the well understood self-interest of Quality Assurance to keep track of the various issues raised and the corrective actions asked for. At times, and depending on the personalities involved, it may not be as easy as it sounds for Quality Assurance to succeed in asserting its point of view. In the worst case, there may be some Study Director who takes the position " I'm a practice-minded person" and "I know jolly well what I'm doing", who therefore would pay no attention whatsoever to the Quality Assurance's admonitions, and who thus would continue in his or her own ways regardless of any encumbrances such as having to update SOPs or having to write amendments before the change in the respective study plan is implemented. In this extreme case, Quality Assurance would certainly need the full support of management in order to prevail with its interpretation of GLP. In other instances, it is, however, probably just the carelessness of the Study Director which leaves such corrections unimplemented, or it is the reluctance to change ingrained ways of doing certain things which impedes rapid action for the required improvement in GLP compliance. Here, the Quality Assurance can ameliorate the situation if it keeps track of the corrections, changes and improvements requested from the different Study Directors and in the different test facilities. Monitoring the implementation of such requests will entail primarily the requirement of back-reporting by the Study Director on the planned or implemented measures, with the concomitant setting of a time limit for this reaction.

> Inspections are not just an exercise which Quality Assurance performs for its own sake. Since it is the task of Quality Assurance to ensure the maintenance of GLP compliance, it is obvious that any deviations from the rules of GLP that are observed in these inspections should be corrected. Inspection reports therefore serve the dual function of permitting test facility management to judge the functioning of the Quality Assurance itself, and of permitting test facility management and Study Directors alike to institute the measures deemed necessary for a full and continued adherence to the GLP Principles.

4.4   Audits of Raw Data and of Final Reports

The review, or audit, of a study's raw data by Quality Assurance can be carried out in a number of ways. During inspections of experimental phases of the study, Quality Assurance may already examine the records existing at that time; also during process-based inspections such data audits will be performed. These data, however, will only be a subset of all raw data belonging to the respective study. For example, during an inspection of the dosing procedures in a toxicology study the Quality Assurance inspector will have to restrict the raw data review to the in-life animal records available at the time and place of inspection (i.e. data on body weights, cage-side observations of clinical symptoms, weighing and preparation of the test item, and dosing records). Even more evident is this problem in the case of field studies, where raw data pertaining to one and the same study might be dispersed over the whole globe. Therefore, raw data will be scrutinised primarily during audits of final reports. Only at that time are the raw data pertaining to one study available in their entirety, being assembled and collected at one location in order to enable the Study Director to write the final report.

An additional point to be considered is the economy in utilising the resources of Quality Assurance which mandate that such audits should be conduc-ted at the final draft stage of the respective report, i.e. when all raw

2.4 The QA Programme

data have been gathered and no major changes are intended to be introduced into the report any more.

There are a number of issues to be addressed in a final report audit which in their entirety would then serve to determine whether the study had indeed been conducted in compliance with the Principles of GLP. Thus, the Quality Assurance inspector performing such an audit should try to determine whether the study was carried out in accordance with the study plan and the applicable SOPs, whether the study has been accurately and completely reported, and finally whether the raw data are complete and have been recorded and compiled in compliance with GLP. There are some points to be addressed which are more of an administrative nature, like the determination of whether the report contains all the elements required by GLP. One important aspect of the report audit will also be the question of whether the report is internally consistent, although this question may be seen to relate more to the scientific side than to the purely GLP aspects of the report; on the other hand, internal consistency is one of the main quality characteristics of a report, a document which is intended to provide the scientific data for the safety assessment of the test item investigated in the respective study.

While in many cases the audit of final reports can be done by accessing the entirety of the raw data, there are possible situations where this may not be feasible for a single Quality Assurance. Consider the case of a multi-site study with one or more Principal Investigators performing their parts of the study and writing their own reports. The Study Director may then possibly rely on these reports and not wish to go through all the corresponding raw data again; in some cases, especially if CROs are involved, these raw data may be deemed (intellectual) property of the Principal Investigator's test facility and might thus not be released to the Study Director and his or her Quality Assurance. Consequently, it could be impossible for the Study Director's Quality Assurance to conduct an audit of all raw data during the final report audit. In such cases, the Quality Assurance of the Principal Investigator's test site would obviously have the obligation to perform an in-depth review and audit of the Principal Investigator's final report, taking into account all the necessary issues that would have to be addressed in such a final report audit.

Until now it has been implicated that the raw data were to be scrutinised in their entirety during a final report audit. This should certainly be done in the case of smaller studies with an easily surveyable amount of raw data. A carcinogenicity study, with data on clinical symptomatology and body weight development for six hundred animals over more than seven hundred days, necropsy records and histopathology data from all these animals make for a huge amount of single data points, which are impossible to scrutinise completely within a reasonable time frame and by a reasonable effort. In the same way as it is done in study-specific inspections, where only representative samples of single activities ("critical phases") are inspected, will the audits on such study reports be confined to the scrutiny of a randomly selected sample of data. Again, the Quality Assurance SOPs have to identify those study types which will undergo a complete data audit, as well as to define the extent of data and the way of (randomly) choosing them in those studies which will undergo a restricted data audit.

At the beginning of section 4.3 the work of Quality Assurance has been likened to the conduct of a study. This likeness may be seen to continue in the recording of the procedures and the progress of a final report audit. Quality Assurance may namely find it helpful for the resolution of later emerging questions and queries to record the audit of the final report in such a form as to enable the respective audit to be reconstructed. This would entail a sufficiently detailed description of all steps taken, and of all findings observed with the accompanying actions to be proposed, in order to enable the reconstruction of the steps leading to the reasons for any changes or corrections that Quality Assurance might have deemed necessary. As already explained above in the preceding section the complete documentation of findings, changes and corrections requested, the answers to these requests, and the records of the measures finally taken, is of special importance in the context of a final study audit. In contrast to the situation with inspections, Quality Assurance has an efficacious instrument for pressing these changes in the possibility to refuse the signing of the Quality Assurance statement, which only will give the study the status of a GLP compliant one. If, however, Quality Assurance would have to resort to this ultimate means of asserting its points, then one could well become of the opinion that "something is rotten in the state of Denmark".

Economy in Quality Assurance resource utilisation has been mentioned above already. Although it is recommended that a final report audit should only be performed on the final draft of a report, there may still be some raw data outstanding at that time, or additional investigations might crop up as necessary. It would be at least annoying for Quality Assurance to have the audit finished only to become aware of some major addition to, or change in, the final report, which would necessitate to do the audit all over again. Therefore procedures must be established which will guarantee that Quality Assurance is made aware, as they occur, of all additions or changes made to the study data and the study report during the audit phase. As a last point, it goes without saying that any correction of, or addition to, a completed final report must be audited by Quality Assurance. The question of how to formally and materially deal with such grave issues, i.e. concerning "non-finality" of the final report, will be considered in the section on the final report (see section 11.4, page 255).

AUTHENTICATION

'I, the undersigned, hereby declare that this work was performed under my direction according to the procedures herein described and this report represents a true and accurate record of the results obtained.'

QUALITY ASSURANCE STATEMENT

Neither the procedures involved in this study nor the report of this study require to be inspected or audited by a Quality Assurance Unit.

Figure 13: This study had not been in need to be conducted to the GLP Principles, therefore the Quality Assurance issued a statement to this effect.

One might possibly be tempted to interpret the economy issue in Quality Assurance resource utilisation also in an additional way, namely to perform report audits for short-term studies in a random fashion, analogous to the inspectional practice. This possibility, however, is precluded by the GLP Principles. They require that Quality Assurance "*inspect the final reports to confirm that the methods, procedures, and observations are accurately and completely described, and that the reported results accurately and completely reflect the raw data of the studies.*" Furthermore, since Quality Assurance has to "*prepare and sign a statement, to be included with the final report*", which should also "*serve to confirm that the final report reflects the raw data*", it follows that all final reports, for which GLP compliance is claimed, have to be audited by Quality Assurance. This requirement holds in an absolute way: For every GLP study the final report has to be audited and raw data checked for completeness and for congruency with the report. There are no special provisions for "sample" or "random" auditing of short-term studies, for which the possibility of performing process-based inspections may lead to some studies not being inspected during the experimental phase. GLP compliance of any study can only be claimed, if the final report has been inspected by Quality Assurance and has been found to reflect the raw data. If a study were not in need of such a final data review because it would not fall under the necessity of full GLP compliance, the Study Director might nevertheless claim that the conduct of the study had followed, if not the letter, then at least the spirit of the GLP Principles, together with a declaration that no report audit had been performed because of the perceived lack of need (see the example of such a statement in figure 13 above).

> Even the best inspection programme cannot ensure in an absolute way the complete, continuous adherence to the GLP rules during a study. The final report audit may, however, be considered to make up in a way for this "deficiency". Through the scrutiny of raw data with respect to their GLP compliant mode of recording, and through the comparison of the original raw data with their representation in the final report, Quality Assurance will be able to obtain a fairly accurate picture of the extent of GLP compliance in the audited study. The audit of the final report, with its detailed assessment of GLP compliance throughout the study and with its concomitant review of all relevant information, records and data, thus serves to ascertain, from another angle, the quality and integrity of the specific study.

4.5 *The Quality Assurance Statement*

The Principles of GLP require Quality Assurance to "*prepare and sign a statement, to be included with the final report, which specifies types of inspections and their dates, including the phase(s) of the study inspected, and the dates inspection results were reported to management and the Study Director and Principal Investigator(s), if applicable*". Thus, after having audited the final report of the study, and after having been satisfied with the way the raw data have been recorded and represented in the final report, Quality Assurance has to look at its own records to identify the respective activities performed in connection with the study in question. Quality Assurance has then to draw up a list of such activities with their dates and to include this list in the Quality Assurance statement.

This may sound rather easy, but the revision of the GLP Principles has brought about a major change in the requirements for the content of the Quality Assurance statement. The earlier OECD guidelines called only for providing the dates when inspections were made and for providing the dates when these inspection reports had been made available to management and to the Study Director. This has led to relatively meaningless Quality Assurance statements like the one presented on the next page as a very minimalistic example (see figure 14). In conformance with the regulations of the US FDA and EPA, the Principles now are calling for a further qualification of the inspections made, namely for the specification of the phase of the study which had been inspected. This requirement poses now a certain difficulty, since the term "phase of a study" is nowhere defined in the GLP Principles; neither do the respective regulations of FDA or EPA define it, short of putting it on equal footing with the term "segment of the study". The problems encountered with this term have been discussed already in section 2.10, where this subject has been exhaustively treated (see page 88).

In the context of the Quality Assurance statement it remains to be discussed which one of the two options, whether the precise description of single activities (see figure 12), or the simple enumeration of wide-ranging test parts, e.g. the distinction between study conduct and data review (see the example in figure 15), would be the most useful one. If one would equate the term study "phase" with a "segment" of the study, then the Quality Assurance statement would include the mention of the dates on which a toxicology study had been inspected in its in-life phase, and in its necropsy and histopathological phase, whereas a Quality Assurance statement of a field study could be restricted to the mention of the field phase and the analytical phase. On the other hand, and what might be considered by the Regulatory Authorities as being much more informative for their assessment work, a Quality Assurance statement of a toxicology study might mention the particular activities monitored, like the preparation for, and execution of, the dosing, the sampling of urine or blood for, and the procedures of, the respective analytical activities, or the necropsy with its accompanying activities (see figure 12). In a field study these activities could also be described in exact detail, such as the preparation of the spraying concentrate and the respective tank mix, the spraying itself, the sampling of the crop and the ensuing analytical activities. Such an interpretation of this requirement may, however, lead to a very long and detailed Quality Assurance statement, covering in the extreme two or three pages.

QUALITY ASSURANCE STATEMENT

Quality assurance inspections of Study   were made and the findings reported to the Study Director and management on the following dates:

01-21-91	07-17-91	12-10-91	05-13-92
01-22-91	07-25-91	01-13-92	05-28-92
01-24-91	07-30-91	01-15-92	06-01-92
01-31-91	08-06-91	01-24-92	06-10-92
02-18-91	09-05-91	01-30-92	11-16-93
02-21-91	09-10-91	02-04-92	
03-13-91	09-11-91	02-11-92	
03-26-91	10-29-91	02-13-92	
04-24-91	11-05-91	03-03-92	
05-07-91	11-12-91	04-07-92	
05-31-91	11-13-91	05-12-92	

This study was conducted in accordance with Good Laboratory Practice for Nonclinical Laboratory Studies regulations (21 CFR Part 58). Data reported were compared to data records and found to be accurate.

Figure 14: Completely uninformative Quality Assurance Statement, where it is not possible to establish what kind of activities within the study had been the subject of the inspections.

To describe in this most exact way the activities which the Quality Assurance had monitored would certainly provide for the highest level of transparency with regard to the activities of the Quality Assurance. Regulatory Authorities might then be better able to check the real value of this Quality Assurance statement. On the other hand, fears might linger that, given that such exact details were provided in the Quality Assurance statement, some Regulatory Authorities would ask nasty questions why this or another activity had not been part of the surveillance programme of the Quality Assurance, and might ask for an official study audit. The use of a more general description of the "inspected phases of the study", rather than a description to the fullest extent feasible, might thus seem a balanced compromise avoiding fruitless discussions, while retaining the relevant information.

There is another point that has to be taken into account by Quality Assurance when writing (or by the Regulatory Authority when reading) a Quality Assurance statement. The format of the Quality Assurance statement will be specific to the nature of the study and thus of the final study report. While it is required for all studies that the statement include on the one hand the full study identification and on the other the dates of relevant Quality Assurance monitoring activities, these latter may be of a variable nature with respect to their connection to the study. For example in the case of short-term studies, where single or repeated inspections for each study are inefficient or impractical, where thus individual study-based inspections have not been part of the Quality Assurance activities, and where therefore the Quality Assurance statement is referring to process-based inspections only, details of the monitoring inspections that did take place must be included. These should demonstrate that Quality Assurance did in fact monitor the critical phases of similar studies in a high enough frequency and inspected them in a time frame

---

QAU INSPECTION STATEMENT

SINGLE DOSE TOXICITY STUDY OF   IN MICE AND RATS

This study has been inspected by the Quality Assurance Unit, and the findings have been reported to Management and the Study Director on the following dates.

Items	Inspected Dates	Reported Dates
Conduct of the Study	8/23, 8/24, 8/28, 9/11, 9/28/1990	9/11/1990
	2/15, 2/26, 2/27, 3/13/1991	3/13/1991
Records and Report	6/26, 6/27, 6/28, 7/23, 10/30/1991	7/23, 10/31/1991

Figure 15: Subdivision of Quality Assurance activities in inspections and audits, without detailing the inspected activities within, and "phases" of, the conduct of the study.

that provided sufficient and relevant coverage of the GLP compliance for the procedures used in the "un-inspected" study to be distinguished, nevertheless, with a Quality Assurance statement. It would certainly not be considered sufficient coverage, if the "critical phase", say the observation of the actual melting point determination, would have been the subject of an inspection more than three years ago, and possibly on an apparatus that is no longer in use. Like the utilisation of historical control data in toxicology studies, which are also only considered supportive, if the data base is from within the last two years, a process-based inspection should be as recent as possible in order to really support the claim of the compliant conduct for an uninspected study. It would also be advisable to list the last few inspections, if possible bracketing the study in question, in order to enhance credibility for GLP compliance.

Although not expressly stated in the Principles, it may be argued - as it has been noted in the Consensus Document on Quality Assurance (OECD No. 4, 1999) - that it should be the responsibility of management to provide policies, guidelines, or procedural descriptions to ensure that this statement reflects Quality Assurance's acceptance of the Study Director's GLP compliance statement. Thus, the responsibilities are clearly distributed between Study Director and Quality Assurance: The Study Director is ultimately responsible for the GLP compliant conduct of a study, and the Study Director has also to ensure that any areas of non-compliance with the GLP Principles are identified in the final report. On the other hand, Quality Assurance has to make sure that the Study Director's claim of GLP compliance can indeed be sustained, at least as far as can be ascertained through inspecting study conduct and auditing the raw data and the final report. The Quality Assurance statement should furthermore indicate that the study report accurately reflects the study's raw data. Before signing the Quality Assurance statement, Quality Assurance should ensure that all issues raised in the Quality Assurance audit, i.e. in the audit report to the Study Director and to management, have been addressed through appropriate changes of the final report, that all agreed actions have been completed, and that no additional changes have been made to the report which would require a further report audit. Through management policy it should certainly be made clear that the Quality Assurance statement would only be completed if the Study Director's claim to GLP compliance can be supported.

The problematic area of "study fragmentation" which has already been discussed at various places in this book has its reverberations also in the question of the Quality Assurance statement. Especially for multi-site studies, where more than one Quality Assurance unit might have been active, or for subcontracted parts of studies conducted at test sites not connected with the Study Director's facility, the question may arise whether there should be multiple Quality Assurance statements appended to the final report, each of them reflecting the involvement of the different Quality Assurance units with the respective, specific parts of the study.

The logical dissection of the various possibilities will, by application of the guiding idea of GLP, provide the unambiguous answers. The Study Director's Quality Assurance unit has to judge and attest to the overall GLP compliance of the study. If this "primary" Quality Assurance cannot perform the necessary inspections itself, then it has to rely, for this judgement to be rendered, on the reports and statements of any "secondary" Quality Assurance, irrespective of whether this "secondary" Quality Assurance were employed by the same company or by another entity. Thus, such inspection reports, or the therefrom derived Quality Assurance statements will constitute the raw data on which the "primary" Quality Assurance will have to base its own, comprehensive statement. The statements of other Quality Assurance units will thus just support the judgement of the ultimately responsible, "primary" unit in those areas or test facilities where this unit could not – for one reason or another – satisfy itself of the GLP compliant conduct of the respective study parts. Since the "primary" Quality Assurance of the Study Director's test facility is responsible for attesting to the GLP compliance, there is no requirement for such "secondary" statements to be regularly appended to the final report or the "primary" Quality Assurance statement. In the case of the Study Director appending the report(s) of Participating Scientists or Principal Investigators to the final report, an inclusion of the respective "secondary" Quality Assurance statements with these reports would be certainly advisable. On the other hand, if the Study Director receives only raw data from these study phases, or utilises the submitted reports as "raw data" for his own report, without appending them, then the inclusion of any "secondary" compliance statements is clearly not warranted. The Quality Assurance should, however, handle these situations in a consequent way, which would preferably also be described in their SOPs.

There may be further special cases connected with study reports and Quality Assurance statements, which will have to be resolved on a case-by-case basis.

One such case may be the question of report amendments. If a report should have to be amended, because either further data from other studies may have invalidated some of the conclusions drawn in the original report, or because an error has been detected which had escaped earlier on the attention of the Study Director and the Quality Assurance, then this amendment to the report will again have to be audited by the Quality Assurance, which will furthermore have to prepare an amendment to the former Quality Assurance statement.

Another such case may involve the treatment of areas of non-compliance. There may be differences in opinion between sponsor and Quality Assurance of the test facility with regard to certain exceptions to be addressed in the Quality Assurance statement. There may even be divergences arising between Study Director and Quality Assurance in their respective judgement of compliance issues; here, again, test facility management would be summoned to arbitrate the case and to decide on the action to be taken with respect to the Quality Assurance statement.

> The Quality Assurance statement has a two-fold function. It firstly serves to demonstrate that Quality Assurance has adequately monitored the conduct and progress of the study, from the first check of the study plan for GLP conformity to the audit of the final report as a "second opinion" on the completeness of the reporting and the adequacy of raw data coverage. It secondly provides the study with the seal of approval by attesting to the GLP compliant conduct. In this sense the Quality Assurance statement carries as much weight for the assessment of the study's integrity and validity as the Study Director's signature.

5. Facilities

A whole section in the OECD Principles is devoted to facilities in general, and to some special ones in particular. Facilities are only a part of the **test facility** which is defined as the sum of *"persons, premises and operational unit(s) that are necessary for conducting the non-clinical health and environmental safety study"*, and it is the management of the test facility which is responsible for ensuring that "... *appropriate facilities, equipment, and materials are available for the timely and proper conduct of the study*". Several types of facilities can be differentiated, each of them having separate requirements for becoming compliant with the GLP Principles. Although being formulated in slightly different ways, these requirements do not differ very much in their basic tenet: The protection of the study against any possibilities of jeopardising its integrity can be regarded as their common key.

5.1   General Requirements

Facilities need to conform to a number of general rules before they can be considered as GLP compliant. These general requirements are not formulated with the purpose of compelling management to provide a kind of facilities that would allow the employees, the technicians, the farm hands or the animal caretakers to enjoy the most spacious, comfortable or recreational workplace. The facilities should be designed for the utmost suitability to the studies that are to be performed within. Some comfort for the employees comes of course with the all-embracing requirement of study quality, which means that the people working in a facility should certainly have sufficient room to move around in order to be able to perform the duties which the study calls for, and to perform them in a manner compatible with the quality, integrity and validity of the study. This is recognised implicitly in the general requirement that a test facility should be of suitable size, construction and location, with the dual purpose to "*meet the requirements of the study*" and to "*minimise disturbance that would interfere with the validity of the study*". The former point would obviously mandate field plots to be large enough for the siting of the entire study or study part, or storage rooms to be spacious enough to accommodate all the necessary materials. The latter requirement, on the other hand, would certainly not be met by an animal room so small that each employee working in it would constantly bump into the cage racks; such a room would tend to maximise rather than to minimise interferences with the test system and the study!

It stands to reason that, depending on the test system used, the availability of an adequate environmental control system has to be considered. Such controls should provide, e.g. for animal rooms, adequate lighting and photoperiod conditions, and an adequate air conditioning system in order for the animals to be kept under standard and well defined conditions.

On the other hand, there may be problems to keep standardised conditions in a field study, since there the facilities to be utilised, e.g. for the preparation of tank mixes, may not be environmentally controlled. The US EPA regulations, which are much more specific in many ways regarding test facilities, address this problem by stating that *"Testing facilities which are not located within an indoor controlled environment shall be of suitable location to facilitate the proper conduct of studies"*. This may mean, e.g. in the case of temperature sensitive test parts or procedures, that it would be advantageous to locate the respective rooms so as to face away from the midday sun, rather than being placed in the most sunny corner of the building. Thus, already the lay-out of the test facility with its rooms and their later destination could be of importance to judge the GLP compliance of the facility as such.

There are more examples for the application of these requirements: If a study involves analytical procedures, the facility has to have an adequate power supply with adequate provisions for the case of power failures or breakdowns. The same provisions have to be taken for the air-conditioning system of the animal rooms. Furthermore, an adequate ventilation system will be needed in order to protect equipment and technicians from noxious or corrosive gases and volatile solvents. The appropriateness of the facility and its construction can be followed down to the small table on which the balance to be used in the study is placed: Does this table have sturdy legs and a special, vibration-proof insert so as to really *"minimise disturbance"*, or will the weighing become perturbed every time when somebody enters the room, or even by the breathing of the person performing the weighing?

Another of these general requirements is the provision that there should be *"an adequate degree of separation of the different activities to assure the proper conduct of each study"*. Only if this is accomplished, a study can be conducted in such a way that the various activities do not interfere with each other and with the test system itself. Thus it is not only the separation of different studies which is envisaged (which of course is also an important aspect to which we will come later), but any function or activity that might have an adverse effect on the test system and thus on the integrity of the study should be performed separate from other such activities or functions. There are activities which are connected with high levels of noise, as e.g. the washing of animal cages or the pelleting of feed, activities which would certainly disturb the animals, if these activities were performed in, or adjacent to, their housing.

With regard to the general provisions for facilities it is important to note that not only the facilities for the test system as such require attention, but also

2.5 Facilities

the facilities in relation to test facility and study logistics, starting from the supply of materials needed for the maintenance of test systems and going through to the final disposal of wastes, need to be considered. Since an adequate supply of such materials as animal feed, animal bedding, or soil for plant test systems should be available at all times, there have to be adequate storage rooms or areas. Again, these rooms or areas need to be separated from rooms or areas housing the test systems in order to avoid disturbances to, or untoward influences on, the test systems. The storage facilities should also provide adequate protection against infestation by insects or rodents or contamination by other environmental influences, and they should allow the preservation of perishable supplies by appropriate means e.g. by providing control of environmental conditions. This latter condition may be exemplified by the storage requirement of Guinea pig: Due to its high content in vitamin C needed to ascertain the adequate nourishment of these animals, their feed has to be kept under conditions of cool storage in order to retain its nutritive value over the required time period.

At the other end of the material flow, facilities need proper provisions for the collection and disposal of all kinds of wastes, e.g. animal waste and refuse or contaminated water, soil, or other spent materials. If such waste cannot be disposed immediately provision has also to be made for safe sanitary storage of such waste before removal, by appropriate transportation procedures, from the test facility. Inherent in these requirements for safe handling and disposal, too, is the question of the integrity of test systems, or as the OECD Principles put it: *"Handling and disposal of wastes should be carried out in such a way as not to jeopardise the integrity of studies."*

There is a final point with regard to the general demands on the test facility, which is not mentioned in any guideline. This point may be summarised under the heading of "tidiness". Not only does a room which is kept in an orderly and tidy condition provide the appearance of being a place of quality work, it indeed makes it much easier to comply with the requirements of a study. When the laboratory bench is cluttered with clean and dirty glassware, when there are instruments lying around, some of which are being used and some are not, then it becomes difficult to locate all the materials needed for a specific activity. It would also not enhance the confidence in the work performed at a place, where the variety of nozzles for the spraying equipment to be used in field tests are being kept in such a fashion as if they had just been tossed into a bucket after use. Tidiness therefore has the dual function of inspiring trust into the quality of the work performed, and of facilitating the performance of the daily activities according to the quality standards aspired to. Last but not least tidiness makes it easier to survive a compliance monitoring inspection, if – even under the stress of being interviewed by the inspector on a specific point of activity performance – the technician can find the folder with the SOPs at once and without having to hunt for it.

> Facilities have to be designed for the utmost suitability to the studies that are to be performed within. The general requirements calling for adequate environmental controls, for proper separation of activities, and for allowing studies to be conducted under conditions of minimal disturbance aim at the protection of the study against many possibilities of jeopardising its integrity.

5.2 Test System Facilities

Of eminent importance are the facilities that are used for the housing or siting of test systems. Some of the requirements dealing with test system facilities have already been addressed above, especially those that are concerning the technical side of the issue. In this section the more organisational matters of test system facilities are to be described. One of the fundamental issues in the requirements for test facilities is the provision for adequate separation. Only if an adequate separation of different activities, various materials, different test systems and, last but not least, different studies is accomplished, the integrity and validity of studies and their results may be ensured.

The sad experience with the conditions at IBT, where multiple studies were run simultaneously in the same room with the concomitant problems of mix-ups of animals and treatment cross-contamination by volatile test substances, have resulted in the requirement of sufficient space to "*assure the isolation of test systems*". Through a sufficient number of rooms or at least sufficiently separable areas, it should become possible to avoid any cross-contaminations or mix-ups of projects, tests or treatments. Also the positioning of test systems used in field tests requires an appropriate degree of separation, as it is specified in the GLP Principles ("*Test systems used in field studies should be located so as to avoid interference in the study from spray drift and from past usage of pesticides*"). In the same sense, isolation of individual projects in aquatic toxicity testing should be applied to the extent necessary, to prevent cross-contamination through spray, mist or overflow.

A second, special aspect of this requirement is the necessity to isolate *"individual projects, involving substances or organisms known to be or suspected of being biohazardous"*. Such isolation involves as much the question of study integrity as it does concern the safety of the study personnel. Biohazard may involve infectious microorganisms, carcinogenic substances, or radioactive compounds, to name just the most blatant examples. Legislation regulating the utilisation of radioactive substances, e.g., will call for a separate "Radio-Lab", if the amount of radioactivity used in, or applied to, the test system is exceeding a certain level; however, even with the application of radioactivity to a test system in such low amounts which *per se* would not necessitate the use of a special, protected laboratory, it might be advisable to separate this test or test system from any other project where no isotopes are used. The same precautionary measures of isolating certain projects from others should be adopted for studies or test systems employing infectious agents, or involving gaseous or readily volatile test, control, and reference items or preparations which might be forming aerosols. In all these cases a test facility should be able to provide adequate separation possibilities.

In many instances there is no need to completely isolate single studies from each other by assigning different rooms to each of them, as long as the actual extent of separation allows for an adequate achievement of the study objectives. What may be needed for toxicity studies may not apply for analytical ones. In the latter case, study separation is obtained on the test system itself, as the samples corresponding to one study will be analysed in one run together, while those of the next study will still be sitting in the refrigerator and wait for their turn at the HPLC. There is most certainly no need to utilise separate rooms for the housing of test systems which are already intrinsically separated, as are plant or aquatic test systems. Proper separation of test systems consisting of fish, Daphnia, algae or higher plants within a room or area can be accomplished without difficulty as the single study groups or individuals will anyway be kept separately in different chambers or aquaria. Finally, in instances of ecotoxicology testing with whole ecosystems or mesocosms, where the protocol specifies the simultaneous exposure of two or more species in the same chamber, aquarium, or housing unit, separation of species would certainly run counter to the study purpose.

Isolation is the keyword to another aspect of facilities for biological test systems. Animals will at one time or another get ill or will suffer from injuries, and there should thus be "*suitable rooms or areas ... for the diagnosis, treatment and control of diseases, in order to ensure that there is no unacceptable degree of deterioration of test systems*". Effective isolation of test system individuals or of whole test systems which are either known to be, or are suspected of being, diseased will be needed to ensure the integrity of other test systems being used at the same time. Infectious diseases may not only be lethal and thus wipe out whole test systems, annihilating whole studies in the process. Even in cases, where a disease would apparently only affect the appetite and thus the bodyweight development of the animals, problems with the scientific validity of such a study could arise due to the probable concomitant changes in physiological or immunological parameters of the test animals, and the validity of the conclusions drawn from such studies might then well be doubted.

Isolation does, however, not mean that access to the test system or their housing or treatment localities generally needs to be restricted, save in some special circumstances, and apart from what would be dictated by sound scientific reasons. An example in case would be the restrictions for the access to certain animal facilities: Because of special requirements, e.g. because of a "specific pathogen-free" or even sterile environment, access to these may have to be limited to authorised persons only.

> A study will yield valid results only, if the respective test system has been properly located or housed. Its integrity will also critically depend on the proper separation from other studies, and the respective requirements aim at minimising not only the potential for mistakes but also the occurrence of reciprocal disturbances. In this sense all conditions or situations which might lead to untoward influences on the study in progress will have to be taken care of through the adequate design and lay-out of the respective facilities.

5.3 Facilities for Handling Test and Reference Items

Even in medium- to small-sized test facilities a multitude of test items in different stages of their "life-cycle" could be present at any one time, some of them being utilised in tests, some awaiting testing, some remaining as samples of items that had been tested, and even some remaining as left-overs awaiting final disposal. This situation calls for a well thought-out concept of logistics for receiving, storing, handling and disposing test items, together with provisions for the adequate documentation of all procedures connected with test item handling. One aspect in this area of test item logistics is the physical location of these activities, and the GLP Principles underline the importance of identifying adequate facilities for them.

We have already seen in the foregoing section (see above) that the GLP Principles are stressing the potential problems of test item mix-ups or contamination, and they are therefore keenly striving to avoid such occurrences by instituting a number of measures intended for ensuring the identity, purity and stability of the test item throughout the study. One important aspect in this chain of measures is the requirement that there should be *"separate rooms or areas for receipt and storage of the test and reference items, and mixing of the test items with a vehicle"*. The separation of receipt and storage rooms or areas from the ones where other activities are performed with test items has to be seen in the light of the above mentioned problem. While receipt and storage involves mainly the handling of closed containers, the opening of such a container, in order e.g. to take out a sample for the preparation of the application form, exposes the test item to the facility environment and leads consequently to the possibility of contamination of either the test item or the environment. Furthermore, the greater the number of different test items that are standing around, the greater the danger that somebody would, in a hurry, and because of the similarity of containers, mistake one test item for another one, and the mix-up would be perfect.

Therefore it is foremost common sense which would dictate that work in the special area where test items are prepared for application has to be carefully organised. For weighing of the test item and its mixing with the vehicle, it should be made mandatory that only one test item (or at most a very few ones) would be present in that area at any one time. Not only could mix-ups be prevented in this way, but also the possibility, actually the necessity, of cleaning this area before the next test item is being weighed and mixed should lead to diminishing the danger of contamination or cross-contamination. By the same reason, GLP mandates that test item storage locations should be available separate from the rooms or areas containing test systems in order to prevent undue exposure of the systems to test items other than the intended one. However, the wording of the respective paragraph (*"Storage rooms or areas for the test items should be separate from rooms or areas containing the test systems. They should be adequate to preserve identity, concentration, purity, and sta-*

bility, and ensure safe storage for hazardous substances.") would seem to leave some room for interpretation, since the separation is not confined to "separate rooms", but "areas" are mentioned as well. Would it thus be admissible to store the test item in one corner of a room, while housing the test system in the opposite corner of the same room? The question may not be relevant for the relatively small amounts of a test item in its pure form, for which a storage area may be found outside the room housing the test system, but it may become highly important when it concerns the storage of test item mixtures with the respective carrier, e.g. in the case of feed admixtures. There, large quantities of feed containing the different dosing concentrations of the test item will have to be stored, and it might pose logistic problems to find adequate space for this purpose.

Certainly, rationality and judgement have to be exercised in such instances, which have to be considered on a case-by-case basis, and according to the known physico-chemical properties of the test item. Furthermore, under a pragmatic approach, a limited supply of the ready-to-use test item/feed mixture could well be kept in the room housing the test system, but it would not normally be acceptable to store longer-term supplies of such mixtures there. In any case it would certainly be necessary to assure on the one hand that the storage of test item or its application mixture under these conditions would not compromise the integrity and homogeneity of the mixture nor the stability of the test item in this mixture, and on the other hand that this storage would not jeopardise the integrity of the study itself, e.g. through the danger of mistake of confusion.

Of course, the storage facilities should provide adequate conditions to preserve the identity, purity and stability of the test items. Normally, storage areas at different temperature levels, for storage at room temperature or in refrigerators and deep freezers would therefore be needed. Also protection from light, humidity or oxygen may be necessary for special cases.

Last but not least there are also the security aspects to be mentioned. Storage provisions don't need to go as far as requiring a lock on each and every cupboard or refrigerator, where test items are maintained. A suitable limitation for access to the test items should, however, be advisable, so that not everybody could just drop in and take some spoonfuls out, or tamper in any other way with test items. It is very important that a good, accurate accounting system should be in place, which could be used to reconstruct the course of test item utilisation. The introduction of such a system would necessitate the nomination of a "test item handler" by the test facility management, through whom all the different test item activities - receipt, distribution, return and disposal - would be conducted. Unlimited access to the test item storage locations with the concomitant loss of control might in this sense be regarded as counter-productive.

> When many activities of a similar kind are performed at one and the same place, and when large numbers of test items or test systems are handled within one room or area, the number of errors, mistakes and mix-ups would probably increase exponentially. A proper separation is therefore necessary to provide an appropriate guarantee for the correct conduct of studies and for their integrity.
>
> In providing guidance on what to separate, how to separate, and to what extent to separate, GLP aims on the one hand at minimising the risk for "annoying" mistakes to happen, and on the other hand at maximising the trust in the correctness of study conduct, when - even in retrospect - it can be assumed that no chance exists for mix-ups to have occurred.

5.4 Archive Facilities

The most important aspect of GLP, the provision for the possibility of a study to become fully reconstructed in order to enable the ascertainment of the study integrity and quality in retrospect, is contingent on the full and secure retention in an archive of the whole documentation connected with the various studies. The term "documentation" should not be regarded to cover written documents only, since the GLP Principles call for the archiving of all material originating from studies; furthermore, also material generally related to the test facility has to be archived, as well as Quality Assurance is obliged to retain the respective records in a special archive. Therefore, management is responsible for providing archive facilities, which have to be adequate "*for the secure storage and retrieval of study plans, raw data, final reports, samples of test items and specimens*".

One of the considerations is that the storage should be "secure", and this aspect is addressed further in the requirement that the design of, and environmental conditions in, these facilities should protect the archived material *"from untimely deterioration"*. It is therefore not sufficient to assign some shelves on the back wall of a humid cellar, or a wooden cupboard under the roof of the building, or some other area, unsuitable for any better use, to hold the GLP archives. It may even mean that, depending on the test facility and the nature of testing being performed there, a number of different archive facilities have to be available, each of them specially constructed and maintained to provide adequate storage conditions for specific materials. While it may be sufficient for the archiving of paper raw data, study plans and final reports to provide the necessary space under dry conditions, protected from fire, water and corrosive gases, more stringent conditions will be necessary for the storage of tissue specimens from toxicology studies. The formalin fixative used to store such specimens would necessitate cooled rooms with adequate ventilation for the reduction of the formaldehyde concentration in the ambient air. Samples of the test and reference items have of course to be stored under the original conditions which were already applied during the testing phase. Specimens from field tests or animal plasma samples from toxicokinetic studies may need storage in freezers in order that their usefulness for further analysis be maintained. Furthermore, a special case may be seen with the problems of archiving electronic storage media, where not only the commonly used precautions for safety will apply, but where also the necessary protection from strong magnetic fields will be an issue.

> Reconstruction of a study is only possible, if all documents, records and materials from this study can be made available in an unadulterated and unspoiled condition for scrutiny and for tracking the course of events. Archive facilities have to satisfy such conditions as to provide an environment where the completeness and the integrity of study-related materials can be continually ensured.

6. Apparatus, Materials and Reagents

During the course of a study a number of apparatus and materials will be utilised for various purposes. Technical equipment, instruments and apparatus will be required either directly for the generation, storage and retrieval of study data, or indirectly for the control and maintenance of suitable environmental conditions. Materials of the most varied kind, like bedding for animals, animal feed, test system containers, bags and jars for the collection of samples and specimens, or protective clothing will have to be utilised in the test facilities and test sites, materials which will also have to satisfy certain standards and specifications. Finally, reagents and solutions will have to be prepared in the test facility for a number of uses, directly or indirectly related to the test system and the study.

In order to satisfy the basic requirements of GLP, chemicals for general uses, like the various substances needed for the preparation of buffer solutions, the wide range of chemicals used in the fixation, storage, embedding and staining of histopathological specimens, or the fungicides, insecticides, herbicides and fertilisers needed in the preparation and maintenance of test plots in the field, should *"be labelled to indicate identity (with concentration if appropriate), expiry date and specific storage instructions. Information concerning source, preparation date and stability should be available."* These requirements are again meant to provide the means for the reconstruction of a study through ascertaining that also these ancillary products have most probably performed to specifications and have therefore not negatively influenced the quality and integrity of the study itself. For a more in-depth discussion of the required information on expiry dates, the reader is referred to section 9.3 (see page 218). Certainly, not every single instance of preparing, e.g., a 70 % ethanol solution or a common phsphate buffer, would require complete documentation in the same way as it would be the case for the preparation of the test item application form. In adhering to the correct labelling of all chemicals, reagents and solutions at all times, the test facility will be able to provide "circumstantial evidence" for the correct preparation and use of such reagents and solutions, while the information on the source of the chemical substances used may be considered as adequate guarantee of the quality and suitability of the original product.

In an analogous way, materials used in studies, and in the test facility in general, will have to be documented with regard to quality and suitability. For instance, periodic analyses will have to be conducted on animal feed, animal bedding or drinking water, in order to ascertain the quality and suitability of these materials. Such analyses will not only provide evidence for the quality of the respective materials, but will also be useful for fulfilling the requirement that "*apparatus and materials used in a study should not interfere adversely with the test systems*". For materials coming into direct contact with the test system, this may mean that they should be free from toxic contaminants which might alter the response of the test system to the test item in a way unrelated to the action of the test item itself and thus possibly leading to wrong conclusions from such adulterated study results.

Two aspects have to be addressed in the context of the apparatus used within the test facility which are of importance for the quality and integrity of the studies in which these are utilised. The first aspect concerns the suitability and adequacy of the respective equipment for the specific task it is expected to perform. Consequently, the GLP Principles require that apparatus should be "*suitably located and of appropriate design and adequate capacity.*" This first point may be regarded as the static prerequisite of the second, dynamic aspect, the adequate maintenance and functional control of the respective technical equipment. In consideration of the ultimate purpose of the GLP requirements, i.e. to provide the means for the possibility for the reconstruction of a study, this second point is of utmost importance. Only through the required documentation proving that the instruments had been "*periodically inspected, cleaned, maintained, and calibrated according to Standard Operating Procedures*" will it be possible to ascertain with confidence, at a later date, the correct functioning of any relevant apparatus during the conduct of a specific study (for an illustrative example see figure 16 on the next page).

The GLP Principles refrain here from suggesting or requiring any specific time intervals for such activities. On the one hand cleaning and maintenance intervals may be different from one type of instrument or apparatus to the other, and such intervals may as well depend on the frequency of use or the workload imposed on the respective equipment. On the other hand the question of the correct frequency of such activities should be considered as a scientific one, calling for the expert judgement of the responsible scientists. In many cases the manufacturer's manuals will provide useful hints or recommendations for cleaning and maintenance intervals. These same aspects are valid also for calibration frequencies, where in some cases calibration is routinely performed before each measurement, while in other cases the respective frequencies may be set in an arbitrary manner. A balance will most probably have to be calibrated before each weighting, while the volume calibration measuring pipettes might well be confined to a check every three to four months. Since these intervals and frequencies will have to be set individually

2.6 Apparatus, Materials, Reagents

CompliantLab, Inc.

Logsheet for (Apparatus)

Internal Instrument No.	Manufacturer Serial No.

Date	Use	Remarks	Initials
17-01-00	study X	calibration ok	JM
18-01-00	study X	calibration ok	JM
19-01-00	check	weekly maintenance: ok	HS
20-01-00	study X	calibration ok	JM
24-01-00	study Y	calibration not linear	AM
29-01-00	service	performed by manufacturer; calibration ok (see service report 00358/00	HS
02-02-00	study Y	calibration ok	AM
03-02-00	check	maintenance: ok	HS

Figure 16: Example of a log-sheet for apparatus showing the entries for the regular maintenance and calibration activities; the log-book or log-sheet has to be kept at the apparatus itself, since external technicians will have to provide records for their maintenance services, too, and the presence of the log-book will thus serve as a reminder.

for single instruments or types of equipment, they should be described and fixed in Standard Operating Procedures for the respective apparatus.

The key point in the consideration of maintenance and calibration frequencies is the necessary assurance of data validity. In certain cases it will additionally be necessary to ensure the traceability of the calibrations performed to *"national or international standards of measurement"*. Although this requirement may be seen to lean more towards the scientific side than to the side of reconstructability, in the sense that this traceability will provide evidence for the degree of exactness in some parameter(s) measured during a study, it can nevertheless be important for certain validity aspects in a study, depending on the nature of this study. In some studies, the experimental results are just the raw material for a scientific interpretation, which will become the ultimate result of the study.

The main example for this kind of studies is the toxicology study, where for instance the determination of animal body weight has no absolute interest in itself. It will not matter whether the weight of the rat no. 634 is precisely 257.9 grams (precision meant in terms of the conformity with the Standard Kilogram), since the body weight of a rat will anyway depend more on a number of additional factors (e.g. defecation or water consumption prior to, or immediately after, the weighting) than on the absolute precision of the balance. What really interests furthermore, in terms of the toxicological assessments of the test item effects, is the relative development of body weights of treated *versus* control groups, which will be used to determine whether the test item evoked some unwanted, noxious effect on the animal. Since these body weights would therefore only be used for a comparison to the respective concurrent control values, i.e. used in a relative way, their true, absolute value could be considered as being not overmuch important. The calibration of the balance used in such a toxicity test with any kind of calibration weight might therefore be considered sufficient, since such a calibration would provide for internal consistency of the weight measurements within the study, and the traceability of the laboratory set of calibration weights to the international mass standard would be considered of lesser importance or even unnecessary. In contrast to this, there are those studies in which the value of the experimentally determined parameter is forming the ultimate test outcome. To this category belong the determinations of physico-chemical parameters, where the absolute values are required as the end result of the study. In the determination of a melting point, it is the "true" value of the melting temperature which is the aim of the study, and therefore in this case, it should be important to be able to judge the reliability of the measurement by being able to trace the calibration of the thermometer back to an acknowledged standard.

However, this requirement may not have to be interpreted in an as stringent way as it would seem at first sight, and it could be fulfilled in some instances in a rather easy way. In utilising calibration weights provided by the manufacturer of the respective balance the test facility could rely on the certification of the manufacturer who had calibrated them against the national standard. For temperature determinations, the respective thermometers or measuring devices might, at intervals, be checked against a standard thermometer in the test facility, which in turn could be calibrated once in a while against the national standard.

If the above mentioned calibrations against an international standard of measurement are representing one end of the spectrum of measurement precision, there may be other measuring devices which are very difficult to calibrate. One of the most obvious examples for the difficulty of calibrating a device to any standard can be seen in the spraying nozzles of field test equipment and in the application of test item in field tests in general. Apart from obvious technical difficulties of collecting a representative sample from the sprayer in action, there are also some other, physical/chemical determinants which make calibration of such equipment a less than exact science. The preparation output by the spraying equipment is determined by the inner diameter of the nozzle as well as by the pressure exerted and the viscosity of the test item preparation. The pressure achieved in the tank containing the test item preparation may be adjustable to only an approximate value, the viscosity of the preparation may change with the outside temperature, and the bore of the nozzle itself may become partially clogged by coarser particles. Any of these events would therefore tend to alter in an indeterminate way the output of the device. Furthermore, if a herbicide were to be dosed in terms of "weight of active ingredient per area", the exact application of this "dose" would depend not only on the output of the spraying device (in terms of "volume or weight delivered per time unit") but also on the exact movement velocity of the device along the field. Under these circumstances, no exact calibrations are possible, and the GLP compliant characterisation of such equipment has to involve other means. In analogy to the possibility of determining the concentration, homogeneity and stability of a test item used for field tests, where also not the actual tank-mix needs to be analysed, but where these parameters may be determined in a laboratory experiment, the calibration of, e.g., spraying equipment could also be performed as a separate experiment in the laboratory using standard conditions of pressure, temperature, viscosity etc.

> The results of a study can be relied on only as far as the study itself is being properly conducted, but also only as far as the circumstances surrounding the study can be ascertained to have been supportive of this reliance. Suitability of apparatus, materials and reagents is thus one of the key points in this judgement.
>
> Suitability of materials may be simply determined by their specifications. Suitability of apparatus and instruments, on the other hand, has to involve the demonstration of their proper functioning through the necessary calibrations which may have to be addressed, however, in a study-related context. For determining the suitability of reagents a further aspect is important. To be suitable for its intended purpose, a reagent has to work in the proper way, a parameter which it might not be possible to ascertain retrospectively. Thus, in order to provide at least indirect reassurance for the proper past usage of reagents, GLP insists on their actual and exact labelling.

7. Computerised Systems

7.1 Introduction

Computerised systems have taken over an ever increasing part of different tasks in various areas within our daily lives, but especially so in the conduct of safety-related, GLP-requiring studies and test facilities: They are used during the planning, conduct and reporting of studies for a variety of purposes, including the direct or indirect data capture from automated instruments, the recording, processing, reporting, general management and storage of data, as well as in controlling and steering functions in numerous kinds of equipment. For these different activities, computerised systems can be of varying complexity from a simple, microchip controlled instrument, an application on a personal computer, a programmable analytical apparatus, up to a complex laboratory information management system (LIMS) with multiple functions. It stands to reason, however, that whatever the scale of computer involvement, the GLP Principles have to be followed. The correct application of the GLP Principles to ensure compliance of computerised systems with the GLP rules may, however, pose some problems, which might be regarded to stem at least in part from the very origins of GLP.

At the time when the principles and regulations of GLP had first been formulated, data acquisition was mainly performed by manual recording of data. Therefore the various prescriptions, rules and procedures were primarily adjusted to the process of manual data recording, and thus the increasing use of electronic data acquisition and management necessitated a new look at the interpretation of these rules with regard to their applicability to computerised systems. Difficulties and problems were encountered not least because the term "computerised system" may involve a vast array of very different equipment and applications of varying complexity. Because of the necessity to apply the GLP Principles to computerised systems, irrespective of the scale of computer involvement, this problem has been studied already in the 1980's, with the UK guidelines for the application of GLP to computerised systems as one of the first documents dealing with this area ((UK, 1989). The US EPA followed suit with the publication of their Good Automated Laboratory Practices (GALPs; EPA, 1995), and the specific problems of their use in the context of the OECD GLP Principles have been addressed also in an OECD Consensus Document (OECD No. 10, 1995). Recently, the US FDA has furthermore published its final rule on "Electronic Records; Electronic Signatures" in the Federal Register (FDA, 21 CFR 11, 1997).

The OECD Principles of Good Laboratory Practice themselves do not deal specifically with the ways in which they have to be applied for ensuring GLP compliance of computerised systems, as they do not describe in detail how GLP compliance of computerised systems is to be ascertained and maintained. They have not even provided a definition of what kind of equipment would be considered under the term of "computerised system". Computerised systems are just being mentioned at the appropriate places, e.g., where the GLP compliance of apparatus in general is discussed. This lack of detailed provisions is the expression of the fact that the field of electronic data capture, data handling, and data storage and retrieval had been, and still must be, considered as changing so rapidly as to preclude a meaningful and at the same time permanently valid regulation. Therefore, only the more general instructions and statements on the GLP compliant introduction and use of computerised systems have been included in the GLP Principles, whereas in the already mentioned Consensus Document a number of additional issues have been further defined, elaborated and discussed. In this way, a certain guidance could be provided without running into the problems associated with fixed regulations.

According to the definition given in the Consensus Document a "computerised system" consists of a group of hardware components equipped with the appropriate software. Hardware on the one hand is defined as the physical parts of the computerised system, including the computer unit itself and the associated peripheral components. Software on the other hand means the program(s) that control the operation of the computerised system. The combination of these two components enables the system to perform a specific function or group of functions. All GLP Principles which apply to equipment therefore apply to both hardware and software. Figure 17 provides for a schematic representation of the interrelations of the various components within the term "computerised system".

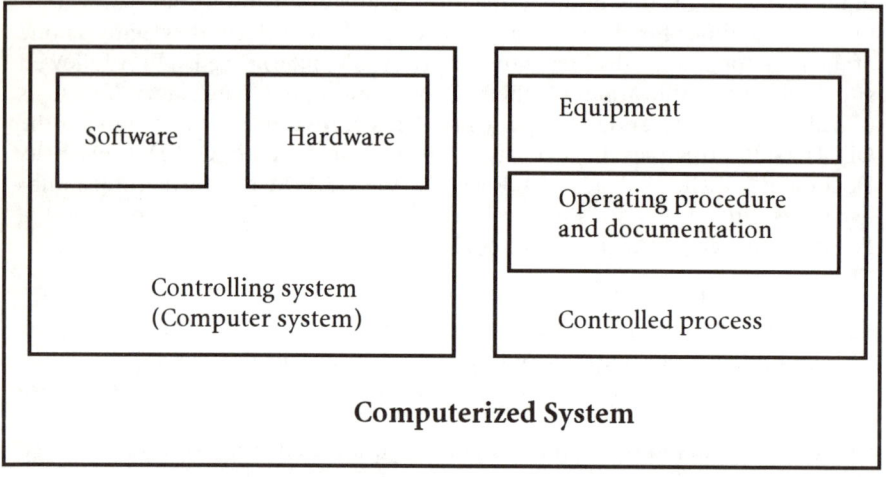

Figure 17: Schematic representation of the various components which, in their combination, form the "computerised system".

One further component has to be taken into consideration, when dealing with computerised systems, namely the nature and extent of the communication system between several computers or between computers and peripheral components. It has to be recognised that all communication links, on whatever level, are potential sources of error, and faulty transmission of data may result in their loss or corruption. This is true of personal communication links, e.g. between Study Director and Principal Investigator, but in the case of communication links between and within computer systems the problem may appear to be aggravated by the perceived lack of transparency of the communication process itself.

In a broader sense it is therefore not only the computerised systems which need consideration with regard to their GLP compliance, but the whole environment of Information Technology (IT) needs to be addressed in order to arrive at sensible solutions for the extended variety of such applications.

The special place of IT with regard to the application of the GLP Principles derives from its very nature. For the non-adept, an IT application is a "black box", where an input is processed in some mysterious way and then presented to the operator in some form of data output. It is like putting some colourful handkerchief into the sorcerer's hat and - "look what's coming out" - waiting for the white rabbit to be pulled out. The average user just assumes - and has to assume! - that the programming people understood the problem, that they were experts in their job and that they did the programming professionally and correctly, so that the result obtained through this "black box" can be trusted. There is, for instance, no obvious means for determining whether the application introduces some kind of systematic error into the data output. A rather simple example in this respect could be the case of rounding figures. In the borderline case of a figure ending with the digit "5", there are two acknowledged possibilities of rounding: Either the figure is generally rounded up, or it is only rounded up when the result would provide an even digit, or down if it were an odd one. Thus, "2.5" could either become "3" with the first method, or "2" with the second one, while "3.5" would yield "4" in both cases. In a computer calculation the average user is not aware of the method which is used by the programme in dealing with this problem. Furthermore, the assignment of a limited number of digits may also affect calculations in still another way. A calculation program might, e.g., just truncate the values at a certain point without bothering to round up or down. In the worst case truncation might then lead to sense- and worthless results as is reportedly the case with one major program for the calculation of statistical parameters which may return P-values of "= 0.0000", or even "< 0.0000" (Finney, 2000) Within a computer application, however, the problem itself may not become obvious, nor might subsequently the solution to the problem be easy.

There are even worse examples of computer failures which may not have been directly caused, but certainly not be alleviated and resolved, by this "black box" problem. As has been described in the example with rounding of figures, there are other algorithms, parameter values and other assumptions which go into a computer programme. If the programmer then assumes that the steering thrust of a rocket will be calculated in pounds per square inch, while the engineer in the control centre is thinking in Newton per square centimetre and entering the respective – for him appropriate - values into the programme, the difference between the assumed and the actual thrust provided by the calculated firing time of the steering rocket suffices to bring a spacecraft to its catastrophic end on Mars.

In manual calculations, such problems could easily be spotted by an independent, critical observer, as one of the points made in the Study Director's statement shown in figure 30 (see page 252) will demonstrate. In an electronic system only an expert programmer could be expected to deal with such issues, and furthermore only so, if he had full access to the source code. Therefore, the basic process for ensuring GLP compliance of an IT application is to demonstrate that the respective application does indeed perform the intended task in a correct way, and that it does so in the actual working environment of computer hardware and connected peripheral components and apparatus. This demonstration of the suitability for its intended purpose is obviously of fundamental importance, and this process is referred to as "computer validation". In essence, such validation should provide a high degree of assurance that the system will meet its pre-determined specifications.

Depending on the situation, this assurance can be obtained in two ways, i.e. by either a quality controlled development and validation of the respective computerised system, or by a (retrospective) verification or qualification of the application in its working environment. There are a number of considerations determining the actual choice of the way in which this assurance is obtained. Firstly, there is the question, whether indeed the producer and vendor of the system has had the GLP compliance of the application(s) in mind when developing it. The software should have been developed within the framework of an acceptable quality system and thus more or less in a GLP compliant manner, but a system validation as it is understood in the context of GLP might be lacking. The end-user of the system may have no direct control over this problem other than the possibility of performing an audit of the manufacturer's or vendor's documentation on the system development. There are considerations, however, which can be addressed fully by the system's user. On the one hand, the level of sophistication of the system may play a role in determining the level of validation needed, and on the other hand the question of the importance of the application may be asked. These areas will be dealt with in the following sections in some more detail.

This section does not, however, intend to provide the "ultimate recipes and solutions" for the GLP compliant development, installation and use of computerised systems. Such a task would be already difficult in itself but would become even more complicated by the actual situation, since, although there is agreement in general terms among the experts in the field, the details for these "recipes and solutions" are still hotly debated issues. Furthermore, many publications have already dealt with this subject, of which the few citations given in the reference list are only a limited sample Thus, the book will rather present some general points on how to develop policies for dealing with IT problems in GLP. In the following, an attempt shall thus be made to address in a general way some of these issues in the context of the GLP compliant validation, installation and application of computerised systems.

7.2 Basic Considerations

With regard to the conformity of computerised systems with the rules and requirements of GLP a number of specific issues are surfacing immediately. There are on the one hand the issues of GLP conformity in the development, validation and operation of complex computer software and its practical applications, and on the other hand there are the problems with the single apparatus equipped with various levels of resident, pre-programmed software applications. There are also various levels of complexity in computerised systems themselves, which necessitate different approaches when considering the ways and means for ensuring GLP compliance. In looking at the relevant parts of the GLP Principles one might be able to perceive in a general way what GLP expects with regard to compliance of computerised systems.

In a top-down approach it has to be emphasised that it is the responsibility of the test facility management to *"establish procedures to ensure that computerised systems are suitable for their intended purpose, and are validated, operated and maintained in accordance with these Principles of Good Laboratory Practice"*. Management should therefore establish policies and procedures dealing with these topics to ensure that this goal will be reached. To this end, it has to choose and designate suitably qualified personnel with relevant experience and appropriate training, and has to charge it with the specific responsibility for the development, validation, operation and maintenance of computerised systems. Management should furthermore ensure that these policies and procedures are understood and followed, which means that an effective monitoring of the adherence to them has to be instituted.

It certainly goes without saying that other general responsibilities, of management, of Study Directors, of personnel and of Quality Assurance, will also have to be applied with regard to computerised systems. Thus, personnel should be appropriately trained also in the operation and maintenance of computerised systems, adequate facilities for the location of such systems (especially for the central, core hardware of the test facility's computer network) have to be provided by management in the same way as for the localisation of other equipment, and Quality Assurance has to ascertain routinely the GLP compliance in the utilisation of computerised systems.

Suitability for the intended use is something else that holds for a number of apparatus, equipment, facilities etc. and is certainly not a specific aspect of computerised systems. Computerised systems, as any other apparatus or instruments, should be of appropriate design, adequate capacity and should be suitable for their intended purposes. In the same way as for other apparatus, procedures have to be developed and documented to control and maintain computerised systems, and they should be developed, validated and operated in a way which is in compliance with the GLP Principles. The key word in this sentence is "validated", and this term returns in the list of responsibilities of the Study Director, who has to "*ensure that computerised systems used in the study have been validated*". This translates into the requirement that only systems that have been proven to be GLP compliant should be used in GLP studies, and the Study Director is consequently held responsible for ensuring that these systems have indeed been validated. In order to be able to fulfil this obligation the Study Director has to be actually aware of all computerised systems that are to be used in a study.

More of the special issues in computerised systems are surfacing in the enumeration of Standard Operating Procedures (SOPs) needed in this area: For computerised systems, there should be SOPs for their "*validation, operation, maintenance, security, change control and back-up*". The most extensive description for the utilisation of computerised systems can be found in the section on the conduct of the study, where for the use of such systems it is required that "*data generated as a direct computer input should be identified at the time of data input by the individual(s) responsible for direct data entries. Computerised system design should always provide for the retention of full audit trails to show all changes to the data without obscuring the original data.*

It should be possible to associate all changes to data with the persons having made those changes, for example, by use of timed and dated (electronic) signatures. Reason for changes should be given." This whole paragraph transcribes the requirements for the manual recording of data into the respective activities connected with electronic recording. It can be seen that again the full traceability and reconstructability of all events and steps leading to the final result is required as the prerequisite to GLP compliance.

The issues relating to facilities, equipment, and security of operation and communication belong to the basic aspects for a test facility. With regard to computerised systems the GLP Principles are listing a number of specific considerations in this area. A Personal Computer as a stand-alone apparatus may tolerate a certain degree of environmental changes with regard to temperature and humidity, and if in these circumstances some "misbehaviour" of the computer is detected, this may not matter overmuch for the single user. The core computer unit, the server for the whole test facility may, on the other hand, not only be much more prone to malfunctions under stress from changing environmental conditions, but a large number of users will be dependent on its proper functioning. Therefore, due consideration has to be given to the physical location of computer hardware, peripheral components, communications equipment and electronic storage media within a test facility. Environmental conditions have already been mentioned, but among these not only humidity and temperature, but possibilities for electromagnetic interference and proximity to high voltage cables has to be taken into account when deciding over the siting of the test facility's central computer.

> The basic principle in the use of computerised systems within regulatory safety studies can be very concisely brought to the point:
>
> *All computerised systems used for the generation, measurement or assessment of data intended for regulatory submission should be developed, validated, operated and maintained in ways which are compliant with the GLP Principles. Appropriate controls for security and system integrity must also be adequately addressed during the whole life cycle of any computerised system.*

> These requirements may not seem to be helpful suggestions, but they provide the basic understanding that data quality, reliability and integrity have to be ensured in ways that are not intrinsically different, whether they are acquired and handled by computerised systems or not.

7.3   Data considerations

Already in the development of an IT application intended for use in a GLP setting, certain special aspects in the area of data handling would have to be taken into consideration. In the first instance, a distinction would have to be made between applications for direct and immediate data capture and primary data processing on the one hand, and applications for secondary processing and storage on the other. The underlying reason for this distinction is the problem of the definition of raw data and their subsequent treatment. Raw data are defined as all primary test facility records and documentation resulting from the original observations and activities in a study. The GLP Principles require that all data generated in the course of a study have to be "*recorded directly, promptly, accurately, and legibly*", and that all data entries "*should be signed or initialled and dated*" by the recording person. Furthermore, changes in raw data have to be made so as not to obscure the original entry, and all such changes have to be justified, dated and signed or initialled by the individual making the change. In the same way as it is described for the manual data acquisition, the GLP Principles require for electronic data capture that "*Data generated as a direct computer input should be identified at the time of data input by the individual(s) responsible for direct data entries. Computerised system design should always provide for the retention of full audit trails to show all changes to the data without obscuring the original data. It should be possible to associate all changes to data with the persons having made those changes, for example, by use of timed and dated (electronic) signatures. Reason for changes should be given.*" Thus, electronic systems used for the recording of original observations, in other words for generating raw data, have to satisfy identical conditions to those that are applicable to manual recording of data.

Raw data may subsequently be processed in a number of different ways, where it might not be necessary anymore to follow this strict regime of controlling any change. The analytical test item residue data from a field test, or the plasma concentrations of the test item from a toxicology study are constituting the raw data of the respective experiment, and they are subject to the strict rules governing raw data corrections and alterations. The residue data may subsequently be used to derive the environmental half-life of the test item by means of one specific procedure. If the procedure for calculating the half-life were to be changed, the results obtained with the "old" formula would not necessarily have to be retained, although in an amendment to the study plan the change and the reason for it would have to be described. The same holds for the plasma concentrations on which additional or changed calculations may be performed as deemed suitable.

In the area of electronic data, these situations have to be considered in an analogous way. If the system is used to capture original observations and thus to generate raw data, they have to satisfy exactly the same conditions and requirements as do the manually recorded data. Thus, a computerised system to be used in a GLP area should include in its design not only the dating and timing of the original entry with identification of the individual having made it, but it also has to provide for the retention of a full audit trail, to show on the one hand all changes to the data without obscuring the original ones, and, on the other hand, to associate these changes with the persons having made them. Furthermore, the system should not accept any changes to data without concomitant entry of a reason.

The other side of the coin would then consist of those systems, where raw data are input for further processing, and where it might be advantageous to have the possibility of first "trying out" an evaluation before the final choice and fixing of fringe parameters. In this situation, again, the requirement for a full audit trail would not be as stringent as in the case of the raw data themselves. As the most obvious example in this area one might consider the writing of the final report, where revisions to the original draft are not retained. Since only the combined result of all the revisions and alterations is of importance, the final report approved and signed by the Study Director, the way in which the approved version had been obtained is of no interest. Another example of allowable "disappearance of original entries" may be encountered in the way, a histopathologist is working when evaluating the tissues from a toxicology study. The first diagnoses to be entered into the system will be only tentative ones, which might become changed with the progression of the work. Since pathological tissue alterations are continuous in nature, from the normal to the highly pathologic state, the grading of such lesions may depend on a synoptic view of the tissues from the whole study, and therefore only with progressive accumulation of experience with the actual study can the histopathologist arrive at a final assessment, which then, and only then, will constitute the raw data of the histopathology evaluation.

With regard to computerised systems, an important conclusion emerges from these various considerations, illustrated by the above examples. It is imperative that for each computerised system the nature of data that are to be processed by the system should be exactly defined. Especially the situation concerning raw data deserves very careful consideration, reflecting on the requirements which have to be imposed on the system. Many test facilities have in the past tried to circumvent the problems with electronic raw data capture, processing, storage and retrieval by producing a print-out of the captured data on paper and defining these hard-copies as the actual raw data. This problem has already been discussed in section 2.9 (see page 85) and need not be expanded here again. It should only be reiterated that with today's technical possibilities there is no reason for having to resort to this apparently expedient solution. Furthermore, based on the publication in the Federal Register of March 20, 1997, of the final rule on "21 CFR Part 11 - Electronic Records; Electronic Signatures" the US FDA has made its intentions clear to exact in the future the electronic retention of data which had been captured electronically in the first place.

> In manual recording the entries made on a sheet of paper can be dated and signed to attest to the validity of data and to accept responsibility; corrections to them remain visible unless the erasure of the superseded data has been done very artistically. These safeguards have to be retained in the use of computers for data capture, processing and storage, since, e.g., the "bits and bytes" in computer memory are invisible, and corrections to them will under normal circumstances leave no trace. GLP therefore wants to ensure that data safety and integrity remains the same in electronically as in manually recorded data, irrespective of how they were recorded, and that reconstruction of the way in which the final results and conclusions were obtained remains fully possible.

7.4 Prospective Validation

When working with a computer one unconsciously assumes that the machine will normally perform the functions for which it is programmed in a completely correct way. Although there are "bugs" in each software program, which may or may not be listed in the "read-me" file, everybody tends to trust the developers and software programmers with the professionality of their work, and the common computer applications generally deserve this trust. Applications like "Word" or "Excel" have undergone careful development (or at least one hopes so) and extensive testing, and their (relatively) faultless operation in the million-fold daily use may be regarded as an indication of their validity.

A computer application which should be used in the context of a GLP study, however, might be a special development for just one or a few test facilities, and therefore any malfunctions, which might occur under special circumstances only, could go undetected during software development and limited program testing. As any other apparatus, computerised systems have to be suitable for their intended use, and the user should be able to expect the output of correct results from the system under all imaginable circumstances. The fundamental importance of the suitability for its intended purpose for the GLP compliance of a computerised system requires that the validation process should be performed, and concluded, prior to the operational use of the system. Computerised systems should thus be validated prospectively, rather than being retrospectively assessed for their suitability. More than with other apparatus or equipment it is important for the validation of a computerised system to define the "suitability for the intended purpose" in terms of clearly documented specifications and quality attributes. This entails the preparation of a detailed documentation of the user requirements and system specifications.

Validation of apparatus, including computerised systems, is sometimes rather simply regarded as being fulfilled by either the expected outcome in the routine use of "quality controls", or at most the by running a test study on the system and to compare the actual outcome with the expected one. Validation in its properly understood sense, however, involves somewhat more than just that. Validation has to be undertaken by means of a formal validation plan which addresses the various aspects of the system's suitability. In other words there should be documented evidence that the system was adequately tested by the test facility for conformance with its pre-defined acceptance criteria, prior to being put into routine use. Formal acceptance testing requires therefore the conduct of tests following a pre-defined plan, the concomitant retention of documented evidence of all test procedures, data and results, a formal summary of testing and a record of formal acceptance. In this respect, the formal acceptance testing or validation may be compared to the GLP compliant conduct of a study, where a pre-formulated, formally adopted, study plan forms the basis of the study, where the test results are the raw data from which to draw the conclusions, and where the study report summarises conduct, results and conclusions of the study. All the specific requirements that are detailed in the GLP Principles for the conduct of a study can also easily be applied to the process of computerised system validation.

The process of validation, and the determination of its necessary extent may be best understood, if the developmental and operational life-cycles of a computer application or system are considered (see figure 18).

Whether for a completely new development, or for the simple purchase of a commercially available computerised system, the user has to define and compile his or her requirements in an as specific manner as possible. The new HPLC system, e.g., should be able to detect the test item at a pre-defined sensitivity and to allow different calculation modes for study-specific conversions of the detector output. A telemetry system should allow to acquire different signals with a given time resolution, and to calculate specific parameters from the recorded values. The more specifically the user can define these requirements, the better will he be able to judge the system's suitability.

However, not only the technical aspects will finally have to be addressed, but the whole range of scientific and business, regulatory, safety, performance, and quality requirements will have to be included, amongst which the existence of an audit trail may be cited as an important part.

At the next level these user requirements will have to be translated into the functional specifications of the system. In the case of a specific new development, this translation will be expected to be performed on a 1:1 basis, i.e. all of the required functions will be specified, while in the case of a commercially available system the user may have to compromise between the maximally required specifications and the ones that are offered by the available systems. Finally, the new system will also have to be integrated into the user's environment, which means that the system design, including its components, the operating system software and the network requirements are to be specified. Testing will therefore have to be performed at all these levels, as the suitability of a computerised system can only be demonstrated convincingly on this basis. As it is the case in many other situations, the whole system has to be regarded as more than just the sum of its parts. To remain with one of the above examples: It is certainly important that the detector for the HPLC system should be designed to allow e.g. the photometric detection at a specified

2.7 Computerised Systems

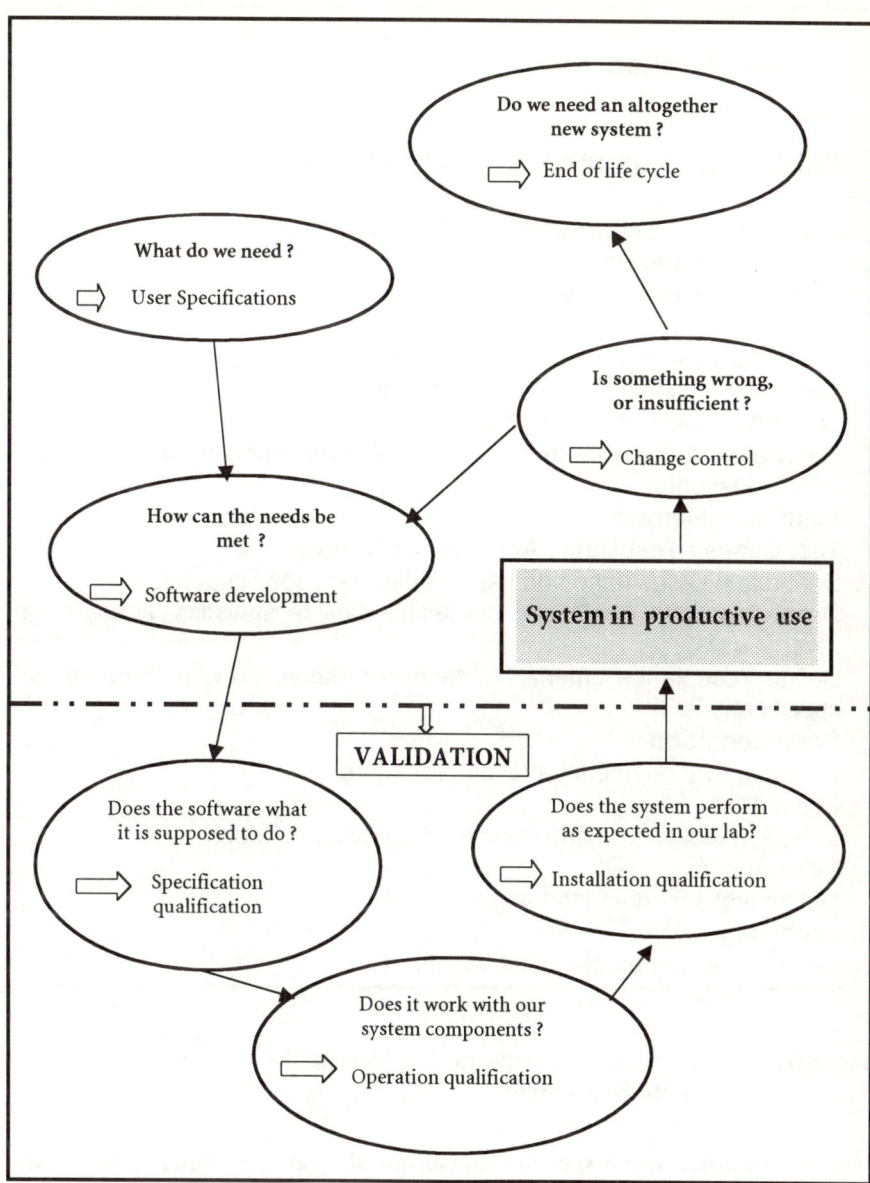

Figure 18: Graphic representation of a computerised system's life-cycle

> **CompliantLab, Inc.**
>
> **Policy Paper on the VALIDATION PLAN**
>
> **Purpose and Background of validation**
> **Scope of validation**
> System parts (individual programs, versions)
> **Responsibilities**
> Test facility management - System owner - IT responsible - Validation manager - Validation team – Quality Assurance
> **Test Environment Description**
> Hardware, all Software levels, network (model, version, further details) - Test location(s)
> **Validation Method**
> **Test Cases / Test Data / Acceptance Criteria**
> Define test cases based on user requirement specifications
> Define test data (real data, borderline data re statistics, stress data, reverse data)
> Define acceptance criteria for each test based on user requirement specifications
> **Documentation**
> Define individual documentation procedure
> **Procedure**
> Establish tests – Implementation – Evaluation - Report
> **Schedule**
> Define test schedule start and end date
> **Archiving**
> Define archiving process and location

Figure 19: General list of points to consider in the validation plan for a computerised system

low level of noise with a specified signal-to-noise ratio, and this property will have to be checked by a specification test. Whether the detector will be able to perform to specifications within the system, will depend, however, on the interplay of the various components, e.g. on the periodic solvent flow-rate changes caused by the specific characteristics of the chosen pump model. Thus the operational qualification will provide assurance of the suitability of the detector within the HPLC system. And finally, the verification of the suitability of the whole system after its installation will be needed to ensure the faultless operation under the actual conditions of use.

The main part of the foregoing paragraphs would probably seem to cover more the demonstration of the "suitability for the intended purposes" of apparatus in general than the prospective validation of computerised systems. While it may be easy to grasp what is meant by the different steps in a validation process for any such "physical" apparatus, it may call for some more imagination to look at an electronic, "computerised" system in the same way. However, one important part of this validation corresponds of course to the validation of the computer software which is needed for the computerised system to perform its functions. There, the validation process follows the same steps and serves the same purposes. The specifications of the software have to be ascertained, the operational qualification has to guarantee that the various software modules will not negatively interfere with each other, and finally the software should run faultlessly in the user's real environment. Thus, the validation of a computerised system has to follow a logical sequence of steps, which have to be defined in the validation plan, executed in the validation test, and assessed in the validation report. Figure 19 can be used as a general guide to any kind of validation.The problems of operation and installation qualification may in general not be insurmountable, but the first step, the investigation of the software specifications might pose some problems. If the test facility does not belong to a company which is large enough to employ a software development group of its own, then it will depend on the professionality and the quality of the software developers. These developers should then, for their part, be able to provide documented evidence for the quality system they used in the development of the different software modules, for the results of the acceptance testing of the single modules and the functional testing of the whole system. The test facility itself may then either rely simply on a declaration by the vendor (for one such example, see figure 20, page 189), or it may, especially for larger systems like a LIMS, perform an audit at the vendor's development site (Segalstad, 1996) in order to ensure the GLP compliant development and testing of the software itself.

> All apparatus used in a GLP context have to be suitable for their intended use, i.e. they have to satisfy the specified requirements of the users. For computerised systems the prospective proof of suitability is provided by the validation procedure. This has to start with the exact definition of the user requirements which have subsequently to be translated into proof of adequate operation of the system in the actual environment. With this prospective validation assurance should be provided that the computerised system will perform the tasks it is designed to execute in a correct, reproducible and reconstructable way.

7.5 Retrospective Evaluation

The awareness for the need of GLP compliance in the development, and functional and operational testing of instruments increasingly equipped with software applications had not held pace with their increased introduction in laboratories of all possible denominations. For many manufacturers of such apparatus the use of their products in a GLP environment represents only a minor share of the market, and therefore any consideration of additional, GLP-based requirements in the development of these applications was either not apparent or deemed to be too expensive to be fulfilled. On the other hand the application of the GLP Principles to computerised systems makes it necessary to deal also with this type of equipment. While the performance of a complete, prospective validation in the sense as detailed in section 7.4 above is most probably impossible for such instruments, their suitability for the intended use has nevertheless to be ensured in an appropriate way. This purpose may be achieved in these situations by a retrospective evaluation of the system's performance relative to the user requirements.

Such a retrospective evaluation will necessitate in the first place the compilation of all available documents on the instrument or system, an evaluation with respect to their extent of coverage of GLP-relevant areas, and a report on the experience with the system. This experience report may contain data on the performance of the system with regard to data obtained on the commonly and regularly performed quality control runs, or the results of positive and negative control samples investigated in the context of GLP studies. This retrospective evaluation summary should thus specify what kind

THERMO SEPARATION PRODUCTS

45757 Northport Loop West
P.O. Box 5116
Fremont, California 94537
Tel. 510-657-1100
Fax. 510-490-8182

DECLARATION OF SYSTEM VALIDATION

RELEASE DATE   May 20, 1994
PRODUCT NAME   PC1000 Software
VERSION NUMBER 2.5.X

THIS DOCUMENT CERTIFIES THAT THE ABOVE NOTED SOFTWARE PRODUCT/SYSTEM WAS DEVELOPED, TESTED AND VALIDATED ACCORDING TO THERMO SEPARATION PRODUCTS' SOFTWARE DEVELOPMENT GUIDELINES. THE PRODUCT HAS MET ALL FUNCTIONAL AND PERFORMANCE SPECIFICATIONS AND PRODUCT RELEASE REQUIREMENTS.

TO SUPPORT THE GLP REQUIREMENTS OF THE PRODUCT USER, WE WILL PROVIDE ACCESS TO SOURCE CODE AND PERTINENT DOCUMENTATION AT OUR CALIFORNIA FACILITY TO AUTHORIZED GOVERNMENTAL OR REGULATORY REPRESENTATIVES ON REQUEST. EXECUTION OF OUR STANDARD NON-DISCLOSURE AGREEMENT MAY BE REQUIRED AND ALL SUCH DOCUMENTS AND THEIR REPRODUCTIONS WILL REMAIN THE POSSESSION OF THERMO SEPARATION PRODUCTS.

ENGINEERING MANAGER

QUALITY SYSTEMS MANAGER

PRESIDENT

Figure 20: Good example of a vendor declaration with acknowledged possibility for verification of source code and other validation documentation by monitoring authorities (reproduced by permission of ThermoQuest, Inc., San Jose, CA)

of validation evidence is available, and it should provide an assessment of the additional efforts needed to ensure the suitability of the computerised system. Defining and listing those functions of the system which would be needed in the conduct of GLP studies will then help to devise an adequate system test for the assessment of the proper functioning of the system under the actual environmental conditions. This latter test should be conceived and conducted in an analogous, albeit possibly somewhat simpler, way to the operational qualification test within a proper prospective validation.

In summary, for every computerised system which has not been validated in a proper, prospective manner, there should be documented justification available for its use, on the one hand in terms of historical records of proper functioning and on the other hand in terms of continued suitability ensured through the conduct of appropriate system tests.

> In every test facility there will be computerised systems which have not been formally validated because of various reasons. Their use in a GLP environment still necessitates, however, clear proof of their suitability which, in these cases, can only be obtained through an evaluation of their historical as well as their actual performance. For the sake of reconstructability and transparency, this proof has to be planned and documented, resulting in a final conclusion on the past, present and future suitability of the respective system.
>
> In this way GLP aims at providing evidence for the correct functioning of the computerised system and for estimating the extent of GLP compliance.

7.6 Maintenance and Change Control

The successful conclusion of a prospective validation or a retrospective evaluation cannot mark the end of the efforts needed to assure the GLP compliant status of a computerised system. A number of support mechanisms have to provide for the continued suitability of a computerised system. System management and technical support have to ensure the day-to-day operational maintenance and trouble-shooting. The correct utilisation may necessitate training and supervision of study personnel. Formal, periodic performance assessments may furthermore be needed to ensure that the system continues to meet its performance criteria in terms of reliability, responsiveness and capacity.

As a necessary prerequisite for the correct operation of maintenance functions as well as for the GLP requirement of traceability it is important that any problems or inconsistencies detected during the daily operation of the system are recorded, and that any consequent remedial actions taken are documented. Changes to the computerised system may thus become necessary through maintenance-related events, but they may also be intentionally introduced through the perceived need for additional or changed application modules. All such changes have to be evaluated in terms of their influence on the validation status of the system. To deal with such situations, change control procedures have to be effective once the computerised system is operational. Change control is defined as the formal approval and documentation of any change to the computerised system during its operational life. Change control, however, cannot be regarded in this narrow sense, but has to include the recording of events which might or should necessitate a change in the system, the evaluation of the change with respect to the necessity of system re-testing, and of documenting the approval to any such measures. The change control procedure should furthermore describe the method of evaluation to determine the extent of retesting necessary to maintain the validated state of the system, and it has also to identify the responsibilities for initiating change control and its approval.

Another aspect of computerised system use may be regarded as belonging in some way to this context of change control. While the events leading to changes may in general be singular defects or "bugs" detected during routine use, major malfunctions up to the complete computer crash will necessitate emergency procedures for "disaster recovery" which also have to be in place from the very moment of the introduction of the system into its operational use. To really serve the intended purpose of retaining continued data integrity, these contingency plans need not only to be well documented and validated, but it has to be ensured that personnel involved in the conduct of GLP studies would be aware of the existence and the content of such contingency plans. While measures may range from planned hardware redundancy to transition back to a paper-based system of data recording, their execution should certainly not interfere with test systems in a way as to compromise the integrity of the studies. Depending on the criticality of the system the procedures for the management of disaster recovery may differ in terms of priority. While a harddisk crash or software malfunction on the central server will certainly range highest on the priority list, an analogous malfunction on the PC of a stand-alone gas chromatograph may not need immediate action, as long as a spare, second GC apparatus were available. In the context of such recovery procedures, it is of self-evident importance that back-up copies of all software and data be maintained to allow for recovery of the system following any failure which compromises the integrity of the system. With regard to study raw data recorded electronically, the implication of such a situation is that the back-up copy may have to replace the "original" and would thus become raw data; therefore, the back-up copy of study data must be treated in the same way as the original raw data. It has furthermore to be recognised that, if recovery procedures would entail changes to hardware or software, it could become necessary to re-validate the respective system.

> Whether any system has been fully and prospectively validated or has just been retrospectively evaluated and qualified, there is a need for continued maintenance of the validation status to ensure continued data validity. This is accomplished through formal procedures that require any changes to the system to be fully documented. Data integrity will, however, not only depend on the validation status of the system, but also, and to a very important extent, on the security measures developed for the utilisation of the system. Through the requirement of documented security procedures for the protection of hardware, software and data from corruption, unauthorised modification, or loss, GLP intends to provide for continuous data integrity.

7.7 Security

In general terms, security issues can be divided into measures of physical security, i.e. measures that can be instituted on the facility and apparatus level, and logical security, i.e. those that are related to software security at the access level.

The former aspect will include protective measures through restrictions in the access. For instance, only authorised personnel should have access to the central server of the test facility and other computer hardware, to communications equipment, and to electronic storage media held within specific computer rooms. While the physical security may not seem to pose real problems, achieving complete logical security may be much more difficult, although its fundamental aspects may be quite straightforward: Logical security measures for the prevention of unauthorised access to the computerised system, as well as to applications and data, may be based on a system of extended password-protection, which would include the unique identification of the respective user. Recent developments in biometric identification of individuals will certainly provide for additional possibilities in this area.

The progress in computer technology with the exponential increase in memory which enables the storage of whole test facility databases on the harddisk of a portable "notebook" and to take it anywhere on the world, or which allows to access such databases remotely through modem links has added another dimension to the question of data security. The utilisation of these and other possibilities will necessitate additional security measures to be taken in order to prevent unauthorised access or changes to the computerised system as well as to the data held within the system. In this respect, and with the global connection of computerised systems through either company networks or more generally the internet the potential for corruption of data by viruses or other agents needs definitely to be addressed. In this respect, any introduction of data or software from external sources has to be controlled. These controls may be provided by the computer operating system software, by specific security routines, routines embedded into the applications or combinations of the above.

Data integrity can thus only be maintained, if adequate security measures are fully implemented, and if everyone associated with a computerised system is aware of the necessity for the above security considerations. It is again the test facility management who is responsible for ensuring that the test facility personnel are aware of the importance of data security, of the procedures and system features that are available to provide appropriate security, and of the consequences of security breaches.

> Data security is a general problem in the area of computer use, and the restricted access to data and databases through the use of personal passwords is nowadays certainly so routine that nobody would spend a thought about this aspect of security. In GLP, however, consideration has not only to be given to the aspect of access restriction, with the purpose of

> allowing only those individuals to view or handle data who are authorised to do so, but the whole issue of data integrity as a primary objective of the GLP Principles has to be taken into account.
>
> Therefore, not only are security measures to be developed which should provide for an optimal protection of data integrity, but the full documentation on, and validation of, these measures will provide the necessary evidence for the effectiveness of the security provisions, which, in turn, will again enhance the confidence in the integrity of the data.

7.8   Levels of Complexity

It has been mentioned already that there are different levels of sophistication with regard to the involvement of computer applications in GLP studies and test facilities. These levels may range from the complex problems involved in the GLP compliant management of computer networks and of laboratory information management systems (LIMS) to the question of whether a simple instrument controlled by a built-in, pre-programmed chip should be treated in the same, extensive way with regard to "software validation". It is certainly self-evident, as these two examples demonstrate, that not all types of IT applications have to be considered as equal with regard to GLP compliance; it may indeed be impossible to do so. As it is commonplace nowadays that the silicon chip penetrates the operation of practically all kinds of work, the elucidation of its involvement in the operations of test facilities becomes an essential part of the implementation of GLP.

In section 7.2 it has already been mentioned that one of the first issues in regulating IT and its part in GLP compliance would be the task of management to formulate an IT policy. This policy document should, amongst other important topics, also address the issue of prioritising the different levels of complexity and sophistication in electronic data management. Only when clear priorities are provided will it become possible to determine which level of attention, with regard to special measures to be taken, these different applications would deserve. Since many IT applications within a company may be of a general nature and thus be equally applied in all test facilities, while others may be confined to one single test facility, this policy document should furthermore define the various levels of responsibility with regard to the management of IT problems. Again, it has to be reiterated that all the provisions and prescriptions given should serve the one main purpose and goal of GLP, namely to provide for the possibility of reconstructing and tracing back of activities and data.

Even under this aspect of data validity, integrity and traceability, there will be distinctions between different levels of system complexity with regard to their need for validation and operational qualification. It stands to reason that a complex system, like a LIMS, serving a whole test facility and providing diverse applications to a large array of test systems and study types will have to be validated to the fullest extent and significance of this term. On the other hand, an instrument or other measuring device, not connected to the test facility computer network, with a built-in microprocessor performing some limited pre-defined functions, like a pH meter or a stand-alone electronic balance, will be in much less need for a full validation of the instrument and its software application. In this latter case, the instrument furthermore lacks the ability to process, transfer or manipulate data, and the respective software can not be modified or altered by the user. Suitability for the intended use will therefore be much easier to determine, and in fact this determination will practically correspond to the one used for a purely mechanical apparatus. Thus in this case the "validation" will consist of compiling the instrument characteristics in an user SOP and of formulating a calibration schedule, with records to be retained in the instrument logbook.

Of course, not every piece of equipment can be placed into one of these two categories, and there are many possibilities in between the two of computerised systems of varying complexity. When apparatus or instruments contain microprocessors enabling the system to process or transfer raw data, are controlled by an external computerised system, or transfer data to a computerised system for processing, then a more elaborate suitability testing may have to take place, in the extreme all the way through to a full prospective validation.

> The GLP Principles require that validation should provide a demonstration of the suitability of the system for its intended purpose. The complexity of computerised systems spans, however, a vast range from a simple analytical instrument with two or three programmable functions, to a complex laboratory information management system (LIMS). Although the same formal process of validation applies in principle to all

> systems, independent of whether they are big or small, and whatever their scale of complexity, the extent of testing and documentation needed to ensure the suitability of the system may, however, be related in a pragmatic way to the complexity of the system.
>
> In order to avoid unnecessary validation efforts it is important that the test facility should develop a well reasoned approach to validation, including the categorisation, with regard to the respective validation needs, of the various apparatus, instruments and systems which may be regarded as "computerised systems".

8. Test Systems

Test systems are the tools with which the purpose of the study, the search for, and the investigation of, effects produced by a test item, or the elucidation of its properties can be fulfilled. Test systems may therefore be "*any biological, chemical or physical system or a combination thereof used in a study*". Since the test systems are the instruments for the generation of the safety data, documented evidence for their adequacy and integrity, as well as their properties has to be provided in order to ensure the scientific validity of the studies conducted.

It is certainly useful, in view of their divergent nature and properties, to distinguish between physical/chemical and biological test systems. The former ones consist of apparatus and instruments, the properties and specifications of which can be described in unequivocal, physical/chemical terms, while the latter ones may need much more intricate descriptions in order to fully characterise them. In consequence, the OECD GLP Principles are following this systematic differentiation.

8.1 Physical/Chemical Test Systems

As can already be guessed from the above sentence on the relative ease involved in the characterisation of physical/chemical test systems, the OECD Principles are very concise in this respect, in that they just express the conditions that: "*Apparatus used for the generation of physical/chemical data should be suitably located and of appropriate design and adequate capacity. The integrity of the physical/chemical test systems should be ensured.*" These four conditions, if met by the physical/chemical test system used in a study, should provide on the one hand the guarantee for the generation of quality data and on the other hand the possibility for study reconstruction. Suitable location is a prerequisite for the correct functioning of the respective apparatus; the reader is reminded of the example given in section 5.1 (see page 157) for the correct positioning of a balance, which needs a stable support in order to return correct values for the items weighed on it. The appropriate design will allow the test system to perform not only according to its own specifications, but to the expectations of the user; to remain with the balance example: the type of balance to be used has to be adapted to the range of weights that are expected to be determined on that specific instrument. To ensure an adequate capacity will be important for the timely conduct of the study. If dozens or hundreds of samples are to be processed an instrument, e.g. on an HPLC apparatus, automation of the chromatographic procedure (from the injection up to the rinsing of the chromatography column to ready it again for the next sample) would enable the system to run practically unattended through nights and weekends. Or, the capacity of an infrared spectrometer would depend on the time it needs to record a full spectrum. Finally, the integrity of the test system is an absolute requirement for the integrity of the study conducted with this test system.

Most of the conditions and prescriptions under which those physical/chemical test systems may be used in a GLP study are the same as for all other apparatus, instruments and technical equipment to be used in a test facility. These have already been described in some detail in section 6 (see page 167); therefore, reiteration of these points and further discussion of them is not needed.

> The quality and reliability of test data depend to a major extent on the state and condition of the test system which is used to produce them. For test systems used to characterise the physical/chemical properties of a

> test item this involves the proper definition and control of a number of technical features and specifications which are needed to ensure the integrity of the system and the quality of the data generated. For compliance with GLP, the most important aspects may be characterised as "suitability", "capacity" and "integrity", which have to be defined in a study-related way.

8.2 Biological Test Systems

Compared to the physical/chemical test systems, the characterisation of biological test systems is quite another kettle of fish (literally!). Such test systems have not only to be described by a few specifications provided by the "manufacturer", but they have to be cared for, properly housed or sited, they may need acclimatisation to the test environment before use, and their characteristics may need to be re-ascertained on a regular basis. They may react very sensitively to disturbances in their environment, and therefore the quality of the data obtained from these test systems can be ensured only through the establishment of the proper conditions. Consequently, the list of points to be observed with biological test systems for ensuring their GLP-compliance is more extensive than the one for the physical/chemical systems.

Physical or analytical chemists, at this point, might object to ascribing such a special position to biological test systems, and they may point out that they, too, are utilising quite a number of very sensitive test systems, i.e. instruments, for which the proper environmental conditions might as well play a decisive role for obtaining quality data as this might hold for any biological system. This may certainly be true for environmental conditions like temperature and humidity, but an instrument may not react to the moods of the technician in the same way as a dog or rat would. If the technician is nervous, distracted or stressed, a test animal may become stressed, too, and it will consequently return abnormal test results, while a computer would, at each erroneous or incorrect command, stolidly return the same message "error 34", or "wrong entry, try again" time after time.

A number of the requirements listed in the GLP Principles are points which relate more to the "good scientific practice" than to GLP proper, although without observing these points conscientiously the quality of the data

2.8 Test Systems

obtained would certainly suffer even under the best conditions of GLP compliance in its purely "administrative" sense. Thus, the Principles describe the conditions under which test systems should be received, observed for any aspects which might negatively influence the future study, maintained, cared for and used. For animal test systems, many of these points are also addressed by the requirements of national animal protection laws, and the GLP Principles are certainly not intended to change, supersede or replace the respective regulations. For field studies, care has to be taken that the respective plots to be used as test systems are situated "*so as to avoid interference in the study from spray drift and from past usage of pesticides*" which again can be seen as a more scientifically dictated requirement. What distinguishes the provisions of the GLP Principles from the "good scientific practice" are again the documentation requirements. Not only should "*records of source, date of arrival, and arrival condition of test systems ... be maintained*", but also other records need to be generated and retained in order again to ensure the reconstructability of the study. The requirement that newly arrived test systems should be evaluated with respect to their health status or their suitability for use, that test systems found to be diseased or in any other way unsuitable should not be used in a study, and that those which become diseased during the study should be isolated and treated is certainly good science and will help to maintain the quality and integrity of the study and of the data generated. For a GLP compliant study, all these aspects of test system status should be documented properly, and the Principles require therefore that "*any diagnosis and treatment of any disease before or during a study should be recorded*". The same holds for any disinfection or pest control agents the use of which might become necessary, and which has to be documented.

An important aspect in the use of biological test systems is their proper identification within a study. In contrast to physical/chemical test systems, on which multiple studies are conducted in a sequential way, one biological test system is used just for one study, and therefore there may be several similar test systems in use at the same time for different studies. As an example: On an HPLC apparatus used as the test system, analytical determinations will be performed in a sequence of individual analyses for one study at a time, and the test system itself may not need to be visibly tagged with special study identification, since this information will have been entered into the system as an identification of the single analyses. If there are, however, different studies to be conducted with rainbow trouts as the test system, the aquaria with their inmates will look very similar if not identical to each other, and therefore the exact identification of the different individual test systems is an absolute necessity. Therefore the GLP Principles require that "*all information needed to properly identify the test systems should appear on their housing or containers*". This sentence does not simply mean that the test system itself should be described on the housing or the container. Everybody would certainly be able to discern a dog test system from a rabbit one, and a test system consisting of green algae is easily distinguishable from one which uses mammalian cells. As soon as a test system is entered into a study, however, supplemental information has to be provided by additional entries. Study identification, treatment group information, experimental start and end dates and other pertinent information will then be required on the appropriate cage or container cards.

As important as the identification of the test systems on their housing or containers is the appropriate identification of test system individuals or test system parts that may need to be temporarily removed. This may constitute no problem, if the test system consists of (larger) animals (or plants), where tags, tattoos, colour codes on fur or tail, implanted microchips or other suitable markers may be used to characterise and identify the single individuals. On the other hand, individual water fleas might have to be removed from the test system container for some observation; if they were to be returned afterwards to the correct place, the problem of an individual identification becomes obvious. Although it constitutes a general principle that any parts or individuals removed from a test system should be appropriately identified so as to allow their correct return to it, the respective requirement is therefore pragmatically mitigated by the addition of the words "*wherever possible*".

There is a final point which may be addressed here, although it is not expressly mentioned in the GLP Principles under the heading of "Test Systems", but is required information in the study plan and is listed amongst the necessary SOPs: Test systems have to be properly characterised in order to ensure their suitability for the respective studies utilising them. On the one hand, such characterisation may be obtained from the supplier of the test system, who will have developed this information under some quality standard to be defined in the supplied documentation. The OECD Consensus Document on "Compliance of Laboratory Suppliers with GLP Principles" (OECD No. 5, 1999) maintains that, where suppliers of GLP test systems belong to some national regulatory or voluntary accreditation scheme, e.g., for laboratory animals, users might be provided with additional evidence for the defined quality of the test system, which would be considered sufficient for the purposes of GLP. On the other hand, test systems may have to undergo periodic characterisation with respect to their sanitary status or their inherent properties. Examples for such periodic verification of test system characteristics may

2.8 Test Systems

be the assessment of the microbiological status of test animals, especially those held under SPF ("specific pathogen free") conditions, the assessment of the strain-specific properties of the various bacteria utilised in mutagenicity studies, or the mycoplasma contamination status of mammalian cell cultures. If such investigations on test system characterisation have to be performed to ensure the continued suitability of the test system, these should certainly be documented in the same way as, e.g., the environmental conditions in the test facility are monitored and documented. It may be disputed, however, whether independent laboratories conducting such investigations on a contractual basis, do really need, or can indeed obtain, a recognised GLP compliance status, since these control tests, especially those concerned with the microbiological status of test animals, do not investigate the influence of a test item on the test system and as such do not constitute GLP studies (Homberger et al., 1999). Whatever the conclusion about the possibility of operation under GLP of such laboratories may be, the evidence for the characterisation of the test system has to be developed in such a way as to ensure the documented quality and suitability of the test system.

> Properties of biological test systems will mostly be more complex and more changeable than the ones of physical/chemical test systems. Therefore biological test systems need very careful characterisation in order to ensure the quality and integrity of the data derived from them. This is also of special importance with regard to the reconstructability of studies, since the actual outcome of a study may have been influenced by the state and condition of the test system at the time of the study.
>
> The GLP Principles, in formulating the requirements for the housing and siting of these systems, for their maintenance and use, and for the concomitant documentation, aims at providing the necessary basis for confidence into the results obtained from biological test systems.

9. Test and Reference Items

The purpose of a "non-clinical human health and environmental safety" study is to investigate the physical/chemical, biological and/or toxicological properties of the test item under study, sometimes in comparison with an appropriate reference item. Conversely, the term test item *"means an article that is the subject of a study"*, as it is expressed in the concise definition of the OECD Principles. This test item may be anything from a pure chemical substance to a complex preparation, a device or an organism, of which the properties with regard to safety aspects are to be evaluated. In consequence, the revised OECD Principles are referring to the entity tested in a study as the test "item" instead of test "substance" as it was called in the original Principles. There are much more elaborated definitions in use in some national GLP guidelines, clearly delineating the physical nature of the test item (test substance, test article) subject to the respective legislation (see the respective paragraphs in the FDA and EPA regulations in appendices II and III to this part).

Since it is the purpose of a study to enable a correlation to be derived between the application of the test item to, and the effects observed on, a test system, i.e. to enable the establishment of a causal relationship between the two, it is imperative that the causative agent can be unambiguously identified. It is certainly not sufficient, nor indeed possible, to rely solely on the name printed on the test item container. In other words, in order to ascribe the properties detected in the study to the item named on the label of the container and in the study plan three conditions have to be fulfilled: First of all the identity of the test item should unambiguously be known at all times. Thus, the test item should, under GLP, leave a reconstructable trail from its exit of the premises of the manufacturer or the sponsor, till its final fate within the study; and even after the termination of the study, there should be some way of ascertaining its identity, i.e. by the retention and archiving of a sample of the respective items. Secondly the identity of the test item should be preserved from the beginning of the study to the end. Stability data and suitable storage conditions should thus be available to cover this requirement. Thirdly, it should in some way be possible to prove with sufficient probability that the test system effectively has been exposed to the nominal test item and not to anything else, e.g. only to a decomposition product. This means that the stability of the test item under the experimental conditions has to be demonstrated. Finally, the amounts used in the test should be documented in such a way as to allow the retrospective assessment of the probability that the test system indeed was treated with the targeted amount, concentration or dose.

Only in this way, and aided by the proper records and documents, can the properties which are derived from the results obtained from the test system be ascribed in reality and with confidence to the influence of the test item itself.

9.1 Handling and Documentation

The logistics of test and reference item receipt, handling and storage have to provide for the possibility of tracing, in retrospect, the complete "life cycle" of any test or reference item. Thus, appropriate procedures have to be defined, adequate records have to be maintained, and proper identification of test and reference items has to be assured.

In the first instance, the adequacy of the rooms where procedures for test item receipt, handling and storage are carried out, should be considered as one of the basic aspects of GLP compliance in this area. Concomitant with the adequacy of the physical locations (which has been described in section 5.3, see page 162) there has to be the necessary, and sufficiently detailed, documentation on all facets of test item handling. The organisation of test item handling and logistics would certainly involve an appropriate system for the orderly storage and easy retrieval of test items, as well as of adequate information on them. Records of receipt containing the necessary information on test item identity and characteristics would allow the identification of the proper storage conditions "*in order that the homogeneity and stability are assured to the degree possible*", while clear instructions and SOPs concerned with the handling of these items would serve to preclude contamination and mix-up. All these, more "administrative", aspects may not give rise to too many problems in the daily work with test and reference items. There are two points, however, which may deserve some more detailed discussion: The first one is the requirement of the GLP Principles that "*Records including ... quantities received and used in studies should be maintained*", while the second one is the requirement that the storage containers should "*carry identification information, expiry date, and specific storage instructions*".

The first of these two points deals with an important consideration in the whole area of test item documentation, namely the accountability of test item and its usage. The GLP Principles are not only concerned about the possibility for checking back the identity of the test item, i.e. that the correct test item had been applied to the test system, but also that it should be possible to reconstruct the probability that the test item had been applied in the correct amounts, concentrations and/or doses at all times. Since with its application the test item disappears in the test system, there is no immediate and direct means of ascertaining in retrospect that the correct amount had been applied. Obviously, it would provide for the highest degree of confidence in the assertion that indeed the target doses had been applied to the test system, if each and every preparation were to be analysed. It is, however, as obvious that such an effort would be stretching the analytical (and financial !) resources to the unbearable, and therefore some compromise will have to be reached. On the whole, it is considered sufficient, if periodic analyses are performed which, in connection with the full documentation of the actual preparation activities, i.e. weighing protocols, dilution calculations and dissolution prescriptions, would constitute satisfactory evidence for constancy in doses or concentrations, providing reassurance for exposure of the test system to protocol-specified quantities of test item. This reassurance, however, is reaching only as far as the respective documentation allows for tracing back the probability of having used the correct amounts all the way through the whole study.

This problem can be tackled, however, by observing a number of documentation steps. First of all, there is the attestation in the records of the technician having applied the test item, who has signed or initialled the respective application sheets, and thus has assumed responsibility for the correct application of the test item preparation. These application records will certainly not suffice for the assurance being sought, since calculation, weighing or dilution errors might have led to incorrect dosage concentrations being prepared and used. Therefore, the weighing protocols together with the dilution recipes will provide for the next step in this string of evidence. But even at this point, human error could lead to an incorrect documentation of the weighing. The last link in this chain will therefore be forged by the recording of the gross weight of the test item container before and after the weighing out of the amount needed, and the subsequent comparison of the actual *versus* the expected weight differences. An example of such an accounting sheet is shown in figure 21.

With this train of information available, it may be possible to exclude calculation and weighing errors at least to a certain extent, if the records of the test item logistics, compared to the application records, would show no discrepancies between these two data sets. The side-by-side comparison of the

2.9 Test and Reference Items

CompliantLab, Inc.

TEST ITEM ACCOUNTING SHEET

Study: CL - 99/304

Test item name: SPO-add-775 09
Batch number: 775 09/02
Container number: 1
Gross weight received: 225.7 g
Net weight: 50.0 g
Date of receipt: 17-02-2000
Storage conditions: RT
Re-Analysis date: 31-10-2000

Date	Remarks	Amount removed	Gross wt. remaining	Diff.	In.
18-02	Anal. sample	998.5 mg	224.65 g	1.05 g	JH
18-02	Dose prep.	12450 mg	212.10 g	12.55 g	JH
18-03	" "	14750 mg	197.30 g	14.8 g	JH
13-04	" "	16500 mg	180.65 g	16.65 g	JH
29-05	Amount removed tot. → corresponds	44698.5 mg		45.0 g	JH

Container back to substance logistics
29-05-2000 JH

Figure 21: Test item accounting sheet with information on amounts removed and remaining, balancing the total amount removed against the difference in container gross weight, and on the fate of the remaining test item / container after study completion. (Note the badly executed correction in the calculation of the total amount removed)

progressive diminution of test item remaining, with the records for the amounts weighed out for the application, will enable the reconstruction of the day-to-day procedures of test item preparation.

This kind of accountability may not be perfect, since in many cases losses of test item will be incurred during the removal from the container, e.g. when syrupy liquids are involved, of which a certain amount would stick to the spatula used for sampling. In other circumstances, the utilisation of such an accounting system may not be possible at all. Especially in the case of reference items used in analytical work, where only a few milligrams may be needed for the preparation of a stock solution, such a scheme could be impossible to institute, or at least become a meaningless exercise. If these few milligrams are, e.g., to be taken out of a 100 g bottle, such accounting might fail because of the lack of precision of the balance involved in the determination of the container gross weight.

An analogous accounting has to be instituted at the level of the whole containers, where the time frame of use and the final fate have to be documented, so that it will at all times be unambiguously known from which one of the possibly many containers the test item had been removed for a certain, defined activity.

The labelling of the test item containers is another bone of contention. The GLP Principles clearly express what has to be stated on the label of a test item container. Thus, while the first example in figure 22 would certainly not satisfy these rules, the second one (figure 23) would fit them better. When this required list of information, would give rise to difficulties, these will have to be resolved by application of common sense in conjunction with the spirit of GLP. While it is certainly not difficult to paste all the necessary information (and much more) onto a 100 kg drum, there might be problems at the other end of test item container sizes. When the respective containers would be too small for permitting the placement of a label, sufficiently large to accommodate the required information, then the question about the correct, GLP-compliant labelling will emerge. The test item may for instance be delivered in small ampoules for injection: Should these now be labelled individually, or could the label, containing the necessary information, simply be put on the box containing the vials? And what should happen in the case of a prolongation of the expiry date which is a mandatory information on the label? To arrive at a correct solution to this problem one has to remember, that the identity of the test item has to be assured at all times, and that for each portion of the test item used in the study the relationship to the remaining

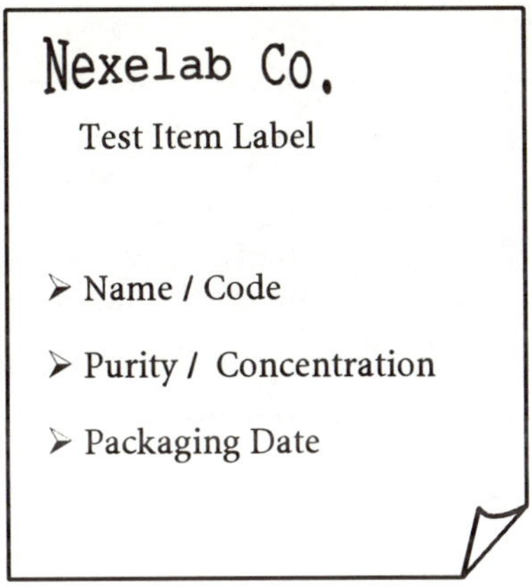

2.9 Test and Reference Items

Nexelab Co.

Test Item Label

➤ Name / Code

➤ Purity / Concentration

➤ Packaging Date

CompliantLab, Inc.

Test Item Label

✓ Name / Code
✓ Batch / Lot No
✓ Purity / Concentration
✓ No Certificate of Analysis
✓ Expiry / Re-analysis Date
✓ Date Received
✓ Amount Received
✓ Gross Weight
✓ Storage Conditions
✓ Date Container Opened
✓ Test Item Logistics OK (Initials)

Figure 22: Test item label with completely insufficient information

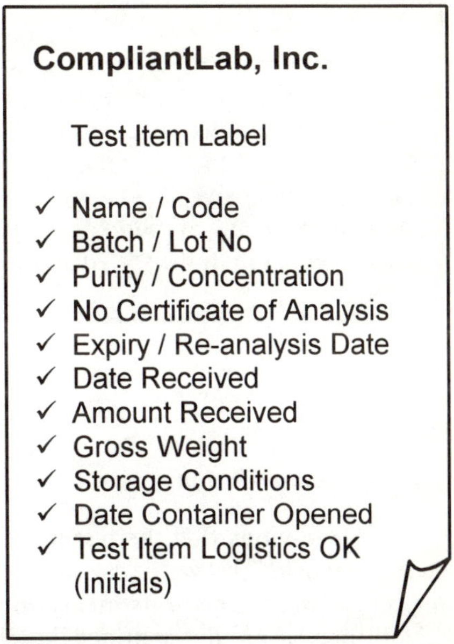

Figure 23: Test item label showing all information needed for GLP compliance

bulk can be ascertained. Therefore, if it can be ensured that the small vials in the above example are rigorously kept in the storage box, that they are taken out only for specified activities, like an administration to the test system, then it can be considered sufficient if only the outer storage container would bear the complete information. Certainly, the vials themselves should be also clearly marked as belonging to the respective container.

> One aspect of traceability in GLP means that there has to be an uninterrupted line of evidence, chaining together the test item (the one which is named in the study title or in the submission for a marketing permit) with the effects exhibited by the test systems. The test item can, in its original state, be characterised by physical/chemical, analytical or

> other means in an unequivocal way. Once it "disappears" into the test system, it may not be possible anymore to ascertain whether in fact the correct test item had been applied. Therefore GLP aims at ensuring as far as possible that the occurrence of mistakes or mix-ups can be minimised through extensive and specific labelling requirements, and that documented information can be provided evidencing the application of the correct item in the stated amounts to the relevant test system.

9.2   Characterisation

It is obvious that the test item has to be "*appropriately identified (e.g., code, Chemical Abstracts Service Registry Number [CAS number], name, biological parameters)*" for its inclusion in a GLP study. The requirements of the GLP Principles are not confined, however, to the identification of the test item, but they call for appropriate characterisation of test and reference items, and of the specific lot or batch of these items as used in a study. This characterisation includes "*batch number, purity, composition, concentrations, or other characteristics to appropriately define each batch of the test or reference items*", parameters which should be known for each single study. It further extends to the necessity of knowing the "*stability of test and reference items under storage and test conditions*" as well as the "*homogeneity, concentration and stability of the test item in (the vehicle in which it is applied to the test system)*". It may be remarked here, that most of these characteristics are derived from the use of chemical substances as test and reference items. In cases, where the test item might represent something else, e.g. a biological entity or a medical device, not all of these characteristics will apply. Thus, for a device that is to be used as a test item, it could certainly not be possible to ascribe to it properties like "purity" or "concentration". However, the requirement for an as full description as possible of the test item's characteristics would remain unchanged ("*other characteristics to appropriately define each batch*"), and it will necessitate a pragmatic approach to judge what kind of characteristics might be used to appropriately and unequivocally describe the test item used.

Since in the large majority of cases, the test item will be a chemical substance or preparation, the following paragraphs will deal exclusively with this kind of test item and thus be confined to the application of the GLP Principles to chemical substances.

Most problems and questions in relation to the requirements for appropriate characterisation of test and reference items are revolving around two pivotal points:

- Who has to generate and provide the appropriate characterisation data, and
- at what time point should these data become available.

Neither of these two questions is directly and expressly regulated by the GLP Principles. While the general data on the test item itself will have to be known (or at least be estimated) before the study actually starts (see the requirement for the labelling with an expiry date), this question becomes less clear with regard to the determination of the respective characteristics of the test item in the application vehicle. Thus, analytical data on the stability, homogeneity and concentration of the test item in the application vehicle may either be generated by the test facility or by the sponsor of the study. The data may be generated before the study actually starts, during the course of study conduct, or even after study termination. All of these cases may be compliant with the GLP Principles, provided they are adequately described in the study plan and the final study report and are appropriately documented, although some national regulations may be more stringent in this respect. As an example of how such issues might be dealt with, figure 24 shows a Study Director's compliance statement which addresses this point succinctly and in a perfectly GLP compliant way.

STATEMENT BY STUDY DIRECTOR

This study was designed in accordance with national and international guidelines, to fulfil the requirements of regulatory authorities, for the toxicity testing of new drugs.

This study was conducted in accordance with FDA Good Laboratory Practice Regulations (FDA, Title 21 CFR Part 58, Federal Register, 22 December 1978 and subsequent Amendments 11 April 1980 and 4 September 1987) except that data concerning the identity, purity and stability of the test material and stability of the test material/carrier mixture were generated within a GMP environment.

Figure 24: Full transparency in this Study Director's statement with regard to test item characterisation under conditions other than GLP.

The least problems, of course, are encountered, when the characteristics of the test item are known to the test facility and the Study Director already before the start of the study. The situation becomes somewhat more problematic, when these data are not immediately available to the Study Director (although presumably existing somewhere, e.g. at the sponsor's). In the former case the test item is fully characterised and complies with all requirements of the GLP Principles. The GLP compliance statement of the Study Director can then, with regard to test item identity, stability, homogeneity and concentration, confidently assure the GLP compliant conduct of this study. In this respect, it would not matter, where these characteristics have been determined: This could have been done directly within the Study Director's test facility and under the direct responsibility of the Study Director, or this information might have been developed in some test facility of the sponsor. In this latter case the data might have been regarded as confidential property of the sponsor and not been disclosed in full to the Study Director (see below). As long as the Study Director is aware of the existence (and nature) of these data, an appropriately worded compliance statement can be issued and signed by the Study Director.

Characterisation issues can, however, become very contentious ones, especially when they relate to the testing of an item by a Contract Research Organisation (CRO) as test facility, to which the sponsor of the contracted study delivers an "off-white powder" in some sort of a container, labelled with "SPO-00115-xyz" or some other such code. The sponsor may possibly not want to disclose the exact identity of this test item, nor to provide the analytical method to be used for ascertaining the "homogeneity, concentration and stability" as it is demanded by the GLP Principles. Thus, the CRO will not be able to confirm, in a GLP compliant manner, these properties; what is even more problematic, the CRO cannot, in such a case, ascertain whether the coded item received at the test facility indeed corresponds to the item that has been intended for testing. If there were no way of ascertaining the identity of a coded substance utilised in a certain study, it would be possible to substitute this study for a study of another test item where an unfavourable outcome had been noted.

The GLP Principles have recognised this problem and are therefore calling for the institution of *"a mechanism, developed in co-operation between the sponsor and the test facility, to verify the identity of the test item subject to the study"*. For the resolution of this general requirement several methods are imaginable. This mechanism might entail the declaration of some easily determined physical/chemical parameters which could be use to "prove" the identity of the item received with the test item that is supposed to be tested; or the test facility might send back to the sponsor a series of coded samples for analysis, among which one sample of the test item in question. The important thing in general is, however, that in this verification mechanism the test item should leave a documented trail from its synthesis and packaging at the sponsor's premises to its receipt, distribution and testing at the facilities of the CRO, until its re-analysis (for identity, homogeneity and concentration) at the sponsor's analytical laboratories.

While this problem of test item identity can thus be resolved to satisfaction by the introduction of such a mechanism, there are the other properties of the test item which have by necessity to be known, in order for the study not only to be GLP compliant, but indeed to be conducted so as to yield scientifically meaningful results. The GLP Principles are requiring that *"the stability of test and reference items under storage and test conditions should be known for all studies. If the test item is administered or applied in a vehicle, the homogeneity, concentration and stability of the test item in that vehicle should be determined."*

There are two aspects to be considered in these requirements, namely stability under storage conditions, and stability under the actual test conditions, both of which serve to ensure that the item studied corresponds to the item that is supposed to be tested.

The stability of the test item under storage conditions is something that the sponsor can address in a sufficiently clear way by specifying exact storage conditions and by providing an expiry date on the label or on the test item data sheet; the disclosure of specific analytical data would not be an absolute necessity. This combination of expiry date with storage conditions to be observed will ensure (or at least make it plausible) that the test item used in the study is still identical to the one that has originally been received, and that it still is of sufficient content and purity for this study. Of course, it has also to be ensured that *"the stability of test, control, and reference substances under storage conditions at the test site shall be known"*, because conditions at test sites might be different from those available at a test facility. In such situations relevant testing should demonstrate stability under the actual storage conditions at the test site (i.e. humidity, temperature, etc.). When the stability of the test item had been determined prior to the study, it is furthermore necessary to ascertain that the conditions which were used in the stability testing apply, or can be extrapolated, to the actual storage conditions at the test site. If this is not the case, or if the stability data are otherwise insufficient, it would certainly be necessary to reaffirm test item stability by appropriate investigations either before, or concomitant with, the respective field study.

A greater problem may be envisaged when the requirement of having to know the stability of this item under test conditions is considered. But again, an indication of stability in aqueous solution or in another appropriate vehicle on the test item data sheet provided by the sponsor would be considered sufficient information to cover this aspect of test item characterisation. Of course, if the purpose of the study is to determine the stability of the test item under these conditions, it is logical and self-evident that the requirement cannot apply, since this property will only be known after the end of the study. This is not only true for studies which are performed for determining the stability of the test item in a vehicle, most notably in feed, but also for those field studies that are conducted to determine the environmental fate of the test item. The biggest problem with these requirements for the availability of test item characteristics lies, however, with the one that calls for the determination of test item homogeneity, concentration and stability in the vehicle in which the item is applied to the test system. This problem, of course need not be addressed to the full extent in the case of true solutions, where homogeneity is guaranteed. Specific stability testing in the vehicle might, on the other hand, not be necessary at all, if the test item were to be applied to the test system immediately after preparation, and if its stability could be estimated to be sufficient to cover this period. Such exemptions from the GLP requirements would certainly need to be discussed in the study report. However, if the test item is insoluble and can only be applied in the form of a suspension or emulsion, these preparations may not stay in the required homogeneous state but will, with time, separate again, so that the intention to apply equal dosages to all individuals of the test system may become jeopardised. This holds as well for toxicity studies, where single individuals may become irregularly dosed, as for field studies, where some parts of the field may receive higher concentrations of the test item than other parts, if the test item starts to settle out in the tank during the spraying operation. Even more crucial becomes this issue in the case of feed admix studies, where the lack of homogeneous distribution of the test item throughout the batch of "spiked" feed may lead to a very unequal exposure of test animals for significant time intervals. These issues, stability, homogeneity and concentration of the test item as it is applied to the test system, are crucial points in terms of the scientific validity of the study, and they will certainly have to be addressed in the final report by the Study Director (see figure 25 a - c). If the sponsor does not, therefore, provide the test facility with the respective analytical method for generating these data, the study

2.9 Test and Reference Items

a)
> The test substance was tested for identity, strength and purity before inclusion in the diet, and was given an analytical reference number. The test article was used within its predicted shelf-life.

b)
> **Analysis of dose preparations**
>
> The results indicate that the achieved concentrations of the dose preparations were within 18% of theoretical concentration (Appendix B). The dose preparations were stated to be stable for 21 days at room temperature; they were used within 21 days of the date of preparation.

c)
> **5.1.1. Homogeneity**
>
> Homogeneity data are summarized below and presented in detail in Appendix B.
>
> The process employed in this study to produce homogeneous dosage formulations was verified. In this study, homogeneity evaluations were conducted at the 0.0250, 0.1000 and 0.4000 mg/ml levels.
>
> The calculated mean concentration of the replicate assay values for dosage formulations ranged from 95-99% of the desired levels and the relative standard deviations (RSDs) were from 0.87 to 1.5%.

Figure 25: a) Simple declaration that the necessary parameters had been determined
b) Declaration of achieved concentrations and of stability in the dosing vehicle
c) Extensive description of homogeneity data in the report body text

might be judged scientifically valueless by the Regulatory Authority. Thus, these data must be provided in some way, and the Study Director has to discuss these issues with the sponsor of the study.

CompliantLab, Inc.

TEST ITEM SHEET

This information is required to carry out the study according to GLP guidelines. Please fill in one form per test item to be tested.

Company: Proposal Number:

Test item name:
Batch number:
Appearance:
Color:
Purity:
Molecular Weight:
Storage conditions (*):
Expiry date:
Solubility (water/vehicle):
Stability in vehicle (+):
Homogeneity in vehicle (+):

Remaining test item shall be
- returned to sponsor after study termination ☐
- stored at test facility until expiry date ☐
- destroyed by test facility ☐

(*)   In the absence of specific instructions, the test item will be kept at 4°C, protected from light
(+)   if applicable

Figure 26:   Example of a test item sheet, submitted by the CRO's Study Director to the sponsor of the study, requesting the necessary information and at the same time drawing the attention of the sponsor to the consequences of a non-compliance with this request.

There are a number of ways out of these situations. In the first instance and with regard to the "simple" stability data for the test item "under storage and test conditions", the Study Director could (and should) exclude these data from his/her Compliance Statement, if they were not provided by the sponsor. This solution may, however, not be liked by the sponsor, since it might be interpreted as the sponsor's failure to comply with the GLP Principles. The most straightforward way of dealing with this issue for a CRO is certainly to draw the sponsor's attention to the fact that the stability of the test item "*under storage and test conditions should be known*" for full compliance with GLP, and that therefore the sponsor would be required to disclose this information. An illustrative example on how a CRO might accomplish this is shown in figure 26. If the respective sponsor really wishes to have the study conducted under GLP, then he would certainly address this issue in some way, at least by ascertaining the availability of this information, if not to the Study Director, then at least to the relevant Regulatory Authorities. In this case the Study Director could, in the final report, indicate that this information would subsequently be provided by the sponsor in the submission package to the Regulatory Authority.

Further, pragmatic solutions may be found in this area. In the case of the determination of the stability, concentration and homogeneity of the test item in the vehicle the Study Director may draw the respective samples from the actual preparations of the test item as used in the study, and send them to the sponsor for analysis. The sponsor may then either release the full results to the Study Director, or state whether there had been deviations from the nominal values, or just acknowledge that the analytical determinations had been undertaken. In all three of these scenarios, the sponsor, or the applicant, respectively, would have to submit these data together with the study results thus permitting the ultimate assessment of the study by the Regulatory Authority.

In some instances, however, pragmatism may even require that these standards should not be fully applied. Consider, e.g., the problem of an assay for sensitisation, where, for the initiation treatment, the test item is suspended in Freund's Complete Adjuvant (FCA). This is a matrix which makes analysis very difficult, mainly because of the sorptive properties of aluminium oxide, one of the constituents of this mixture, and only with huge efforts might a method be developed for the analysis of any single substance in this vehicle. Since the mixture of test item and FCA will be administered to the test animals within a very short time from its preparation it could, if the test item is known to be relatively stable otherwise, be assumed that there would be no major deviation of the actual dosage received by the animals from the nominal dose of the test item through encountering stability problems.

There is still another problem to be considered in the context of test item characteristics, which is most prominent in the field of non-clinical safety testing of drugs. During the development of a substance, some toxicity studies (e.g. genotoxicity studies) may already be performed very early on, i.e. at a time, when a definite, and GLP-compliant, analytical method for the determination of this test item would not yet be available. Thus, although the "*stability of test and reference items under storage and test conditions*" should be known for these studies, too, this information might not be available from a GLP compliant analytical study. It would seem logical from a scientific point of view that such information should, however, be available before the start of a study even in this early stage of development, and therefore, any kind of stability information will be deemed useful to assess the correct exposure of the test system to the test item.

To ensure the validity of these determinations, sampling techniques are obviously also of great importance. Normally, the samples should be taken from the real test preparations. In true solutions, where different concentrations are prepared by sequential dilutions from a stock solution, determination of the test item concentration in the lowest dilution would not only ensure the correctness of the stock concentration but would at the same time confirm the precision of the dilution process itself. Similarly, samples for the determination of homogeneity in suspensions or in feed admixes will normally be taken from the top, the middle and the bottom of the respective preparation during the dosing process, in order to account not only for the spatial distribution of the test item but also for the time factor involved. There is only one acknowledged deviation from the general way of sampling directly from the test preparation, and this is in the case of field tests, where it may be impractical or even impossible to draw the respective samples in a proper way. The preparation of large volumes of spraying emulsions or suspensions in the form of tank mixes, their mode of application, and the geographical localisation of the test site would render the collection of samples for these determinations a rather difficult task. For this special situation, therefore, the GLP Principles state that "*for test items used in field studies (e.g., tank mixes), these may be determined through separate laboratory experiments*".

Another point which is quite frequently debated is a question that stems from the various possibilities of handling the analytical work of test item characterisation: Has all analytical work, even remotely connected with a GLP study, to be performed in every instance under GLP, or under what circumstances would GLP not be required? There are two different sides to this question. The first derives again from the fact that, in some circumstances (e.g. early on in the development of a drug substance), an analytical method may not yet be available which can be used under the provisions of GLP. The second is the question whether the sponsor, to whom the samples for the analytical determination of concentration, stability and homogeneity have been sent, does indeed perform these studies in compliance with GLP. To both of these questions a general answer may be given.

In the first instance, when the item to be tested is in the early stages of development, it may be perfectly admissible to utilise data from analyses that have been conducted under the terms of other quality systems; if the substance has been produced and analysed in a GMP environment, this might constitute sufficient evidence for the test item characteristics. In general, it may be stated that data on the basic characteristics of a test item, e.g. structure and molecular composition, melting point, solubility, or other general physical-chemical data, can be generated under the requirements of any quality system. The Study Director should, however, acknowledge this fact in some adequate way in the compliance statement (see figure 17). Data which are specifically generated for, and within, a GLP study have, however, to be developed in full conformity with the GLP Principles.

In the second case, the question of whether the sponsor performing the analyses would indeed be in a position to generate these data under the provisions of the GLP Principles, the sponsor obviously has to be made aware of this requirement by the Study Director. The GLP Principles do burden the Study Director with the responsibility of ascertaining the GLP compliant conduct of the whole study. This responsibility implicitly includes - as has been clearly stated in the respective OECD Consensus Document (OECD No 8, 1999) - that the Study Director should be aware of the GLP compliance status of any facilities or sites involved in the study. If a contract facility were not GLP compliant, the Study Director would have to indicate this in the final report. Therefore, in the event of an unsatisfactory answer from the part of the sponsor, the Study Director would have to exclude the performance of this analysis from the GLP statement, and it is then again up to the Receiving Authority to judge whether these data can be considered acceptable or not.

> What is applied to a test system should certainly conform to the test item as it is named in the study plan. It should not only be the correct substance, or other article, it should also conform to specifications about purity, concentration and stability. Since the effects produced in a test system will be interpreted as having been induced by the test item, it is of major importance to ascertain the preservation of these specifications throughout the whole study period. Only if the required information on these specifications is present, it can be reasonably concluded that the effects observed indeed derive from the influence of the test item.
>
> The characterisation requirements of GLP have therefore to be viewed in the sense that for study reconstruction it is necessary to have documented evidence for the actual nature of the test item which had been used in the study. In the chain of information needed for a proper reconstruction, the characteristics of the test item play an important role, and GLP wants to ensure that there are no flaws or weak links in this chain.

9.3 Expiry Dates

The requirement to provide an expiry date not only in the test item documentation, i.e. on the analysis sheet, but to print it plainly on the label of the test item container is a very important issue. It has to be seen in the line of all the other requirements for test item characterisation, namely to provide assurance that in the study the correct, unchanged test item has been applied to the test system in its original state. Since study personnel may not be able to consult the analysis sheet before each use of the test item, the requirement to print this information on the label of the test item container should thus ensure that in no instance a test item would be applied to a test system, the stability, purity and decomposition status of which would be in doubt.

There are two points which need to be addressed in this context. In the first instance, the requirement for an expiry date is not confined to the labelling of test and reference items, but is also applicable to any chemicals, reagents and solutions used in a GLP compliant test facility. This issue has already be touched upon briefly in section 6 (see page 167). While there is no question about the utility and feasibility of providing an expiry date for test items, there are regularly questions being asked about the value of such information for simple chemical substances, where, for the most part, no stability data are known or provided by the supplier, and where "infinite stability" may be assumed. Sodium chloride may be considered the proverbial case in point: There is absolutely no question about its chemical stability, thus the need for putting an expiry date on the container with sodium chloride may seem ludicrous. Therefore, failure of the supplier to provide such an expiry date cannot be criticised. On the other hand, once a container has been opened and samples have been taken out repeatedly, the exposure to humidity and the laboratory atmosphere may lead to a progressive contamination of the remaining substance. In order to avoid possible conse-quences from such contaminations it will be advisable to define, not an "expiry date" in the narrow sense of the analytical chemist, but a "usability period" starting from the opening of the container. Such "usability periods" may be laid down in an SOP, which may even differentiate between various reagents, chemicals and solutions with respect to their allowable duration of use.

The second issue is connected with the term "expiry date" as such. Some scientists interpret this term as a "guillotine" date, after which the remaining amount of the test item has to be discarded, because its time limit of usability has "expired", regardless of the possibility that it might still be non-deterio-rated, unspoilt and usable. This interpretation has caused a certain reluctance, especially by analytical chemists, who are charged with developing the necessary stability data for test items, to provide such "absolute" endpoints of usability. However, it is certainly not the intention of GLP to require unnecessary destruction or disposal of valuable materials, and indeed, in the paragraph on the expiry date of reagents, the Principles allow for the extension of the expiry date, if warranted by *"documented evaluation or analysis"*. Although this possibility is not mentioned in the context of test and reference item expiry date, there is no reason to assume that the expiry date of these items could not also be *"extended on the basis of documented evaluation or analysis."* Therefore, and in order to clarify the situation, it has become customary not to speak of an "expiry date", but rather to provide "re-analysis dates". However, it has to be stressed that also the use of the term "re-analysis date" (instead of "expiry date") would not mean a fuzzier endpoint with regard to the usability of the test item. This date should be fully respected as an endpoint, and it would still require that test and reference items could not be used past this date in any study, unless the necessary attestation for quality, purity and usability being still unchanged had been provided by the responsible department or scientist.

> A test item should be used in studies only as long as it can with confidence be regarded as being in its pristine state, pure, unadulterated and not decomposed. Any decomposition or other change in the properties of the test item may lead to spurious and erroneous results, and to wrong interpretations of the effects the test item is supposed to have produced. Stability testing will lead to the definition of a time interval within which the test item will stay in this state, and the resulting "expiry" or "re-analysis" date has to be clearly indicated on the label of the test item container. With this requirement - defining in unequivocal terms the period of time in which decomposition can be ruled out - GLP aims at reducing the possibility that an article will be used in a study which does no longer correspond to the item that had been intended for testing.

9.4 Sample Retention

The underlying principle of GLP requires that the retention of records, other documentation, samples and specimens should provide, wherever possible, the means for full study reconstruction. Thus, it follows logically that also samples from each batch of test and reference item should be collected and retained. In this way it can be ensured that any questions regarding the quality, purity, stability and identity of the test item, that might turn up during the Quality Assurance audit or the scientific assessment of the study, could be resolved by an independent analysis of the reserve sample, without necessitating the repetition of the study itself in case of major doubts about the test item. The requirement that *"a sample for analytical purposes from each batch of test item should be retained"* would therefore not seem to pose major problems of interpretation and implementation.

Two points might still be discussed in this context. The first one concerns the potential utilisation of the sample, while the second one may be seen as an interpretational issue. A third, potentially important point has been dealt with already under section 2.6 (see page 70) and has been qualified in the GLP

2.9 Test and Reference Items

Principles themselves, in that samples of the test item need not be retained for short-term studies.

The GLP Principles, in the paragraph cited above, do already restrict the possible uses of this reserve sample in the sense that this sample should be retained "*for analytical purposes*". Thus, there is no need to retain a very large sample, e.g. of a size allowing the complete repetition of the study. On the other hand, the sample size should be chosen so as to allow multiple analyses. A situation could be imagined, where in the course of a study the development of a refined analytical methodology would allow the detection and quantification of an impurity or decomposition product, not quantified in earlier batches of the test item. In such a case, it may become necessary to perform an analysis of an already used batch in order to ascertain the comparability of test item quality across batches. Or, there might be the case of the test item sent to the test facility in the form of pre-filled vials, where questions of the actual concentration could arise in the wake of, e.g., batch-to-batch variations in the reactions of the test system to the application of this item. In all of these cases, it should be possible to perform an analysis on an aliquot of this reserve sample in order to answer such questions. However, it would be very important that even after an additional analysis, a sufficient amount would be left over for a final verification, if needed.

The second point can be considered a semantic one, and it relates to the definition of the test and reference item. In all those cases, where the test item is applied in a vehicle, the respective vehicle may also be applied to the test system in the property of the reference (or control) item, and therefore, a sample of the vehicle should then also be retained. In the case of toxicity studies involving the admixture of the test item to the feed, the feed therefore becomes the reference item, and thus samples of the respective feed batches would have to be retained "*for analytical purposes*", e.g. for later ascertainment that the "reference feed" did neither contain traces of the test item, nor was it contaminated in any other way which could have influenced the study.

> The purpose of any safety testing is to investigate possible effects of the test item on the test system. To achieve this purpose, it is essential that the effects observed in any test system should be traceable to the application of the item which was the intended subject of the study. In other words, GLP wants to ensure that the effects observed were indeed

the consequence of the application of the test item described in the report or the application. In order to ascertain this even retrospectively, after the conduct of the respective safety test, the documentation on the test item has to fulfil a number of requirements:

- There has to be reassurance, that indeed the alleged test item had been delivered to the test facility for testing;
- there has to be reassurance that the item delivered to the test facility had indeed been used in the studies;
- there has to be reassurance that the test item received and stored at the test facility did not deteriorate during storage, so that it had indeed been the original test item in its documented strength that the test system actually had been exposed to;
- there has to be reassurance that the test item had retained its identity and concentration in all the vehicles it had been admixed to, and
- there has to be reassurance that it retained, during the study, its identity and intended concentration (or dose) during those time intervals after admixture, for which it had been stored before application to the test system.

In summary, for the test item, there must be documented proof that the one item that had been intended to be tested indeed reached the sensitive parts of the test system warranting that the effects observed had really been initiated by the test item, and that the application of this item to man or the environment would therefore not be expected to result in any effects other than those which can be extrapolated from the observed ones in the test systems utilised.

10. Standard Operating Procedures

10.1 Introduction

As their name implies, Standard Operating Procedures (SOPs) are intended to describe procedures that are routinely employed in the performance of test facility operations. Indeed they are defined as *"documented procedures which describe how to perform tests or activities normally not specified in detail in study plans or test guidelines."* The definition furthermore implies that SOPs should describe all steps in the performance of an activity in such a detailed way that somebody not absolutely familiar with this activity might be able to perform it correctly and without having to resort to outside help. In contrast to a guideline, which generally is regarded as something like a "guiding principle", from which reasoned deviations are possible or even necessary, an SOP should be followed faithfully and to the letter. If it is not, this has to be counted among the deviations and should be appropriately addressed in the final study report.

It is counted among the responsibilities of test facility management to ensure that *"appropriate and technically valid Standard Operating Procedures are ... followed"*, even though Study Directors and Quality Assurance personnel would certainly be in a better position to judge the observance of SOPs by study personnel. The real importance of this responsibility lies thus less in the detailed remonstrances with single persons, or admonitions of individual errors, but in the stipulation that management has to act and to implement adequate measures if Quality Assurance reports a less than optimal observance of SOPs.

By helping to ensure that all personnel will use exactly the same procedures for the operations described therein, the SOPs may be looked at as an instrument to minimise the introduction of random error, due to individually varied procedures, into a study. On the other hand, the SOPs need to be written by persons who are experienced in the procedures to be described, and thus the introduction of systematic error into studies should also become minimised.

One very important aspect of the Standard Operating Procedure has to be addressed here, although this issue has already been alluded to in the first paragraph. It must be emphasised that these documents are prescriptive for standard situations, activities and procedures only. The reluctance of research scientists to accept GLP as a valuable way for improving data quality and study reliability and integrity, which has been mentioned in the first part of this book (see section 1, page 4), may stem from the misconception that all and every activities have to be governed by an SOP, thus threatening to introduce an element of rigidity into the conduct of investigations. This misconception can be denounced on two counts: Firstly, an SOP has to be available only for those activities which are "*normally not specified in detail in study plans or test guidelines*", which means that any procedures or activities that must be regarded as singular, or which may be in need of constant adaptation, would not need to be described in such a standard way. Secondly, even from the most "rigid" SOPs deviations are possible, if scientifically or procedurally justified. A very good example can be provided by the (probably world-wide standardised) way of placing tissue sections on the histology slides. The respective SOP will thus show (see figure 27) how the standard arrangement of the sections will look like, while allowing at the same time for special arrangements and additional slides if a specific need should arise. Thus, neither the existence of SOPs as such, nor their application would be averse to flexible approaches where necessary.

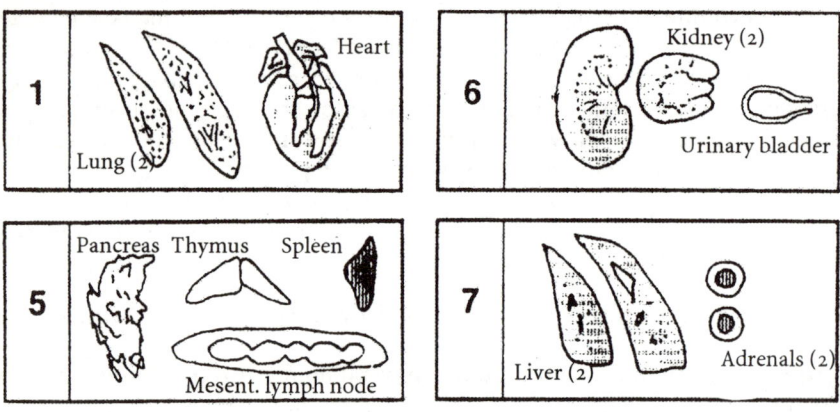

Figure 27: Four illustrative examples for the standardised way in which tissue sections are arranged on the specifically numbered slides; non-standard tissues may also be taken and processed, and these will necessitate additional slide numbers and an exact description in the study plan.

10.2   The Format

One of the most important Standard Operating Procedures which, however, isn't even mentioned in the OECD GLP Principles, is the one which describes the basic processes and procedures for the writing, approving and revising SOPs, the "SOP on SOPs". In this document the general format of the test facility's SOPs should be laid down, and it should describe when and how revisions of SOPs should be done.

Even though each test facility may develop its own format of SOPs, there are a number of general points to be observed when considering the format and layout of an SOP. First of all, an SOP should be recognisable as such, i.e. it should be headed by the words "Standard Operating Procedure". It is furthermore highly advisable to stick exclusively to this expression and not to invent individual terminology such as "Standard Working Procedure", since this may give rise to misinterpretations and deviations in the way in which the instructions in such documents are followed. It would also be advisable to create a template of the front page ensuring that all the necessary information will be provided on this page. While the GLP Principles advise on the content of the study plan and the final report, they do not state which information is considered valuable or necessary on the cover page or in the heading lines of an SOP. These informations can only be derived from the general principles of GLP, namely to ensure traceability and reconstructability. Therefore it is certainly necessary to provide on the front page the following information:

- Date of entry in force;
- descriptive title;
- code number of the SOP, or
- abbreviation of the title used for the indexing in the list of SOPs;
- version number;
- the signatures of the author, possibly also of the Quality Assurance; and
- the approval by dated signature of test facility management.

Helpful is also information on the reasons for the revision, if any has been performed; and the archived copy of an SOP which has been superseded by a revised version should be stamped with the "expiry" date (for an illustrative example, see figure 28).

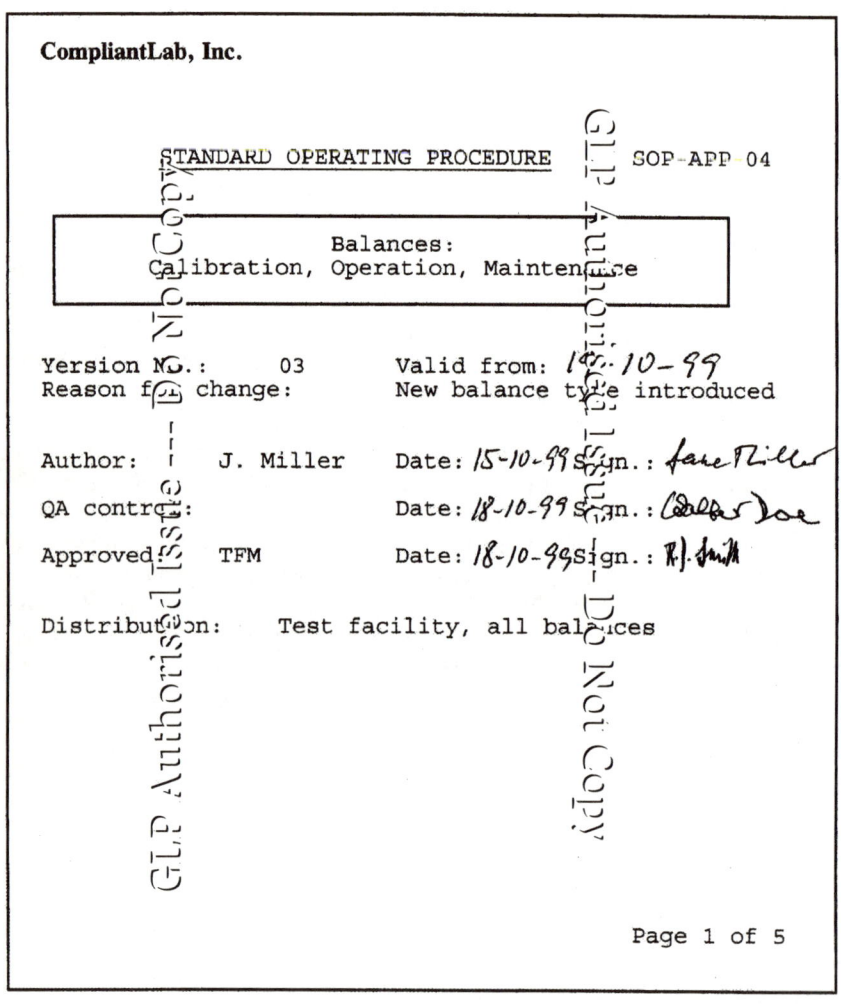

Figure 28: Model cover page for an SOP, containing the minimally necessary information

The format of an SOP may also include appendices, especially in the case where textbooks, publications or instrument manuals are referenced in the SOP. The GLP Principles allow the utilisation of such publicly available information, although only on a secondary level, in that they state that they "*may be used as supplements to these Standard Operating Procedures.*" This means that formally an SOP has to be issued, consisting simply of the cover page and providing a reference to the appended document. The SOP for conducting some analytical determination by means of a commercially available kit - which is supplied together with an exact description of the test procedure - may therefore state in the section dealing with the method, that the description enclosed with each kit (and an example of which should be appended to the SOP) should be used for conducting the experiment. As a further example we may consider the case where a special maintenance of an apparatus is performed annually by the manufacturer according to its own procedures. The test facility's SOP for the maintenance of this apparatus then needs only to describe the routine procedures to be used by the test facility personnel, while referring to the manufacturer's procedures with regard to the special maintenance work. Also check lists, recording forms or other aids for the performance of the operations described in the respective SOP may be appended. It is important then, that the pagination of the SOP would reflect also the appendices, and indeed it is to be recommended that SOPs should show the total number of pages besides the actual page number.

Another consequence in the area of SOP use needs to be addressed in this "SOP on SOPs", and it might probably need some consequence with regard to the formal aspects of SOP generation. It is connected with the issue of reconstructability which is central to the GLP Principles. To realise this reconstructability it has to be ensured that work has - in all probability - been performed according to the instructions provided by the respective SOP. This aspect can only be ensured, if only the current SOP version is exclusively available at the respective workplace. In consequence it can then be assumed that the activities indeed have followed the instructions as specified in this document. This assumption would, however, be invalid, if SOP copies other than the "officially" distributed ones, possibly adorned with personal, individual annotations, were allowed to be in use. Therefore, it is of great importance that this "SOP on SOPs" should address the question of copying SOPs for the personal use. While such copying is not prohibited by the GLP Principles, certain precautions have to be exercised when contemplating the utilisation of such "personalised" SOPs. Two ways are possible to cope with this question, both of which would conform to the intentions of GLP: Copying of SOPs may either be completely prohibited, or it may be permitted under very restrictive, pre-defined conditions.

The former approach might be considered easier to control, and it should in general be the preferred way of dealing with this question. One of the best ways to achieve this goal and to discourage personal copying might be to print the "official" copies on specially marked paper, e.g. "SOP sheets" displaying a colour band with a remark such as "official version, do not copy". As a further precautionary measure, such stationery should be available only to the person (or in the office) responsible for the SOP administration and distribution. Certainly also other, similar means to facilitate the control over the existence of SOP copies may be found. In this way, illicit copying could be efficiently detected and controlled.

Although the exclusive use of the officially distributed SOP copies would in general be preferable, there may be reasons for the latter approach of permitting additional, personal copies to be made and used. In order to retain the possibility for study reconstruction, however, all such personal copies should then be collected at the end of the study in which they have been used and be added to, and archived with, the study raw data. Any alterations having been made to the "official, current" version should furthermore be acknowledged by the Study Director as mandated by the GLP Principles ("*Deviations from Standard Operating Procedures related to the study should be documented and should be acknowledged by the Study Director and the Principal Investigator(s), as applicable.*"). This question gains in importance, as the problem of "on-line" SOPs becomes more and more prominent. This special aspect of "personalised" SOPs will be addressed further in section 10.4 (see page 231).

10.3 Issue, Approval and Distribution

In this area test facility management has again to play a central role: Test facility management is responsible for ensuring that "*appropriate and technically valid Standard Operating Procedures are established*". Management does not have to write or to issue "technically valid" SOPs, as for this task it may simply not have the necessary expertise. In many cases a Study Director would be the most capable person for describing all the standard procedures to be utilised during the course of a study. The SOPs on apparatus, on the use and maintenance of instruments, might be written by somebody among the technical personnel who are most knowledgeable about the functional testing and maintenance of technical equipment. And last but not least, the SOPs of the Quality Assurance Programme most certainly will have to be written by Quality Assurance personnel, since only they have the expert knowledge in the tasks and processes which form the Quality Assurance Programme. However, it is a managerial function to approve the SOPs written by adequately qualified personnel.

It is essential that, once written and approved, these SOPs are not regarded as "commandments, hewn in stone" but that they are constantly reviewed for their actuality, and that they should be revised as soon as such a need would appear. The various people who were responsible for writing the respective SOPs are also the ones who will consequently be best situated to judge the possible necessity of changes in the SOPs, either because of scientific progress, because of a change in the equipment or because of a shift in the test system use, or simply because a better way of performing some task has been developed. In all these cases, SOPs need to be revised in order to reflect the new standard ways of performing tasks and operations. In order to ensure the technical validity of SOPs there has to also be a mechanism, whereby the need for a revision should be judged in regular intervals even without apparent reasons for doing so. In consequence to the requirement for approval by the test facility management, the so revised SOP cannot be acknowledged or approved by any person other than the responsible test facility management. This is also explicitly mentioned in the Principles by requiring that test facilities not only should *"have written Standard Operating Procedures approved by test facility management"*, but also that *"revisions to Standard Operating Procedures should be approved by test facility management."* Although Quality Assurance personnel are not normally involved in the writing of SOPs - with the exception of their own, see above - it may, however, be desirable that they review SOPs before approval in order to assess their compliance with the GLP Principles and the prescriptions set forth in the "SOP on SOPs".

This again documents the importance of separating the various responsibility levels and the respective tasks in the operations of a test facility. As much as the Quality Assurance personnel have to be independent of the studies the quality of which they have to assure, the persons approving SOPs should be different from those writing and utilising them. Therefore, a Study Director cannot be allowed to perform test facility management functions enabling him to approve his own SOPs. While this may not be a problem in large test facilities, it may become one in smaller facilities with a limited number of personnel, and it will call for judicious balancing of GLP requirements and pragmatic approaches to the organisational facts in order to arrive at a solution which satisfies both sides of the coin.

Once written and approved, SOPs need to be distributed and made available to the study personnel and to any other personnel needing such instructions for the performance of their daily tasks. The distribution of SOPs is on the one hand governed by the requirement that the relevant SOPs should be immediately available at the workplace, and on the other hand, that work should be performed only according to approved and current SOPs. This latter point precludes the utilisation of an "old" SOP in any single part of the test facility once a revised version has been approved and made available. There are a number of possible solutions to this problem of adequate and proper distribution. Test facility management may, e.g., appoint one single person to bear responsibility for the administration of all SOPs, which may include tasks like keeping account of versions, and consequently alerting authors of SOPs on the "age" of their documents, keeping the archives and maintaining the historical file of all SOPs, and last but not least the overviewing of the distribution of new SOPs (or new versions thereof). This distribution should be handled concomitant with the reassurance that the old versions are removed from their location (and destroyed). Instead of charging a single person with this distribution task, it may be possible to send the required number of SOP copies to the respective laboratory heads, or other responsible persons, with the request of inserting the new (or revised) SOPs in the SOP collection, of removing the old version and of acknowledging the correct accomplishment of the SOP exchange (or insertion, if it is a completely new one) by dated signature on a special receipt form. The distribution and removal of SOPs can be greatly facilitated, if there is a clearly defined recipient's list, either printed directly on the SOP, or at least available to the person responsible for the SOP distribution.

The GLP Principles are considering it important that SOPs are "*immediately available*". Only in this way can it be assured that there is at least the possibility for (study) personnel to consult SOPs immediately, should the need arise. If an employee would have to go to the other end of the building in order to look up some forgotten detail in the instruction for the procedure to be performed, he might just go ahead with the task as he thinks fit and not bother about the way prescribed in the SOP. Of course, the immediate availability does not guarantee that personnel would indeed consult the SOP at every single step of the activity. In performing really routine procedures the consulting of the respective SOP might be deemed unnecessary by experienced personnel. But even so, slight changes in the habit of doing things might creep in, and these may finally change the whole way of performing the activity. At this point, however, the Quality Assurance would have to step in and either lead the way back to performance according to the SOP, or point to the necessity for a revision of this document.

Another point which has to be taken care of in the context of SOP distribution is the extent of coverage needed at the individual workplaces. Not all SOPs are needed everywhere: An SOP on the manner and frequency of changing the bedding for rat cages will be needed in the animal rooms, but will have no applicability in an analytical laboratory; an SOP on microscope maintenance, while important in the histopathology laboratory, will be of only spurious interest in the farming equipment's shed. Even closely related activities may not be covered by the same SOPs, and therefore even adjacent test facilities may need different make-ups of their SOP collection. Therefore, the GLP Principles expressly ask for the availability of SOPs "*relevant to the activities being performed*" in the respective test facility. Each working place or work station should thus have access to the SOPs applicable to the work performed at the respective place, but should not be burdened or inundated with documents that are completely irrelevant to their specific operations and activities.

10.4   On-line SOPs

Normally, SOPs are still provided in hard-copy form and are assembled in binders or other kinds of collecting gear. With the increase in importance of the "paperless office" there is certainly the question of whether SOPs could not also be provided in electronic form only. There are advantages as well as disadvantages to the introduction of electronic SOPs. It could be imagined that distribution and retraction of SOPs would become much easier, because it could indeed be done just by pressing a key on some central computer, instead of having to go through the different steps described in the preceding section. Furthermore, the utilisation of SOPs in "dirty" areas might be easier, when they could be read on-line in electronic form, since there would be no paper which could become soiled, crumpled, torn and rendered illegible, and therefore inutilisable in the long run through any kind of environmental influences. On the other hand, computers may not be usable under certain environmental conditions, where paper SOPs could still be utilised.

Problem areas on the other hand are also manifold. They may concern the question of documented approval: Although electronic signatures are coming of age (Coombes, 2000), other solutions to this problem may have to be found. This could include the printing of one original copy bearing the necessary dates and signatures which can then be properly archived, or the possibility of utilising a "signature sheet" only in hard-copy format. A more difficult problem area can be derived from the habit of people who are used to printing their own hard copies from any electronic document. This habit might give rise to illicit copies floating around, which are not registered anywhere and the existence of which therefore may escape the attention of Quality Assurance. Revisions to the "official" SOP, or changes made on these hard copies might go unnoticed and could possibly influence the integrity, and certainly jeopardise the recon-structability, of the study. As discussed in the previous section, there are two possibilities: Either it will be totally prohibited to print hard-copies, or these copies will have to be submitted to very strict rules with regard to their use. In the former case, it would be the task of the central information technology people to ensure that personnel would have no possibilities to print such documents (i.e. to assign a "read only" status to these documents), whereas the latter would call again for an involvement of the Study Director and Quality Assurance in the control of such individual hard-copies, in their filing with the raw data of every study so conducted, and in the judgement of their influence on study conduct and outcome.

Other problems may not be specific to electronic SOPs, but may be related to problems of electronic systems in general. If anything happens to a paper SOP, there are many more copies of it around, and the lost or destroyed individual copy can easily be replaced again. If, on the other hand, something happens to the SOP file on the central server of the IT system, then all SOPs would be irretrievably lost, if no back-up copies were to be kept. Therefore, back-up and disaster recovery systems will have to be in place as for any other IT applications, and the respective issues of validation and maintenance will have to be observed (see section 7, page 172).

10.5 The Content

It might be considered as a self-evident prerequisite that an SOP should be "technically valid". If this were not so, then the quality and validity of data generated under the provisions of a "technically invalid" procedure would certainly be jeopardised. Therefore, great attention has to be paid to the content of these procedures and their description. It is certainly not possible in the context of this section to provide detailed advice on the specific content of all imaginable SOPs. For this to do, the area of procedures possibly to be governed by an SOP is much too large. However, a number of general considerations can be presented which should be helpful for the generation of any "technically valid" SOPs.

Before going into a somewhat more detailed description of the necessary elements of SOP contents, a general issue should first be discussed. This issue is connected with the question of the necessary or desirable level of detail with regard to the organisation of SOPs. Two ways can be envisaged in this respect: Either each single activity can be fully described in an SOP, or a general description of similar activities could be combined in one SOP. This issue becomes most obvious when considering SOPs for technical equipment, where it could either be possible to issue one general SOP for one kind of instruments, or to write specific SOPs for each and every single piece of apparatus of this kind. The analogy may be drawn between a study plan, and the "general study plan" for short-term studies with its study-specific extensions.

As the most pointed example for these two possibilities, the divergent ways for dealing with SOPs on balances, their maintenance and calibration, can be cited: On the one hand, it is certainly possible to issue specific SOPs for each and every balance in a test facility. It could be even imagined that there might be different SOPs for use, for maintenance and for calibration of each of the balances. This possibility might be regarded as having certain advantages, since there could be no question that only the really correct SOP would be placed alongside the respective balance, and there could be no misunderstanding about the procedures to be used with this specific instrument. On the other hand, a large test facility can be equipped with a multitude of balances, of which a great number may be of identical make and type. If the SOPs were to be dealing with these balances individually, the inflation in their number would inevitably lead to loss of control, especially with regard to possible revisions of these SOPs. It would become difficult, if not impossible, to keep track of the various revisions made to them in the different laboratories at different time points, and consequently, there might be different procedures applied in different places at the same time. Furthermore, the general list of SOPs would become difficult to survey due to the very large number of SOPs with very similar titles and identical areas of application.

On the other hand, one single SOP could be generated, which would deal with balances in general. In this case, the SOP would not only describe in general terms the procedures for using, calibrating and maintaining the instruments, but the more specialised points to be observed with specific types or models of balances could be described in special sections. Moreover, the SOP should contain a list of all balances to which it applies, together with all the necessary information about types and locations. In this way, one SOP might cover all activities connected with all kinds of balances throughout a test facility. Both of these ways to deal with the problem of "levels of detail" are certainly to be regarded as the extremes in a continuum of possibilities, and it might lie again in the responsibility of test facility management to issue guidelines on the preferred direction in which to proceed.

But let us turn now to the actual topic of this section, the content of the SOP proper.

Basically, an SOP should contain, besides the more administrative parts like title, version number, author, and all the approval dates and signatures, two logical parts: The first one should provide the reason for, or the purpose of, the SOP, while the second one should describe the activity to be regulated. Simple as this advice may seem, it is not so easy to live up to it, since the question, already alluded to above, immediately arises at what level of detail these activities should be described.

The GLP Principles are silent about the amount of detail to be included in the SOPs, but it is a management responsibility to ensure that "*personnel clearly understand the functions they are to perform*" and that the "*technically valid Standard Operating Procedures are ... followed*", which must be read to mean that an SOP should be sufficiently detailed that trained laboratory personnel would not only understand it, but could perform the tasks described therein in a uniform way. Therefore, all the important steps to be performed should be described in such a way as to allow the unequivocal reconstruction of the performance of these activities. Achieving this purpose can necessitate different approaches, however. On the one hand it would be important to generate a very detailed guidance for a complex activity or procedure which had been developed especially in the test facility or for the specific type of study. On the other hand, for more general activities of a rather routine nature, the description needs not be more extensive than just an enumeration of steps to be followed.

We might consider the topic of "calibration" as a good example on which these distinctions can be demonstrated. Calibration of a balance may not involve more than just the placement of the correct calibration weight on the balance and to read off the respective value indicated. Thus, the respective SOP can be kept rather simple. Calibration of an HPLC apparatus for the quantitative determination of test item in a biological matrix will involve, however, more complex and delicate manipulations, so that a much more detailed description of the whole procedure should certainly be advisable.

There is another aspect in the description of activities that is very often overlooked in the drafting of SOPs. In many cases, the SOP can describe a procedure in a completely straightforward way, because it involves an activity that produces a definitive result. In other cases, however, the situation is different, in that the procedure results in an "either / or" situation, where a choice will have to be made. This is especially the case with calibrations of instruments, the results of which may lie either within or outside their specifications. It is of utmost importance that such SOPs do not only describe the actual procedure for calibrating the instrument, but that they also exactly advise on two additional points:

1) the admissible magnitude of deviations from the expected calibration value, and

2) the necessary steps to be taken, if the calibration results in an unacceptable value.

As another example the measurement of a number of biological or physico-chemical parameters might be cited, where it is customary to repeat measurements, if the results from the first readings give rise to suspicion about their validity. In the same spirit as described above, decisions on repetition of measurements should be based on clear criteria, which have to be fully described in the respective SOPs. In connection with this problem of measurement repetitions, the question ensues which one(s) of the different values obtained in the original and repeat determinations should subsequently be used in the final calculations or finally be reported. Again, clear directives for the application of distinct criteria will help in assuring the integrity of the whole study, as it can be clearly seen that only in this way it will become possible to fully reconstruct a study. Without written and approved standard procedures completely and exhaustively delineating the approach to be taken for dealing with these situations, it becomes impossible to address and judge the reasons behind the actions taken at such decision points in a meaningful and conclusive way. It will even appear paradoxical, if for each simple correction of a misspelt word or erroneous date, the person making the correction has to date and initial this change and to provide the reason for it, while in the case of choosing amongst two or three results of repeat measurements, no argument has to be provided for the final choice of the one value which is finally used or reported. Since such choices should be made in a standardised, reconstructable way, however, then at least the standard principles, if not the exact steps, governing the approach to be taken should be delineated in the respective SOP.

Earlier in this section, we have mentioned the two logical parts of an SOP, namely the descriptions of its purpose and of the procedure. The foregoing paragraphs have shown, however, that there is more to the content of an SOP than could be imagined at first glance, since in a number of cases, a third element has to be added: A decision tree for situations where choices have to be made, together with a description of the ensuing standard procedures for the various possibilities, or any other similar description of the applicability of the respective SOP under various circumstances.

This latter point might best be illustrated by the case of the Quality Assurance SOP on inspections. Quality Assurance has the possibility to perform facility-based, study-based and process-based inspections, as has been described in detail in section 4.2 (see page 128). In writing the respective SOPs, it will be insufficient, however, to provide only a description on how such inspections will be performed. This is to say that describing the standard way of performing these inspections will only represent one part of the necessary information to be given in these SOPs. As important as the "how" should be the "when". Thus, the respective SOPs of the Quality Assurance will have to consist of the following parts:

- Purpose: "This SOP covers the way facility-/study-/process-based inspections are to be performed in order to ascertain the GLP compliance in these areas."
- Applicability: "Facility-based inspections will be performed at regular intervals of about x months independent of specific studies. Study-based inspections will be performed within the time limits specified by the master schedule and the single inspections will be set according to the defined critical phases of the studies. Process-based inspections will be conducted at regular intervals (of about x months, or for each yth study) for those study types that fulfil the following conditions: ..."
- Procedure: "Inspections will be carried out according to the following standard procedures and with the help of the relevant check-lists appended to this SOP."

From the various points, issues and problems addressed in this section it can be clearly seen that writing an SOP may not be as easy as writing a recipe for a cookbook, where it is only necessary to list the ingredients and then to describe the way of assembling them into the final dish. The writing of an SOP involves first of all the laying-out of all the necessary points and issues to be addressed. The elements identified as the necessary ingredients have then to be combined in a logical way, and finally it has to be checked, whether indeed all of the specific information needed to attain the goal of the SOP in question has been included and dealt with in a exhaustive manner.

10.6 Where are SOPs required?

It is certainly difficult to define exactly and comprehensively all areas where SOPs may be needed. In general, as has already been mentioned, SOPs do describe activities of a repetitive nature, and therefore all GLP-related areas where such activities are performed, need such standardised descriptions. Furthermore, SOPs are such helpful instruments to define the proper conduct of any activities, that they will also be used outside of the realm of GLP, e.g. in administrative areas where they may describe the procedures for dealing with sponsors at all stages of the contractual relationship, from obtaining a commission until the final invoicing of services rendered.

The OECD Principles of GLP provide a list of areas and topics for which SOPs should be written; the list is expressly said to be non-exhaustive, but the various, illustrative examples provide test facilities (and SOP authors) with an idea about the topics to be regulated in the form of SOPs. Thus, they state that *"Standard Operating Procedures should be available for, but not be limited to, the following categories of test facility activities. The details given under each heading are to be considered as illustrative examples"*.

Five main areas of test facility operations are identified in these Principles, where SOPs should be available. They are centred around the standardised activities connected with the test item, the apparatus, the test system, the study conduct, and the Quality Assurance, as shown in Appendix II.I under the respective paragraphs (see page 272). Insofar as the details of the areas of SOP applicability are concerned, there is thus no need to reiterate the examples given in the test of the GLP Principles. Since the Principles are providing examples only, a few thoughts should be spent, however, on the general aspects of SOP applicability in order to clarify some of these general points.

Before going into these general issues of SOP philosophy and policy, however, one area should be mentioned, which has not been directly addressed in the GLP Principles, but which may be regarded as one that is preparing the ground for the GLP compliance of a test facility. This area concerns the various activities centred around the responsibilities of management and the administrative processes connected with it. The GLP Principles require that management has to ensure, e.g., *"the maintenance of a record of the qualifications, training, experience and job description for each professional and technical individual"*; how these records are to be collected, formatted, updated and retained may best be described in an SOP. The same holds for a number of other management responsibilities, such as the replacement of Study Directors and Principal Investigators (*"Replacement of a Study Director should be done according to established procedures, and should be documented"*), the maintenance of the historical file of SOPs and of the master schedule, or the provision, for multi-site studies, of *"clear lines of communication ... between the Study Director, Principal Investigator(s), the Quality Assurance Programme(s) and study personnel"*. In order to ascertain a clear regulation of all these activities, the existence of, and adherence to, SOPs relevant to these areas, should certainly be regarded as a prerequisite of GLP compliance. In this way, another point which has already been mentioned (see section 3.1, page 94) may be addressed by management in an indirect way, namely the "real" standpoint of management with regard to GLP adherence. If management itself is respecting SOPs in its activities like all other test facility personnel, it will be seen as placed on the same level with the "common folk" with regard to its obeisance of GLP rules. Thus, the area of "management SOPs" may constitute an as important part of the whole collection of SOPs as any other area listed in the GLP Principles.

But let us return to the general, guiding principles for an answer to the question of where SOPs should be required.

It has already been stated that SOPs should be available for all areas, where activities or procedures of a repetitive nature are being conducted. There, the respective, specific SOP has to ensure that the activity described will be conducted in a reproducible, and thus reconstructable, manner. In the same sense, it has to be ensured that all such activities and procedures are covered by the respective SOPs. This means that for each part of test facility operations, a "life cycle" consideration could be applied for determining the extent of SOP coverage. This can well be illustrated by the examples given in the GLP Principles in the various areas. When the extent of SOP coverage is described for, e.g., computerised systems, the examples involve *"validation, operation,*

2.10 Standard Operating Procedures

maintenance, security, change control and back-up." Thus, before a computerised system comes into operation, it has to be validated, at its introduction its performance has to be determined in the working environment, then its operation has to be described, and finally the standard procedures to be followed when its software changes into a new version may be taken figuratively for the end of the life-cycle.

If this way of looking at activity areas in an all-embracing manner is consequently followed, then it may be easy to determine the extent of SOP coverage needed for the respective areas, and no apparent deficiencies will arise. In a way, this may again be considered as a policy matter, which would necessitate, for each of the areas concerned, a well delineated charting-out of the respective activities in their entirety, from the initial, basic requirements, over the activities proper, up to the final dispositions necessary.

> The goal of GLP, the ability to fully reconstruct any study, mandates that it will be possible to determine exactly how a specific activity had been executed on any specific, single day in the operations of a test facility. From this goal, the GLP Principles derive the requirement of Standard Operating Procedures to be written, approved, distributed, used, maintained, revised, and finally archived in chronological order, and GLP wants also to ensure that all activity areas are properly covered by SOPs.
>
> An SOP is therefore a prescription which has normally to be followed to the letter, since it is declared to be the standard way in which the activities described therein are to be executed; exceptions to this rule are possible only if appropriately justified and documented. The emphasis placed on these instructions is demonstrated by the fact that GLP charges the test facility management with the approval of these "technically valid" descriptions of activities. In view of the necessity for an unequivocal determination of the way activities were executed, measures have to be instituted which should make it impossible to utilise more than one version of one and the same activity description at any one time.

> In summary, it can be stated that with the instrument of the Standard Operating Procedure GLP wants to ensure the reconstructability of all activities and events around a study. While it does not intend to curb flexibility in activities, processes and procedures, it aims, however, at eliminating such instances of "flexibility" which are only the result of sloppiness and lack of planning and forethought.

11. Study Performance and Reporting

The performance of a study is governed by a whole set of rules and not exclusively by the GLP Principles, since it is here, where the scientific background of a study is meeting with the managerial quality tools of GLP. In other words, it is here where the scientific reasons for, and the test guidelines applicable to, the use of the test system, have to be merged on the one hand with the SOPs ruling the conduct of standard activities, and on the other hand with the study plan delineating the chronological course and the experimental details as applicable to the actual study, as well as the recording requirements of the Principles. For the GLP compliant conduct of the study, the study plan is constituting a central document and its elements are therefore described in detail in the GLP Principles.

11.1 The Study Plan

The GLP Principles are adamant in the requirement that a study has to be well planned in advance, and to this end the Study Director has to compile a study plan, which has to be approved by dated signature before the study itself can be initiated. A study without a proper study plan can in no way be regarded as a GLP compliant study, even if all other rules of the GLP Principles were to be respected during its conduct. Besides being approved through dated signature of the Study Director, a study plan should also be *"verified for GLP compliance by Quality Assurance personnel"*. This verification should logically be performed already at the draft stage of the study plan, since any change due to some aspect of non-compliance discovered during this process could then be easily incorporated into the final study plan, while any necessary corrections after approval would have to be made through an amendment to the study plan.

The GLP Principles provide furthermore for the possibility of approval of the study plan by test facility management and by the sponsor, without generally requiring it (*"if required by national regulation or legislation in the country where the study is being performed"*). Indeed, since the Study Director has the full and undivided responsibility for the GLP compliant conduct of the study, approval of the study plan by other persons may seem unnecessary from a GLP viewpoint. The requirement that the study plan be additionally approved by the test facility management may be considered to serve the GLP relevant purpose of demonstrating that the test facility management is aware of the study and that with its approval by dated signature it acknowledges its responsibility of providing adequate facilities and resources for this study and at the same time appoints the Study Director. On the other hand, the approval of the study plan by the sponsor may be required by some authorities due to legal considerations related to responsibility for the validity of test data. Notwithstanding the legal aspects, contractual arrangements between sponsor and test facility may also call for the approval of study plans by the former. It would certainly be "good business practice" to have the sponsor agree to the study plan in order to avoid later discussions about study conduct, but from a GLP point of view, it is the signature of the Study Director which would mark the initiation of the study in any of these cases.

One important aspect of the study plan has already been discussed in section 2.8 (see page 80), namely the fact that even the best conceived plan may be in need of alterations as dictated by study events, and that therefore mechanisms have to be defined by which such changes can be introduced. While the definitions of the terms "amendment" and "deviation" have been treated in the above mentioned section, it remains at this point to draw the attention to a small but nevertheless important difference between the ways the study plan and any amendments to it are to be treated with regard to their approval. While the study plan has to be approved not only by the Study Director, but, as described above, (*"if required by national regulation or legislation"*) also by the test facility management and / or the sponsor, an amendment has only to be approved by the Study Director, since the GLP

Principles require only that they "*should be justified and approved by dated signature of the Study Director and maintained with the study plan*".

There is a practical reason for this procedural difference. Just like the study plan has to be written and approved before study initiation, any amendment to it should preferably be approved before the respective change is introduced. This may involve very narrow time windows between the acknowledgement that a change is needed and the necessity for its actual implementation, which in turn might make it impossible to get timely approval by test facility management or, possibly even worse, by the sponsor. Thus, in terms of GLP compliance, the dated signature of the Study Director will mark the time of approval of an amendment. On the other hand, it would certainly be admissible, or even advisable, to let test facility management or the sponsor - whoever approved the study plan in the first instance - also acknowledge these amendments. Since the GLP Principles do not require these additional signatures, these procedures might be described either in SOPs or could be the subject of the contractual agreement between sponsor and test facility.

In this regard, a small problem, which may lie more on the psychological side, could be posed by the reluctance of sponsors to have to acknowledge great numbers of amendments. The sponsors may prefer, or actually request from the Study Director, that amendments would be bundled, so that e.g. the Amendment No. 1 would be the only one, and would document all the single deviations, precisions or additional informations that have become known or necessary within the whole study, instead of having to deal with these changes one at a time as they are turning up. There is nothing in the GLP Principles that would prohibit such a procedure, provided that every single amendment would have been approved by dated signature of the Study Director in a timely manner. After this, the Study Director would be free to retain these amendments and deviations until the end of the study, and submit them subsequently in one bundle to the sponsor, which might then become signed together with the final report.

The format of the study plan is laid out in a more or less definitive manner by the GLP Principles, while its contents cannot be conclusively described, since they will depend on the nature of the study to be reported. Therefore, the GLP Principles state that the study plan "*should contain, but not be limited to the following information*", whereupon the main points to be addressed in the study plan are listed. Generally speaking, two parts may be distinguished: The study plan should deal in its first part with the more administrative information, while in its second, and main, part the chronology and the scientific conduct of the study have to be described.

The first, administrative part of the study plan can be considered as a simple listing of information necessary to identify the study and the various individuals and entities connected with its conduct. Every study needs a descriptive title, so that it may easily be recognised and identified in a list of studies conducted at a test facility. This requirement will be evident to anybody who has tried to search for a specific document in a list of computer files all of which are bearing similar and not very illuminating names like "memo-xx" or "document-yy". Such a search can become a very tedious task, and thus the necessity of having a descriptive title, further on accompanied by a statement revealing the purpose and nature of the study, becomes really obvious. In the same way, test and reference or control items have to be unequivocally identified by adequate descriptors.

In the next place, information concerning the test facility and the sponsor is an obvious necessity. As there might be more than one test facility and / or test site involved in a single study, an enumeration of all the relevant locations where the study is to be conducted has to be provided. The study plan has also to identify the Study Director by name and address. If the study plan has to be approved by the test facility management, this could at the same time be considered as the actual appointment of the Study Director by the test facility management. Careful consideration has also to be given to the study subdivisions, since already at this point the Principal Investigator(s) have to be named, together with the specification of *"the phase(s) of the study delegated by the Study Director and under the responsibility of the Principal Investigator(s)"*. At times it may be difficult to address this point in a definitive manner, e.g. when it is not possible at the time of finalising the study plan to choose between different PIs. In such a case, the study plan may remain "open" in this respect, and address the issue by stating that "the PI will be named by test site management and will be reported in an amendment". The Study Director will then have to be responsible for "timely amending" the study plan with the "missing" information on the PI and the respective test site.

Another important information has to be provided in the study plan by specifying the different "cornerstone dates", which will not only influence the planning of activities at the test facility and at the associated test sites, but will have to be used in the master schedule for the planning of the Quality Assurance activities in the context of the study. The one obvious date to be reported is certainly the approval date, but the planning of the study furthermore has to result in more or less definitive proposals for the experimental starting and completion dates. These dates have already been discussed in a previous section (see section 2.7, page 76) with regard to their definition. Therefore, it shall only be reiterated here that it may be advantageous to provide not only a simple calendar date for the two limits of the experimental part of the study, but to identify, if necessary, the specific experimental activities with which they can be equated.

Finally, the study plan has to define and to list the records which will be collected during and retained at the end of the study.

The second part of the study plan is concerned with the conduct of the study and as such is dealing with its scientific aspects. The GLP Principles require that not only reference is made in the study plan "*to the OECD Test Guideline or other test guideline or method to be used*", but they require the study plan to address a number of scientific issues. These issues may be sub-divided into those dealing with the test system as such and those related to the design of the study.

With respect to the test system there are some obvious points to be addressed, like the justification for the choice of the test system, its characterisation, and the various aspects of the test item administration or application. The most exacting part of the study plan is certainly the one dealing with the experimental design of the study. Although there are internationally accepted guidelines for the conduct of a number of study types, where the necessary "*analyses, measurements, observations and examinations to be performed*" are described, there are nevertheless the issues of the "*chronological procedure of the study*" as well as of "*all methods, materials and conditions*" which have to be described in a detailed manner. Especially in this part of the study plan it should never be forgotten, that this document will be used not only as the actual guide through the study, but that it will provide an important element for the possible reconstruction of a study. If this document succeeds in transparently conveying the intentions of the Study Director with regard to the purpose of the study and its possible meaning, then it will not only become easier to follow the Study Director's reasoning, when deviations from the study plan have to be judged for their relevance with regard to GLP compliance, but the scientific assessment and conclusions from the study may also be improved.

With respect to study plans two additional aspects may be of importance in some instances. The issue of the possibility of employing "general" study plans for short-term studies has been already addressed in section 2.6 (see page 70). Such a general study plan would thus include those of the above points which are connected with the test facility and the Study Director on the administrative side, and those which relate to the test system and the experimental conduct, while all the issues relating to the test item and the dates relative to study conduct would then have to be placed in the "study-specific supplement". For each of the various short-term study types conducted at a test facility, this distribution of information to these two parts may be different, and it will be a case-by-case decision on how to structure the general part of such study plans, and on what information would be included where.

Much more tricky than the issue of short-term study plans is the problem that may arise in multi-site studies. For a multi-site study, for example the one illustrated in figure 3 (see page 59), the Study Director will have to prepare a study plan which will be intelligible to all the test site personnel all over the different places. It may be taken for granted that probably none of the personnel involved in this study may be able to understand the Study Director's native language, in this instance Swedish. It may not even be admissible to assume that all the field hands to be employed in the study at the various test sites would understand an English study plan. Consequently, this specific study plan, or at least the relevant parts of it, would have to be translated into Portuguese, Spanish, Arabic, Urdu and Japanese. For these translations, however, there could be a problem in the extent of their accuracy, since the single, specific activities should be conducted, and the instructions understood, in the same way at all the different test sites. It should again be the responsibility of the test facility management to address such questions and problems in a respective policy or position paper, with the view to arrive at a GLP compliant solution. Certainly this question calls for a very close co-operation and extremely good communication between Study Director, Quality Assurance(s) and Principal Investigator(s) to make sure that every single instruction is interpreted correctly and identically at all test sites.

> For the ultimate goal of GLP, the reconstructability of a study, it is important that it is known in a clear-cut and definitely fixed manner, how the study had been intended to be run. The original design of the study, the reasons for it, the intended investigations and their time frames of

> conduct, the proposed start and end of the study, all these and more details have to be fixed beforehand, so that the actual study conduct can be checked against the intentions of the Study Director. Only good planning will really turn a simple "study" into a "quality study", fit for use in the determination and assessment of product safety.

11.2 Study Conduct

A number of the various issues addressed in the GLP Principles under the heading "Conduct of the Study" have already been described and discussed repeatedly. The requirement that "*all data generated during the conduct of the study should be recorded directly, promptly, accurately, and legibly*", that the entries should be dated and signed or initialled, and that "*any change in the raw data should be made so as not to obscure the previous entry, should indicate the reason for change and should be dated and signed or initialled by the individual making the change*" is the prerequisite for the reconstructability of the study and should guarantee the integrity of the data. It is of course standard practice nowadays how changes to records and data have to be made, and that manual recording has to be made in an indelible way (the US regulations are - as usual - more specific in this respect and mandate that such data "*shall be recorded directly, promptly, and legibly in ink*"). As this requirement is meant to ensure the maintenance and later accessibility of these original entries, it may additionally be mentioned that from a GLP point of view it would not be allowed to use the popular "Post-It" stickers in any way connected with the recording of raw data.

There is no actual distinction being made between manually recorded data and data generated as direct computer input in IT systems, as has already been elaborated in section 7.3 (see page 180). Also the requirement that a "*study should be conducted in accordance with the study plan*" is self-evident and deserves no further discussion, especially since the measures to be taken if the study plan cannot be followed, the documenting of deviations and the issue of amendments, have been already described in full detail (see section 2.8, page 80).

There is one specific point in the GLP Principles with regard to study conduct which needs some more detailed consideration. The GLP Principles require that a "*unique identification should be given to each study*". All single data, records, documents and other items connected with the study have to carry this unique identification, in order to ascertain the traceability of events and activities. Also specimens from the study have to carry this identification to allow traceability from the specimen back to the study. The respective phrase in the GLP Principles, however, shows something of a Janus face, since it cannot only be interpreted as meaning that "no two studies may bear the same code", but also in the opposite way, i.e. that "no single study may bear two different codes". In the first of these situations, no special problems might be foreseen, if these identification codes were centrally managed and distributed, and indeed every test facility will have devised its own general way to systematically apportion such identification "tags" to the studies conducted. Still, there might be problems with this requirement, especially in multi-site or otherwise subcontracted parts of studies.

In a multi-site study, one of the test sites may have in place its own identification system for all studies or study parts conducted there. This may on the one hand even have its advantages, since in the very improbable, but nevertheless possible, event that a test site would have to conduct field tests commissioned by two different sponsors which, however, use an analogous identification system and which, by unfortunate coincidence, happen to have identical identifiers attached to their studies, unequivocal identification would still be possible. On the other hand, it has nevertheless to be ascertained that all these study parts still can be related to the original study through the concomitant labelling with the original identifier. As another example we may consider the situation of a toxicity study, conducted at a CRO, and marked with this test facility's identifier. The determination of the test item concentrations for the toxicokinetic evaluation is being performed, however, in the analytical facility of the sponsor, where another identification system is used. Although only part of the whole study, and conducted under the responsibility of a PI, the analytical records, data and documents will be labelled by the sponsor's identification system, since it may not be possible to introduce a "foreign" identifier into the computer system of this facility. In this case, a note has to be included with the raw data which explains this situation and relates the analytical identification code to the study identifier.

There is also the other side of the coin to be considered. Since it has to be categorically excluded that the same identifier can be used twice for two different studies, a situation may arise, where two such codes need to be used for a single study. This may happen, when similar studies are individually defined in the computer system of a test facility in a rigid manner. If, for instance, the computer-based template of a field study allows for the identification of one test item only, then a study with a mixture of two different test items becomes difficult to define within that system by one single study identification code, and the results for the second test item would have to be generated under a code different from the one for the first test item. As another example the conduct of reproductive toxicity studies may be cited, where for one study type the template will ask for the sacrifice of all pregnant females at a specific time point; if only half of the females were to be sacrificed with the other half allowed to litter, e.g. through an additional requirement of a sponsor, the system might not recognise the littering animals as belonging to the same study, and another identifier would thus have to be used to accommodate the second study part. If such a situation is not very well explained in the final report of the "composite" study, the Regulatory Authority might question the validity and integrity of this study and claim that the results obtained were derived from different studies. Again, an explanatory note, describing the situation in detail and retained with the raw data of the study would serve to unequivocally clarify the GLP compliance and the validity and integrity of the study.

> Study conduct is guided mainly by its scientific purpose and the respective guidelines as substantiated in the study plan. The prescriptions of GLP with regard to study conduct therefore should not be regarded as interfering with these aspects, but they set out the framework needed for making a later reconstruction of the study possible. At the same time GLP intends to prevent the occurrence of instances and situations where the possibility of errors and doubt could not be ruled out. Important considerations in this area are therefore the requirements for a unique study identification and for the direct, prompt, accurate, legible and inchangeable recording of observations and data.
>
> In this respect GLP certainly represents nothing else than common sense, as only through the faithful observation of these "administrative" aspects of study conduct and data recording can transparency be achieved and the reliability of a study be fully ensured.

11.3 The Final Report

The "human health and environmental safety study" is intended to be submitted to a Regulatory Authority with the purpose of supporting the company's application for a marketing permit of the product. This purpose may be reached if the study can be presented in a form that clearly states the reasons for conducting it, the methods and test systems used, the results obtained and the conclusions drawn therefrom. Although, with respect to the purely scientific evaluation, this goal might be achieved by just submitting the raw data to the Regulatory Authority, where the assessors could then try to navigate through the study and to draw their own conclusions, this purpose is better and easier served by the submission of a final report. The final report, which the GLP Principles require to *"be prepared for each study"*, does, however, serve an additional purpose. With the signature under the final report the Study Director assumes the ultimate responsibility for the GLP compliant conduct of the study. At the same time, the inclusion of the GLP Statement by the Quality Assurance provides for the final recognition that the report reflects the raw data and that the pertinent rules of GLP have been observed during the conduct of the study.

In a way, the final report may be regarded as a mirror image of the study plan. All the relevant information that had been provided in the study plan has to be given again in the study report, this time, however, in its definitive form as a true account of what had been, instead of the "statement of intent" form of the study plan. This difference can already be demonstrated in the "administrative section": While, with regard to the test item, the study plan just calls for *"identification ... by code or name"*, the final report has to include, besides the identification of the test item (which, by the way, should be identical to the one given in the study plan), the *"characterisation of the test item including purity, stability and homogeneity"*. The range of individuals connected to the study is also broadened, in that the final report has to name not only the Study Director and the Principal Investigator(s), but has to provide in addition *"name and address of scientists having contributed reports to the final report"*. Furthermore, while the study plan just lists the records to be retained, the final report has now to indicate *"the location(s) where the study plan, samples of test and reference items, specimens, raw data and the final report are to be stored"* in order to allow the report to be traced back to the respective raw data, records and documents.

In the description of materials and test methods care has to be exercised when utilising the study plan as template for the report. Since the report has to describe the materials and methods which had actually been used in the study, there is at least a grammatical difference to be made between the two documents, in that the past tense should be used in the final report, instead of the future tense which in the study plan indicated the (future) intentions of the Study Director.

For the main, scientific part of the final report, the GLP Principles are refraining from specifying the contents in a very detailed manner. The scientific test guidelines do already, for each type of test, present this information, and the GLP Principles are therefore providing only very general guidance in this respect. One important aspect, however, deserves special mention, and this is the requirement, that the final report should contain "*all information and data required by the study plan*". This specific requirement serves again the purpose of traceability and reconstructability, since it will provide the possibility of keeping track of all the various study elements and thus to provide a guarantee that no information has been lost in the course of the study.

While this part with results and conclusions has to be compiled and written by the Study Director, there may be a number of different contributions by other scientists or by Principal Investigators. When the contributors only deliver raw data, or bare results, to the Study Director, these data will be utilised in the report like any other data generated by the Study Director, and they will not obtain a special mention. On the other hand, PIs or other contributing scientists like analytical chemists or histopathologists will frequently deliver the information obtained from their part of the study also in report form. These reports have to be treated in the same way as the final report of the Study Director itself, which means that these "*reports of Principal Investigators or scientists involved in the study should be signed and dated by them*". Consequently they should not only serve as "raw material" to help the Study Director compile the final report, but they should additionally be appended to it in their original form. Appending these partial reports to the final report will have an additional advantage in terms of GLP compliance which has been discussed in detail already in section 4.5 (see page 150). Especially in the case of PI reports from independent test sites, the accompanying Quality Assurance statement will then serve to indicate the GLP compliant conduct of the respective study parts or phases, which had not been inspected by the Study Director's Quality Assurance.

Once the study report has been checked by Quality Assurance and has been finalised by the Study Director, it has to be dated and signed "*to indicate acceptance of responsibility for the validity of the data*". The date of the signature of the Study Director is the decisive moment in the whole study. With this signature, the study is declared to be closed, its results and conclusions to be final, and nothing in this study and in its report can be changed anymore, except in certain, special cases, which will be discussed in the next section below. The signature of the Study Director has, however, to serve still another purpose. It not only sets the final "seal of approval" under the results and conclusions of the study, but the Study Director should also indicate *"the extent of compliance with these Principles of Good Laboratory Practice"*. The Study Director, with this sentence, is required to formally and explicitly indicate whether the study had been conducted fully in accordance with the GLP Principles, or whether deviations from the GLP rules had been present. This requirement may be interpreted in different ways, from providing a simple statement like "The study has been conducted in accordance with (the applicable national or international) GLP regulations", to more elaborate statements listing all occurrences and circumstances which might have affected the validity of the study as well as any deviations from SOPs and study plan together with an evaluation of these points with regard to the GLP compliance and the integrity and validity of the study. It need not be a typhoon (see figure 29), but also small events may be addressed and assessed (see figure 30) in this way. The FDA regulations, e.g. do specify this by requiring the Study Director

---

3. Deviation from the protocol

Dosing, clinical observation and control of environmental conditions in the animal room could not be performed according to the protocol on September 28, 1991 (week 19), owing to the stoppage of electric power caused by a typhoon (No.19) which hit this area the previous evening. However, these deficits were not considered to affect the toxicological evaluation in this study as they occurred on only one day of the 26-week period.

Figure 29: Inadvertent deviation from protocol due to external conditions

to provide in the final report a *"description of all circumstances that may have affected the quality or integrity of the data"*. The inclusion of such a detailed list of critical circumstances encountered during study conduct in the Study Director's GLP compliance statement certainly serves to enhance the transparency of the study with regard to its performance and validity, and the regular

> **CIRCUMSTANCES AFFECTING THE STUDY**
>
> The study director is not aware of any circumstances during the study which could have adversely affected the quality or integrity of the data.
>
> The following represent incidental deviations from the approved protocol:
>
> - The age ranges listed in the protocol were 10 to 11 months for young dogs and 43 to 89 months for old dogs; the actual ranges at dose initiation were 10.2 to 11.2 months for young dogs and 44.8 to 89.1 months for old dogs.
>
> - The dose of Evans blue dye solution was incorrectly listed in the protocol as 0.1 mg/kg; the 5 mg/mL stock solution was dosed at 0.1 mL/kg, resulting in an Evans blue dye dose of 0.5 mg/kg.
>
> - The temperament of dog 39 made repeat blood sampling impossible.
>
> - The euthanasia agent used on study, Socumb, was dosed at the manufacturer's recommended dose level of 1 mL/10 lb; this dose of the 6 grain/mL solution resulted in a pentobarbital dose of 86 mg/kg, which was above the pentobarbital dose of 30 to 60 mg/kg listed in the protocol.
>
> - The room environment records indicate the relative humidity ranged from 43% to 66%; this exceeded the protocol specified humidity of approximately 50% by more than 10%.

Figure 30: Detailed enumeration of all observed deviations from the study plan with an assessment of their impact on study validity and integrity (Note the difficulties encountered with the correct calculation of dose, when the concentration is given in units other than the scientifically used SI-units).

inclusion of such an assessment is therefore to be warmly recommended. Some examples of such declarations are presented in figures 31 a – c..

Finally, in the case of short-term studies, there is again the possibility of preparing a *"standardised final report accompanied by a study specific extension"*. The example of a melting point determination may be used to illustrate this point. All circumstances of study conduct can be considered as absolute routine and every single such determination will follow exactly the

2.11 Study Performance and Reporting

a)

> To the best of my knowledge, there were no significant deviations from the Good Laboratory Practice Regulations which affected the quality or integrity of the study. This study was conducted in conformance with the Good Laboratory Practice Regulations. Comments on study data, including any protocol deviations, are presented in Appendix N. This report accurately reflects the raw data obtained during the performance of the study.

b)

CERTIFICATION OF GOOD LABORATORY PRACTICE

The enclosed report for Study ▮▮▮▮▮▮▮▮▮▮ accurately describes methods and procedures used in the study and accurately reflects the raw data obtained. The study was conducted in compliance with Good Laboratory Practice for Nonclinical Laboratory Studies as described in the Federal Register: 21 CFR 58. There were no differences, with the exception of a few instances of incorrect data entries/corrections, discovered between practices used in conducting the study and those required by Good Laboratory Practice regulations.

c)

PROTOCOL ADHERENCE

The study was conducted in accordance with the agreed definitive protocol, and two amendments. Minor deviations from the protocol are outlined in the body of the report. There was one major deviation from the protocol, this was the incorrect preparation of the control article and vehicle for dosing to animals in Groups 1, 2 and 5 in Weeks 2 to 4. At the Sponsor's request, supplementary groups were added to the study, with correctly prepared vehicle. There were no effects in animals that received the incorrectly prepared control article or vehicle, and this deviation was considered not to have affected the outcome or integrity of the study.

Figure 31 a) – c): Examples for correct, GLP compliant statements resulting in full transparency regarding study conduct and deviations from study plans.

same experimental course, which is why the study conduct can be described in a general study plan. Thus, there will be no deviations possible from the general study plan, the study report will only reiterate the description of study conduct given therein, and the only difference between the individual studies will be the result of the determination consisting of a single figure, the actual temperature of the melting point. The result would not need any interpretation, and no conclusions would be drawn from it; therefore the Study Director may well sign the standardised final report once and for all, and the study specific extension, i.e. the numeric value of the melting point, can be appended to this pre-signed general report form. This provision in the GLP Principles can thus relieve the Study Director from the obligation to sign dozens of nearly identical study reports a day.

A slightly different situation might occur, when large numbers of short-term studies would be conducted at a single test facility which, for some reason or another, would refrain from utilising the facilitations for short-term studies offered by the GLP Principles. Under these circumstances it might become difficult for the Study Director to sign all these final reports in a timely manner. The test facility management might then consider to empower a Deputy Study Director to sign such reports in the absence of the actual Study Director. This question of a Deputy acting for the Study Director by signing the final report could also be posed in the event that the sponsor should press for the release of the report in the temporary absence the actual Study Director. In such situations, a case-by-case solution will have to be found which should, however, satisfy the GLP requirements of acceptance of responsibility for the GLP compliant conduct of the whole study, and such situations would have to be clearly described in a relevant SOP. Further questions and problems concerning the replacement of the Study Director have already been discussed (see section 3.1, page 94).

> The study report is the window through which the Receiving Authority will be able to look into the room behind, i.e. at the study, enabling it to assess the safety of the submitted product. If this window is clean and completely transparent, the study can be assessed without problems on its merits. On the other hand, if this window is dirty, with blind spots, so that the contents of the "room behind" may not be clearly visible, or may be interpreted in more than one way, the assessment of this study will suffer in consequence.

> Apart from providing the more "administrative" requirements of title, names and addresses, which would serve to unequivocally fix the various responsibilities, the rules of GLP intend to ensure that this window is as transparent as possible. Through an enumeration of all aspects, occurrences and circumstances which might have had an influence on the quality and integrity of the study, this transparency will be enhanced. The Study Director has finally to acknowledge responsibility for the GLP-compliant conduct of the study through the dated and signed statement
>
> GLP aims thus to ensure that, as far as it is possible, the contents and the conclusions of a study report can be trusted and that they can be used confidently in the assessment of product safety.

11.4   Re-opening and Amending a Study

It has been repeatedly stated in the foregoing section that once the study report is signed by the Study Director, the study is closed and there is no immediate possibility for any changes to be introduced afterwards. As its name indicates, the final report has to be considered the final and conclusive document with regard to a study. This is the general principle and rule, but there are no rules without exceptions! The GLP Principles recognise that there can be instances, where corrections or additions to a final report should become possible. In the following paragraphs we will therefore look at these possibilities and try to delineate the GLP compliant ways to deal with studies and reports that suddenly become "unfinished" once again.

The GLP Principles do not allow a simple re-opening of a study or a final report. Anything that has to be changed in a final report has to be done in the form of an amendment to it, in the same way as the study plan may be changed during the study. Thus, *"corrections and additions to a final report should be in the form of amendments. Amendments should clearly specify the reason for the corrections or additions and should be signed and dated by the Study Director"*. The wording of this paragraph in the Principles leaves no doubt that even typographical errors have to be corrected by amendment rather than by the simple exchange of the original page with the corrected one. It may, however, be questionable, whether it would indeed help the intelligibility of the report, if the corrections of such errors were to be printed on separate sheets and added to the unaltered report. The purpose of the GLP Principles might as well be served by an exchange of the corrected pages, accompanied by an amendment which states the reasons for the listed corrections, and by the subsequent archiving of the original report pages.

Remaining at the issue of corrections, the GLP Principles provide furthermore for an "exception to the exception". National Regulatory Authorities may differ in their requirements for the formal aspects of study reports. Some might ask for tables to be interspersed with the text of the report, while others might prefer to receive them as appendices. There could be requirements for signature pages, or for additional statements by the sponsor. It might thus be necessary for a sponsor to submit a study report in two or three different formats, although the alterations that would be necessary would not touch upon the actual content of the final report. In these instances, the GLP Principles clearly state that such a *"reformatting of the final report to comply with the submission requirements of a national registration or regulatory authority does not constitute a correction, addition or amendment to the final report"*. It has to be stated very clearly again, that this possibility is open only to those formal adjustments which do not encroach on the actual, scientific and GLP relevant content of the report.

Finally, a number of questions may arise with respect to additional data which may become available, or may become necessary to develop, after the finalisation of a study report and after the closing of the respective study.

Let us consider as a first example the situation in a chronic toxicity study, in which the histopathology evaluation has brought an unexpected but serious effect to light. It might then be desirable to check the histopathological slides of other, already terminated toxicity studies again, or even to prepare additional slides from organs which originally had not been investigated, in order to ascertain the presence or absence of this particular lesion in these other studies. If the results of such additional evaluations were to have an impact on the original conclusions of these former studies, this should logically be reported in an amendment to the respective study reports.

The second example shows that it might even become necessary, under certain circumstances, to extend the experimental phase of an already terminated study. In a plant and soil metabolism study, conducted early in the development of a product, not all possible metabolites of the test item will be known, let alone be available as reference items. The study will therefore be restricted to those metabolites which can be identified at this time and it will be terminated, once the original goals of the study have been reached. Later in the development, other, additional metabolites could be detected, some even of major importance. Instead of repeating the original, large field trial, it might make more sense to re-analyse the specimens from the original study. Since such an additional analysis for the presence, and amounts or concentrations, of the further metabolites will not have been described in the original study plan, it will necessitate either the extension of the original study or the generation of a new one. If, in the first instance, field trial and analytical work had been parts of one study, then it would be difficult from a GLP point of view to generate an individual study out of the supplementary analyses of material from the original study. In this case, the original study plan would have to be amended first to cover the additional analytical investigations, methods and reference items, and then the results from these analyses would be used to amend the original final study report. Only if in the first instance already the field part as well as the analytical investigations would have been conceived as separate studies, could such an additional investigation be planned as an individual study in analogy to the earlier ones. Since this splitting-up of field studies is not to be recommended, the first option of re-opening the original study by a study plan amendment and a report amendment seems to be the major GLP compliant way to deal with these situations.

> While under normal circumstances "final" is to be regarded as a term denoting the absolute end, there may be situations where it will turn out that this term has been precipitately conferred to a study report. In addressing exceptions to the rule of inchangeability of a "final report" GLP wants to assure that also in such instances the complete traceability of the respective decisions remains guaranteed. This entails therefore measures which should ensure that no material alterations of a study report will be possible without the proper acknowledgement and assumption of responsibility, while on the other hand alleviating the respective requirements for purely formal changes.

11.5 Discontinued Studies

Under this heading we will consider shortly two situations, which may sometimes occur in a test facility or at a sponsor, and which may be judged as problematic with regard to their GLP compliant handling.

For some reason, whether explicitly stated or not, a sponsor may suddenly determine that the further development of a test item would be stopped, and therefore all studies still running at this time point would be abandoned and terminated. Since this decision would entail the fact that the respective product would not be submitted anymore to any Regulatory Authority for marketing approval, and GLP compliance of this study would not be at stake anymore, it might be perfectly possible to scrap the study altogether and to forget about it. On the other hand, the study had originally been entered in the master schedule, and it had left other trails of its existence, like records of animal room occupation and other general documentation of the test facility, and therefore it would be advisable to retain the study plan and the raw data collected up to this moment in the archives. An explanatory amendment to the study plan would additionally serve to confirm the fate of the respective study. Obviously no final report would be written, and no Quality Assurance statement could be issued, and the unfinished study would thus not be in compliance with GLP.

The second such situation may occur, when a study for which a final report has already been issued might be superseded by a more recent one. The first study would therefore not be submitted to a Regulatory Authority as part of a submission package, and the question could be raised what to do with this first study. This question relates not to the GLP compliance status of the first study, since it had been conducted and reported to GLP, but to the archiving requirements for such a study which might be questioned. The GLP Principles do not specify that only such studies need to be retained in the archives which have been submitted to a Regulatory Authority. Studies do fall under GLP, if they are intended to be submitted to a Regulatory Authority, or even if there exists just a possibility that some day the report from the study might be used in a submission package. The decision by the sponsor not to use the results of the first study for the actual submission does not alter the fact that the study might still, under other circumstances, become part of a submission at a later date. Therefore, any GLP study needs to be retained in the test facility's archives, even if it is not immediately used.

12. The Archives

It stands to reason, that all the records, data, specimens, samples and documents which are produced and compiled in the context of GLP studies, and of GLP test facilities, including documents and records of the Quality Assurance have to be stored somewhere for possible future examination. In order to allow for a later reconstruction of studies from this documentation, this storage cannot consist of simply creating a pile of all study-related and test facility-related material in a dusty attic or a dank cellar. The first consideration in archiving is that all this material should be stored under the proper conditions suitable to protect the contents of the archive "*from untimely deterioration*". This technical aspect of the archive facilities proper has already been dealt with in section 5.4 (see page 165) and need not be taken up again here.

However, the archiving of study materials and of supporting documentation cannot be regarded only under this restricted viewpoint of the necessary facilities. There are a number of other important issues that need regulation and clarification, and they are addressed therefore in the last section of the GLP Principles. These issues relate to the length of time for which storage should be mandatory. But there is still more to archiving than just to take care that the material may stay in good condition for a specified period of time; there are problems of archive organisation as well as of security to be considered. The storage has to be organised in an orderly form so as to facilitate the retrieval of any document which may be needed. And on the other hand, the material should be secure, not only from unwanted, untimely deterioration, but also from intentional changes and wilful destruction. These additional safety aspects are quite different from the ones being addressed through technical archive facility standards.

12.1 Storage Period

It may be debatable for how long it is useful and thus necessary to retain records, samples and specimens from studies, or check-lists from Quality Assurance inspections, whose results may long have been superseded by new data or by simple experience from the use of the test item. In the one extreme, one might argument like this: Once the product has been approved by the Licensing or Regulatory Authority, one could reasonably assume that the studies, which had been used in the submission, have been accepted as valid, and that therefore further retention might be unnecessary.

Such a standpoint is of course untenable and short-sighted, since it may well be that with increasing use of the product a problem might surface that would necessitate the critical assessment of the GLP compliance of some pivotal study. At the other end of the spectrum of possibilities a complex arrangement could be foreseen in which the supporting documentation for a marketing permit should be retained until, say, five years after the expiry date of the last batch of the product produced anywhere on the world. This solution may in practice not be feasible, and therefore a fixed time limit for the retention of GLP records is the much preferred solution. However, these time limit requirements may vary considerably from country to country, and even from product type to product type. In consequence to this situation, the OECD GLP Principles had to refrain from providing a concrete time limit, and thus they have to refer to national requirements and legislation ("... *for the period specified by the appropriate authorities*") which have to be followed in this respect. Also the obvious question defining the start point of the retention period cannot be answered conclusively, since it depends on national legislation, too, although in general one may assume that the starting point for the retention period would be the date of the Study Director's signature under the final report.

The general time limits for archiving may, however, not be appropriate for every kind of material that has to be stored and retained. There will be some deterioration of certain materials which, under the best of storage conditions, cannot be held back. Consider, for example, the case of a test item that has been applied to the test system in radioactively labelled form. The documentary value of a specimen or sample derived from this study will then relate to its radioactivity, which in turn is remorselessly decaying according to the laws of physics and the half-life of the isotope used. This will not matter for ^{14}C with a half-life of about 5700 years, but if the label had been ^{32}P with a half-life of barely fourteen days, it has to be expected that after fourteen weeks of storage (i.e. after a time spanning seven half-lives, reducing the radioactivity to less than 1% of the starting value) no useful measurements could be made anymore on the sample or specimen. Chemical degradation is another possibility, which even under conditions of storage at very low temperatures may, after a shorter or longer time period, render the respective samples or specimens useless for the purpose of verification of study results. Even though it may sometimes be possible to circumvent this problem of deterioration and degradation - the reader is reminded of the copying of light-sensitive prints - it would be regarded as unreasonable to require storage of materials over and above the limits of their usefulness for analysis, evaluation and verification. Therefore, the GLP Principles concede that in these instances "*samples of test and reference items and specimens should be retained only as long as the quality of the preparation permits evaluation*".

This concession for disposal cannot, however, be read as permission to simply and thoughtlessly destroy study materials. On the contrary, even in this instance the guiding principle of the GLP rules mandates that the final fate of these materials be documented and that, where appropriate, the reasons for the premature disposal have to be provided. Any missing material that should be in the archives, and the fate of which is not documented, would automatically jeopardise the GLP compliance of the test facility: Questions about the GLP compliance of either the archive organisation, including the work of the person responsible for the archives, or even the GLP compliance of the whole study touched by this loss would arise. Therefore, the GLP Principles firmly mandate that "... *the final disposition of any study materials should be documented. When samples of test and reference items and specimens are disposed of before the expiry of the required retention period for any reason, this should be justified and documented*".

There is one big problem with this provision which remains to be solved: The question of who may be allowed to take the decision to destroy a specimen or some other material. Could it be the person responsible for the archives, who may give as his reason that he needed the space? Or should it rather be management, who is ultimately responsible for the GLP compliance in the test facility? Or would it not better be the respective Study Director, who should know best which limitations may be applied to the evaluability of the test material? It might also be the Quality Assurance who should be in the best position to judge whether the further retention of the material would serve a valid purpose in preserving study integrity. Additionally, this problem may be expanded to include not only materials that are stored in an archive and which would not last for the required ten or thirty years of storage, but it may also be encountered in test situations, where "transient" specimens may have to be dealt with.

As one example the case of a bacterial mutagenicity test may be cited, where hundreds of petri dishes with the bacterial colonies grown as a result of the test, and which may be considered to constitute the test system as well as the primary data. After the end of the incubation period these petri dishes cannot be stored for too long a time, since either the bacteria (and any contaminant germs) will continue to grow, or the layer of agar growth medium will dry out, and both of these events will render the plates inevaluable after some time. The petri dishes will therefore have to be discarded shortly after the end of the experiment. The actual raw data of the study to be archived are, however, the bacterial colony counts, whether they have been manually determined or automatically recorded. Thus, if it can be ascertained that the records of the bacterial colony counts truly reflect the actual colony numbers on the plates, the petri dishes may be discarded without any consequences to the integrity of the study. This situation may, however, be regarded as a borderline case, since it will be obvious after some days to weeks at most, that the plates are no more usable for evaluation, unless they were to be stored under very special conditions (e.g. air-tight packing and storage in a freezer), which, however, may not be feasible due to the sheer number of such test specimens. On the other hand, a very similar test system like a mammalian cell gene mutation test, or a cell transformation test, where also colonies are grown on petri dishes, may be treated very differently, since in these cases the cells are adhering to the petri dish surface itself, the growth medium is not an agar gel but a liquid, and the cell colonies are fixed at the end of the experiment, stained and dried for examination. Such dishes have then to be regarded as specimens and may therefore need archiving, although even in these cases, the useful life-span of these specimens may be limited.

There are other situations, e.g. in field tests, where soil or crop samples have to be analysed for their content of test item and its metabolites. Some of these compounds may be relatively short-lived, even under the most optimal storage conditions, and in these instances, the disappearance of the most labile of the analytes present would then dictate the useful life-span or storage period of these samples. Such samples may therefore be discarded, upon the assessment of the analytical chemist, as soon as a re-analysis cannot be used anymore to confirm the original results. As another, similarly obvious situation, the determination of cellular parameters in blood may be considered: White blood cells will disintegrate, lose their specific form or stainability, and thus whole blood may after a short time become inutilisable for the original purpose. In this instance, too, the blood sample remaining after the haematological evaluation may thus be discarded after verification of the data without jeopardising study quality.

The problem may be considered to become more difficult for specimens and samples which have actually to be regarded as raw data, or where re-analysis might be possible for some time even after the conclusion of the respective study. But on the other hand, there might be clear criteria to determine the end of usefulness of these archived materials. Let us consider the case of the test item sample the archiving of which is mandatory under GLP. At intervals, this sample might be analysed, and at some point of time the analytical chemist will determine that the sample is not useful anymore for its analytical purposes. This assessment will certainly mark the end of its required storage period, since the sample would have outlived its purpose. In this case, the required reason for the disposal of the test item sample would be obvious, given the respective analytical data, and the documentation of the decision could be very straightforward. The same can be said of other samples or specimens, e.g. of wet tissues from a toxicity study, where the histopathologist might arrive at the conclusion that no evaluable slides could be made anymore from the preserved material. Analogous cases could be made for soil or crop samples originating from field studies, where also the decision for the disposal could be based on rational arguments.

Although the GLP Principles allow the disposal of specimens and samples, when their condition precludes further meaningful evaluation, the individual responsible for the archives has to pay attention also to the opposite clause of the Principles, namely the requirement that the storage conditions should preclude untimely deterioration. If for example, the jars containing preserved tissues cannot be sufficiently sealed, so that the preservative slowly evaporates, it lies in the responsibility of the archivist to periodically check these jars and to refill them with the respective preservative, if the need arises. Consequently, it would be considered a violation of the GLP Principles, if such specimens were just left to dry out, and the deteriorated specimens were then to be destroyed.

Another aspect of storage and its termination concerns the involvement of Quality Assurance. In order to ascertain the correct application of the provisions given by the GLP Principles, Quality Assurance should be called in to help determine the GLP compliant way of dealing with these different situations. Thus, Quality Assurance should be involved in any decision about the removal of archived materials and their disposal by verifying the condition of these materials and thus the reasons for the final disposal.

In summary, it will be the specialist, who will determine the end of usefulness of any sample or specimen, but it will also be the Quality Assurance who, through inspection and verification, will have to acknowledge the GLP compliance of the disposal procedures in general, and of the single processes and instances of discarding study-related material in particular.

> The archive serves to store the complete materials, records and other documentation specifically related to individual studies, and to the test facility in general. In this regard GLP sets out the general conditions under which archives should be operated in order to ensure the continuous availability and evaluability of any such materials. GLP does not provide for a specific storage period, since different countries do have different legal requirements for storage length, and GLP therefore just refers to these national rules. Since it would make no sense to continue with storing materials which are no longer usable for investigation or evaluation, GLP provides for the possibility of early disposal which, however, has to satisfy again the documentary requirements of full traceability.

12.2 Indexing and Retrieval

The second aspect of archiving is the one of its organisation. An archive is only useful as long as any specific piece of material it contains can be located and retrieved within a reasonable time. While it may be regarded as unproblematic with respect to single studies and the documents related to them - remember that a study and all documents and materials immediately related to it should bear a single and unique identification - this may not be as easy as it looks on the first glimpse, since there are a number of additional points to be observed. Therefore, the GLP Principles require that "*Material retained in the archives should be indexed so as to facilitate orderly storage and retrieval.*"

First of all the physical location of the various study materials may be different, as has been described in the section on archive facilities (see page 165), and this has to be reflected in the archive's indexing and retrieval system. It has therefore to be ascertained that on looking up the identification tag of a specific study one would be led to the location of all the various documents and materials related to this study, and which have been identified as "materials to be archived" in the study plan and the final report.

Secondly, there are not only documents around that are directly related to any single study, like study plan, raw data and study report, but there are the general data and common information that might be pertinent to the assessment of the validity of the study, like environmental condition records, documents on calibration and maintenance of apparatus, temperature records of freezers, or cleaning and decontamination/sanitation records of animal rooms, and not to forget the validation and qualification documentation relating to physical/chemical test systems and computerised systems. Rather than being stored in a study-related manner, they would be archived on a chronological basis, and they should thus be retrievable in this way.

In the third instance, it might become desirable or necessary to be able to locate and retrieve all documents, belonging not to a single study only, but to find all studies and the respective documents pertaining to a specific test item. Or the necessity could arise to check for the fate of a specific batch of test item, and all records of its distribution, use and final disposition in studies or in the test item logistics department should be retrievable in order to find the looked for quantitative answers. Therefore, the organisation of the archive indexing system should allow for multiple searching, either according to study identification, or according to test item.

In what might be called the "olden days" these various search functions did necessitate an elaborate system of two or three different indexing systems, with the respective entries to be made simultaneously on a number of index cards. Nowadays, the computer has made the task easier, but it needs still some good planning to create an indexing system which would allow all the possible, necessary and desirable localisations to be made. The archivist, who will be the one to handle the system and to satisfy the various requests, would therefore be well advised to take all these considerations and possibilities into account when devising, or helping to devise, the indexing and retrieval system. By no means, however, can there be one single, universally applicable solution to this problem, since much depends on the nature and diversity of materials that are retained in the archives. Therefore it is important, that the archiving procedures and the indexing system be described in pertinent SOPs, allowing in this area, too, to reconstruct the pathways of all GLP-relevant materials into and out of the archives.

12.3 Security

Security is a further and final aspect in archiving. One side of archive security has already been dealt with in the section on archive facilities (see page 165), namely the safety from physical destruction through environmental influences, be they slow acting, like the yellowing of paper records or the outright destruction of light-sensitive materials through too intense illumination, or rather dramatic, like the destruction through fire or flooding. There is another side, however, which has to do with the "intellectual" integrity of the archived records, documents, samples and specimens.

As soon as a study is completed and the report finalised, all the study documentation has to be "frozen". The final study report cannot be altered anymore after the date of the Study Director's signature, save by amendment; consequently, raw data, samples and specimens should be preserved unchanged, and no tampering should be possible anymore with any of them. While the Study Director is responsible for the integrity of the whole study, including its raw data and other records up to the time of handing them over for archiving, it is the individual responsible for the archives who will take over at that time point. In order to achieve the necessary control over the integrity of the study documentation in an unchanged state, access to the archive facilities has to be limited to specially authorised personnel only. For this, again the test facility management is responsible, as the Principles assign to management the task of ensuring that *"an individual is identified as responsible for the management of the archive(s)"*. It is to be noted that it is not possible to designate an organisational subunit (e.g. Quality Assurance) as generically responsible for the archives, but that a single individual must be identified for this function. This focusing on a single individual indeed lies in the general line of GLP, where responsibilities are clearly defined in terms of pivotal points of control. The archivist is in this respect placed on the same level as the Study Director: As the latter is the single point of study control while the study is being conducted, so is the former the pivotal point of control for the archived material which forms the documentary evidence for the quality and integrity of completed studies.

There is one exception to this rule, and this is the archives of the Quality Assurance. Quality Assurance is also required not only to retain the documents pertaining to its own activities, but also to retain copies of approved study plans and of test facility SOPs in use. Although it would not be impossible, and certainly not forbidden, to archive these records and documents in the general archive facilities and under the care of the person responsible for the general archives, Quality Assurance may nevertheless choose to store these documents in its own, special facility, to which only Quality Assurance personnel could be given access. For this case, the GLP Principles do not spell out a special requirement for singling out a specific individual.

It is not only the limited physical access to the archive facilities which is of importance for keeping the study documentation unchanged, addressed in the GLP Principles in a very strict and straightforward manner ("*Only personnel authorised by management should have access to the archives.*") but there has also to be a well-designed system for keeping track of any material that leaves the archives for one reason or another, and for ascertaining the integrity of the material when it is delivered back to the archives. This latter point is certainly one that poses the greatest difficulties in its implementation.

At the delivery point of the materials for archiving, the Study Director is responsible for the completeness of the study-related material. The archivist will therefore have only to register the material, to assign it an indexing designation and to integrate it into the proper storage place without having to care for its completeness, although he might still do some checking in this respect. As long as this material rests in the archives the situation regarding completeness remains unchanged. If, however, some of the material will have to be retrieved from the archives, either because the Study Director would want to check some data, or another person might need some material for comparisons within another study, the problem of the completeness upon returning the material to the archives will become critical.

The GLP Principles require therefore that "*movement of material in and out of the archives should be properly recorded*", but just recording these movements may not be enough to secure the integrity of the returned study material. There are a number of points to be observed in this respect, which will, if properly implemented, lead to enhanced security of the study-related materials.

The first of these measures is the already mentioned limited access to the archives. Only those individuals "*authorised by management*" should have access to all archive locations. This translates into the requirement that only a limited number of keys should be available for the archive facilities, all of which should be securely kept, allowing only the authorised persons to use them. It has to be recognised, however, that for cases of an emergency, like fire or a leaking water pipe, threatening the archive and its contents, a reserve key should be placed with the technical department in the test facility.

As a further security measure, a documentation system has to be in place where retrieval of any material from the archives can be recorded. These records will have to show the exact nature and designation of the material to be retrieved, together with the date of retrieval and the signatures of both archivist and retrieving person. A controlling system would subsequently allow to keep track of such material and to ascertain the logistic means for its timely return to the archives. It will furthermore be advantageous to limit the range of persons who may legitimately ask for documents or other materials to be retrieved from the archives.

All these measures allow for the control of the material "flux" in and out of the archives. They do not quite by themselves guarantee for the unchanged integrity of the material at its return to the archives. It need not be by bad intentions that this integrity could be jeopardised, but already the insertion of some raw data at the wrong place in the whole documentation might later on give rise to suspicions of tampering with them. There are a number of ways to deal with this problem, and the decision on which solution to prefer would certainly depend again on the type of the test facility and on the nature of the materials to be controlled. For documents and raw data on paper, it might be possible to consecutively number the sheets through the whole documentation, or some sort of seal might be applied making the removal of single sheets impossible without breaking it. Small documentations might also just be controlled page by page by the archivist upon receipt for return into the archive. Whatever the solution ultimately taken, the application of GLP would mandate that the exact way to ensure the integrity of the archived material should be described in the respective SOPs.

All the foregoing discussions have dealt with physical materials going into, retrieved from, and returned to, the archives. The secure archiving and retrieval of electronic data is posing different problems. Of course, back-up tapes may be stored in physical archives, where they are protected from untoward influences like electromagnetic fields and other destructive processes. Retrieval of a specific part of such archived material. and the subsequent control over it, may be facilitated in electronic storage, since the material need not leave the archive physically, i.e. in the form of the stored diskette or tape, but can just be uploaded into the system once again. On the other hand, the technical development driving up exponentially the limits of storage in central computers to terabytes and more, thus making it possible to store ("archive") practically limitless amounts of information in the central computer unit, may pose special problems of security. If a test facility decides to use the internal storage power of the central computer unit for the archiving of electronically captured and generated data, then it will certainly have to devise also special procedures and safeguards to ensure the absolute integrity of these data and of the respective study records. In this area, the requirements for IT applications in general, and especially for computer validation, operation and safety should be very carefully considered and applied to the largest extent possible.

> During the conduct of a study every precaution is being taken that true and faithful records are taken and that no alterations of data and records are possible, unless the change is justified and completely documented. If the materials stored in the archives should be of any value for the reconstruction of a study, this inchangeability of records, documents and materials has to be fully preserved. The technical conditions in the archive as well as the security measures to be taken should help to accomplish this.
>
> With the formulation of the respective requirements for the archives GLP wants to create a situation which enables an authority to examine, evaluate and judge, at any time after the completion of a study, the quality, reliability and integrity of data, records, documents and the whole study.

12.4 Archive location, merging and dissolution

The question of where to store all the material from studies and from test facility activities which needs to be retained can give rise to some uncertainties, which on the whole are mainly unfounded. The GLP Principles do not address the question of the specific requirements for the location of an archive, except that it should, like any other facility, be "*of suitable size, construction and location to meet ... requirements*". Therefore, there is complete freedom for every test facility to define the location of its archives and to designate the proper locations for each type of materials to be stored.

It follows, that it is not necessary from the point of view of GLP to have a single, central archive, where all materials originating from the test facility's activities should be stored. On the contrary, it might be considered to present an advantage to the pathologist's work, if all histology slides from all studies ever evaluated by him were to be stored in an archive facility very near to his own facility, and to which he might be given immediate access. For small test facilities, the archive might consist of two or three suitably located and securely locked cupboards. Test sites could also have their own, albeit restricted, archive facilities, where general documentation pertaining to the test site's daily operations would be retained, while in this case the study-specific records would be sent to the Study Director and archived at the central test facility archives. On the other hand it might, e.g., be considered a good idea to retain the back-up computer files in a bank vault for security reasons.

The question may seem to become a somewhat more contentious one, however, in the relations between contract test facilities and their sponsors. Contract test facilities certainly have to have their own archive facilities and would thus generally be able to guarantee for a GLP compliant storage for all study-related materiel. Usually, sponsors will agree to have the study raw data retained and archived by the contract test facility, where also the general facility-related data will be available for scrutiny, should the necessity for a study audit arise. Single sponsors might nevertheless require that all raw data, samples and specimens originating from the commissioned studies should be returned to them for archiving. This request can doubtlessly be met, although the contract test facility might wish in such a case to additionally retain copies of the study documentation in their own archives.

The GLP Principles do also direct attention the one fact of business life, the event that a test facility might go out of business, while giving no advice on the way to handle the archive situation in the case of a company merger. In this second case, there is probably no need for overmuch regulation. In the event of a merger, the already archived material may either stay at the present archive location, or it may be moved to the archive of the "buyer". In the former case no special measures have to be taken, while in the latter one some documentation will be needed. The reason for this necessity is that it has to be mentioned in all final reports where the different materials, raw data, samples and specimens are to be stored. Since with the move to new archive locations this statement will not be true anymore, the new location of the various materials has to be given in an amendment to each study report, which can, however, take the form of a general statement that can be declared valid for all studies of a certain time period or of a certain denomination.

Things get more complicated, however, when a test facility ceases to exist. Since there will be study material archived at such a test facility that has still stay archived for a number of years according to the regulations of the country where this facility is situated, the problem arises of how to deal with this situation. The GLP Principles require that in the case "*a test facility or an archive contracting facility goes out of business and has no legal successor"* then its *"archive should be transferred to the archives of the sponsor(s) of the study(s)"*. This may be possible without any special efforts or thoughts for the study-specific raw data, samples and specimens, which can easily be attributed to the single sponsors, and which the respective sponsors will also be well prepared to take back and to have them archived in facilities of their own choosing. The complexity of the situation originates in the general, facility-related data, like the animal room environmental data, or the meteorological data at a field test site. If they cannot be allocated to single studies, then they will have to be made available to all sponsors in a general form. Two solutions are imaginable for this case: Either all the data are copied multiply and the whole set of, e.g., ten year's worth of data given to each sponsor, or these data may be moved to a contract archive, for which the various sponsors will then have to pay in proportion to the number of studies the contractor had performed for them.

In any case, it has to be stressed that it should be possible for all material for which the GLP Principles require retention, to be retrieved and investigated during the whole period of time which the country, where the test facility is located, stipulates as the minimum time of storage. Every change in the conditions of archiving has therefore to be fully documented. Only in this way will it be possible to trace the fate of documents, records, samples and specimens even after their placement into storage.

Appendix II.I

The Revised OECD Principles of Good Laboratory Practice (reprinted by permission of OECD)*

1. Scope

These Principles of Good Laboratory Practice should be applied to the non-clinical safety testing of test items contained in pharmaceutical products, pesticide products, cosmetic products, veterinary drugs as well as food additives, feed additives, and industrial chemicals. These test items are frequently synthetic chemicals, but may be of natural or biological origin and, in some circumstances, may be living organisms. The purpose of testing these test items is to obtain data on their properties and/or their safety with respect to human health and/or the environment.

Non-clinical health and environmental safety studies covered by the Principles of Good Laboratory Practice include work conducted in the laboratory, in greenhouses, and in the field.

Unless specifically exempted by national legislation, these Principles of Good Laboratory Practice apply to all non-clinical health and environmental safety studies required by regulations for the purpose of registering or licensing pharmaceuticals, pesticides, food and feed additives, cosmetic products, veterinary drug products and similar products, and for the regulation of industrial chemicals.

2. Definitions of Terms

2.1 Good Laboratory Practice
   1. Good Laboratory Practice (GLP) is a quality system concerned with the organisational process and the conditions under which non-clinical health and environmental safety studies are planned, performed, monitored, recorded, archived and reported.

---

* OECD Principles of Good Laboratory Practice (as revised in 1997). Copyright OECD Paris, 1998. Material available on OECD website at http:\\www.oecd.org/ehs/ehsmono/index.htm#GLP

Appendix I: The OECD GLP Principles

2.2 Terms Concerning the Organisation of a Test Facility

1. Test facility means the persons, premises and operational unit(s) that are necessary for conducting the non-clinical health and environmental safety study. For multi-site studies, those which are conducted at more than one site, the test facility comprises the site at which the Study Director is located and all individual test sites, which individually or collectively can be considered to be test facilities.
2. Test site means the location(s) at which a phase(s) of a study is conducted.
3. Test facility management means the person(s) who has the authority and formal responsibility for the organisation and functioning of the test facility according to these Principles of Good Laboratory Practice.
4. Test site management (if appointed) means the person(s) responsible for ensuring that the phase(s) of the study, for which he is responsible, are conducted according to these Principles of Good Laboratory Practice.
5. Sponsor means an entity which commissions, supports and/or submits a non-clinical health and environmental safety study.
6. Study Director means the individual responsible for the overall conduct of the non-clinical health and environmental safety study.
7. Principal Investigator means an individual who, for a multi-site study, acts on behalf of the Study Director and has defined responsibility for delegated phases of the study. The Study Director's responsibility for the overall conduct of the study cannot be delegated to the Principal Investigator(s); this includes approval of the study plan and its amendments, approval of the final report, and ensuring that all applicable Principles of Good Laboratory Practice are followed.
8. Quality Assurance Programme means a defined system, including personnel, which is independent of study conduct and is designed to assure test facility management of compliance with these Principles of Good Laboratory Practice.
9. Standard Operating Procedures (SOPs) means documented procedures which describe how to perform tests or activities normally not specified in detail in study plans or test guidelines.
10. Master schedule means a compilation of information to assist in the assessment of workload and for the tracking of studies at a test facility.

2.3 Terms Concerning the Non-Clinical Health and Environmental Safety Study

1. Non-clinical health and environmental safety study, henceforth referred to simply as "study", means an experiment or set of experiments in which a test item is examined under laboratory conditions or in the environment to obtain data on its properties and/or its safety, intended for submission to appropriate Regulatory Authorities.
2. Short-term study means a study of short duration with widely used, routine techniques.
3. Study plan means a document which defines the objectives and experimental design for the conduct of the study, and includes any amendments.
4. Study plan amendment means an intended change to the study plan after the study initiation date.
5. Study plan deviation means an unintended departure from the study plan after the study initiation date.
6. Test system means any biological, chemical or physical system or a combination thereof used in a study.
7. Raw data means all original test facility records and documentation, or verified copies thereof, which are the result of the original observations and activities in a study. Raw data also may include, for example, photographs, microfilm or microfiche copies, computer readable media, dictated observations, recorded data from automated instruments, or any other data storage medium that has been recognised as capable of providing secure storage of information for a time period as stated in section 10, below.
8. Specimen means any material derived from a test system for examination, analysis, or retention.
9. Experimental starting date means the date on which the first study specific data are collected.
10. Experimental completion date means the last date on which data are collected from the study.
11. Study initiation date means the date the Study Director signs the study plan.
12. Study completion date means the date the Study Director signs the final report.

2.4 Terms Concerning the Test Item
1. Test item means an article that is the subject of a study.
2. Reference item ("control item") means any article used to provide a basis for comparison with the test item.
3. Batch means a specific quantity or lot of a test item or reference item produced during a defined cycle of manufacture in such a way that it could be expected to be of a uniform character and should be designated as such.
4. Vehicle means any agent which serves as a carrier used to mix, disperse, or solubilise the test item or reference item to facilitate the administration/application to the test system.

GOOD LABORATORY PRACTICE PRINCIPLES

1. **Test Facility Organisation and Personnel**

    1.1 Test Facility Management's Responsibilities
    1. Each test facility management should ensure that these Principles of Good Laboratory Practice are complied with, in its test facility.
    2. At a minimum it should:
        a) ensure that a statement exists which identifies the individual(s) within a test facility who fulfil the responsibilities of management as defined by these Principles of Good Laboratory Practice;
        b) ensure that a sufficient number of qualified personnel, appropriate facilities, equipment, and materials are available for the timely and proper conduct of the study;
        c) ensure the maintenance of a record of the qualifications, training, experience and job description for each professional and technical individual;
        d) ensure that personnel clearly understand the functions they are to perform and, where necessary, provide training for these functions;
        e) ensure that appropriate and technically valid Standard Operating Procedures are established and followed, and

approve all original and revised Standard Operating Procedures;

f) ensure that there is a Quality Assurance Programme with designated personnel and assure that the Quality Assurance responsibility is being performed in accordance with these Principles of Good Laboratory Practice;

g) ensure that for each study an individual with the appropriate qualifications, training, and experience is designated by the management as the Study Director before the study is initiated. Replacement of a Study Director should be done according to established procedures, and should be documented.

h) ensure, in the event of a multi-site study, that, if needed, a Principal Investigator is designated, who is appropriately trained, qualified and experienced to supervise the delegated phase(s) of the study. Replacement of a Principal Investigator should be done according to established procedures, and should be documented.

i) ensure documented approval of the study plan by the Study Director;

j) ensure that the Study Director has made the approved study plan available to the Quality Assurance personnel;

k) ensure the maintenance of an historical file of all Standard Operating Procedures;

l) ensure that an individual is identified as responsible for the management of the archive(s);

m) ensure the maintenance of a master schedule;

n) ensure that test facility supplies meet requirements appropriate to their use in a study;

o) ensure for a multi-site study that clear lines of communication exist between the Study Director, Principal Investigator(s), the Quality Assurance Programme(s) and study personnel;

p) ensure that test and reference items are appropriately characterised;

q) establish procedures to ensure that computerised systems are suitable for their intended purpose, and are validated, operated and maintained in accordance with these Principles of Good Laboratory Practice.

3. When a phase(s) of a study is conducted at a test site, test site management (if appointed) will have the responsibilities as

defined above with the following exceptions: 1.1.2 g), i), j) and o).

1.2 Study Director's Responsibilities

1. The Study Director is the single point of study control and has the responsibility for the overall conduct of the study and for its final report.
2. These responsibilities should include, but not be limited to, the following functions. The Study Director should:
    a) approve the study plan and any amendments to the study plan by dated signature;
    b) ensure that the Quality Assurance personnel have a copy of the study plan and any amendments in a timely manner and communicate effectively with the Quality Assurance personnel as required during the conduct of the study;
    c) ensure that study plans and amendments and Standard Operating Procedures are available to study personnel;
    d) ensure that the study plan and the final report for a multi-site study identify and define the role of any Principal investigator(s) and any test facilities and test sites involved in the conduct of the study;
    e) ensure that the procedures specified in the study plan are followed, and assess and document the impact of any deviations from the study plan on the quality and integrity of the study, and take appropriate corrective action if necessary; acknowledge deviations from Standard Operating Procedures during the conduct of the study;
    f) ensure that all raw data generated are fully documented and recorded;
    g) ensure that computerised systems used in the study have been validated;
    h) sign and date the final report to indicate acceptance of responsibility for the validity of the data and to indicate the extent to which the study complies with these Principles of Good Laboratory Practice;
    i) ensure that after completion (including termination) of the study, the study plan, the final report, raw data and supporting material are archived.

1.3 Principal Investigator's Responsibilities

The Principal Investigator will ensure that the delegated phases of the study are conducted in accordance with the applicable Principles of Good Laboratory Practice.

1.4 Study Personnel's Responsibilities
1. All personnel involved in the conduct of the study must be knowledgeable in those parts of the Principles of Good Laboratory Practice which are applicable to their involvement in the study.
2. Study personnel will have access to the study plan and appropriate Standard Operating Procedures applicable to their involvement in the study. It is their responsibility to comply with the instructions given in these documents. Any deviation from these instructions should be documented and communicated directly to the Study Director, and/or if appropriate, the Principal Investigator(s).
3. All study personnel are responsible for recording raw data promptly and accurately and in compliance with these Principles of Good Laboratory Practice, and are responsible for the quality of their data.
4. Study personnel should exercise health precautions to minimise risk to themselves and to ensure the integrity of the study. They should communicate to the appropriate person any relevant known health or medical condition in order that they can be excluded from operations that may affect the study.

2. Quality Assurance Programme

2.1 General
1. The test facility should have a documented Quality Assurance Programme to assure that studies performed are in compliance with these Principles of Good Laboratory Practice.
2. The Quality Assurance Programme should be carried out by an individual or by individuals designated by and directly responsible to management and who are familiar with the test procedures.
3. This individual(s) should not be involved in the conduct of the study being assured.

2.2 Responsibilities of the Quality Assurance Personnel
1. The responsibilities of the Quality Assurance personnel include, but are not limited to, the following functions. They should:
   a) maintain copies of all approved study plans and Standard Operating Procedures in use in the test facility and have access to an up-to-date copy of the master schedule;
   b) verify that the study plan contains the information required for compliance with these Principles of Good Laboratory Practice. This verification should be documented;
   c) conduct inspections to determine if all studies are conducted in accordance with these Principles of Good Laboratory Practice. Inspections should also determine that study plans and Standard Operating Procedures have been made available to study personnel and are being followed.
   Inspections can be of three types as specified by Quality Assurance Programme Standard Operating Procedures:
   - Study-based inspections,
   - Facility-based inspections,
   - Process-based inspections.
   Records of such inspections should be retained.
   d) inspect the final reports to confirm that the methods, procedures, and observations are accurately and completely described, and that the reported results accurately and completely reflect the raw data of the studies;
   e) promptly report any inspection results in writing to management and to the Study Director, and to the Principal Investigator(s) and the respective management, when applicable;
   f) prepare and sign a statement, to be included with the final report, which specifies types of inspections and their dates, including the phase(s) of the study inspected, and the dates inspection results were reported to management and the Study Director and Principal Investigator(s), if applicable. This statement would also serve to confirm that the final report reflects the raw data.

3. **Facilities**

   3.1 General
   1. The test facility should be of suitable size, construction and location to meet the requirements of the study and to minimise disturbance that would interfere with the validity of the study.
   2. The design of the test facility should provide an adequate degree of separation of the different activities to assure the proper conduct of each study.

   3.2 Test System Facilities
   1. The test facility should have a sufficient number of rooms or areas to assure the isolation of test systems and the isolation of individual projects, involving substances or organisms known to be or suspected of being biohazardous.
   2. Suitable rooms or areas should be available for the diagnosis, treatment and control of diseases, in order to ensure that there is no unacceptable degree of deterioration of test systems.
   3. There should be storage rooms or areas as needed for supplies and equipment. Storage rooms or areas should be separated from rooms or areas housing the test systems and should provide adequate protection against infestation, contamination, and/or deterioration.

   3.3 Facilities for Handling Test and Reference Items
   1. To prevent contamination or mix-ups, there should be separate rooms or areas for receipt and storage of the test and reference items, and mixing of the test items with a vehicle.
   2. Storage rooms or areas for the test items should be separate from rooms or areas containing the test systems. They should be adequate to preserve identity, concentration, purity, and stability, and ensure safe storage for hazardous substances.

   3.4 Archive Facilities

   Archive facilities should be provided for the secure storage and retrieval of study plans, raw data, final reports, samples of test items and specimens. Archive design and archive conditions should protect contents from untimely deterioration.

3.5 Waste Disposal

Handling and disposal of wastes should be carried out in such a way as not to jeopardise the integrity of studies. This includes provision for appropriate collection, storage and disposal facilities, and decontamination and transportation procedures.

4. **Apparatus, Material, and Reagents**

   1. Apparatus, including validated computerised systems, used for the generation, storage and retrieval of data, and for controlling environmental factors relevant to the study should be suitably located and of appropriate design and adequate capacity.
   2. Apparatus used in a study should be periodically inspected, cleaned, maintained, and calibrated according to Standard Operating Procedures. Records of these activities should be maintained. Calibration should, where appropriate, be traceable to national or international standards of measurement.
   3. Apparatus and materials used in a study should not interfere adversely with the test systems.
   4. Chemicals, reagents, and solutions should be labelled to indicate identity (with concentration if appropriate), expiry date and specific storage instructions. Information concerning source, preparation date and stability should be available. The expiry date may be extended on the basis of documented evaluation or analysis.

5. **Test Systems**

   5.1 Physical/Chemical
   1. Apparatus used for the generation of physical/chemical data should be suitably located and of appropriate design and adequate capacity.
   2. The integrity of the physical/chemical test systems should be ensured.

5.2 Biological
   1. Proper conditions should be established and maintained for the storage, housing, handling and care of biological test systems, in order to ensure the quality of the data.
   2. Newly received animal and plant test systems should be isolated until their health status has been evaluated. If any unusual mortality or morbidity occurs, this lot should not be used in studies and, when appropriate, should be humanely destroyed. At the experimental starting date of a study, test systems should be free of any disease or condition that might interfere with the purpose or conduct of the study. Test systems that become diseased or injured during the course of a study should be isolated and treated, if necessary to maintain the integrity of the study. Any diagnosis and treatment of any disease before or during a study should be recorded.
   3. Records of source, date of arrival, and arrival condition of test systems should be maintained.
   4. Biological test systems should be acclimatised to the test environment for an adequate period before the first administration/application of the test or reference item.
   5. All information needed to properly identify the test systems should appear on their housing or containers. Individual test systems that are to be removed from their housing or containers during the conduct of the study should bear appropriate identification, wherever possible.
   6. During use, housing or containers for test systems should be cleaned and sanitised at appropriate intervals. Any material that comes into contact with the test system should be free of contaminants at levels that would interfere with the study. Bedding for animals should be changed as required by sound husbandry practice. Use of pest control agents should be documented.
   7. Test systems used in field studies should be located so as to avoid interference in the study from spray drift and from past usage of pesticides.

6. **Test and Reference Items**

   6.1 Receipt, Handling, Sampling and Storage

1. Records including test item and reference item characterisation, date of receipt, expiry date, quantities received and used in studies should be maintained.
2. Handling, sampling, and storage procedures should be identified in order that the homogeneity and stability are assured to the degree possible and contamination or mix-up are precluded.
3. Storage container(s) should carry identification information, expiry date, and specific storage instructions.

6.2 Characterisation

1. Each test and reference item should be appropriately identified (e.g., code, Chemical Abstracts Service Registry Number [CAS number], name, biological parameters).
2. For each study, the identity, including batch number, purity, composition, concentrations, or other characteristics to appropriately define each batch of the test or reference items should be known.
3. In cases where the test item is supplied by the sponsor, there should be a mechanism, developed in co-operation between the sponsor and the test facility, to verify the identity of the test item subject to the study.
4. The stability of test and reference items under storage and test conditions should be known for all studies.
5. If the test item is administered or applied in a vehicle, the homogeneity, concentration and stability of the test item in that vehicle should be determined. For test items used in field studies (e.g., tank mixes), these may be determined through separate laboratory experiments.
6. A sample for analytical purposes from each batch of test item should be retained for all studies except short-term studies.

7. Standard Operating Procedures

7.1. A test facility should have written Standard Operating Procedures approved by test facility management that are intended to ensure the quality and integrity of the data generated by that test facility. Revisions to Standard Operating Procedures should be approved by test facility management.

7.2. Each separate test facility unit or area should have immediately available current Standard Operating Procedures relevant to the activities being performed therein. Published text books, analytical methods, articles and manuals may be used as supplements to these Standard Operating Procedures.

7.3. Deviations from Standard Operating Procedures related to the study should be documented and should be acknowledged by the Study Director and the Principal Investigator(s), as applicable.

7.4. Standard Operating Procedures should be available for, but not be limited to, the following categories of test facility activities. The details given under each heading are to be considered as illustrative examples.
   1. Test and Reference Items
      Receipt, identification, labelling, handling, sampling and storage.
   2. Apparatus, Materials and Reagents
      a) Apparatus
          Use, maintenance, cleaning and calibration.
      b) Computerised Systems
          Validation, operation, maintenance, security, change control and back-up.
      c) Materials, Reagents and Solutions
          Preparation and labelling.
   3. Record Keeping, Reporting, Storage, and Retrieval
      Coding of studies, data collection, preparation of reports, indexing systems, handling of data, including the use of computerised systems.
   4. Test System (where appropriate)
      a) Room preparation and environmental room conditions for the test system.
      b) Procedures for receipt, transfer, proper placement, characterisation, identification and care of the test system.
      c) Test system preparation, observations and examinations, before, during and at the conclusion of the study.
      d) Handling of test system individuals found moribund or dead during the study.
      e) Collection, identification and handling of specimens including necropsy and histopathology.
      f) Siting and placement of test systems in test plots.

Appendix I: The OECD GLP Principles

5. Quality Assurance Procedures
   Operation of Quality Assurance personnel in planning, scheduling, performing, documenting and reporting inspections.

8. Performance of the Study

8.1 Study Plan

1. For each study, a written plan should exist prior to the initiation of the study. The study plan should be approved by dated signature of the Study Director and verified for GLP compliance by Quality Assurance personnel as specified in Section 2.2.1.b., above. The study plan should also be approved by the test facility management and the sponsor, if required by national regulation or legislation in the country where the study is being performed.
2. a) Amendments to the study plan should be justified and approved by dated signature of the Study Director and maintained with the study plan.
   b) Deviations from the study plan should be described, explained, acknowledged and dated in a timely fashion by the Study Director and/or Principal Investigator(s) and maintained with the study raw data.
3. For short-term studies, a general study plan accompanied by a study specific supplement may be used.

8.2 Content of the Study Plan

The study plan should contain, but not be limited to the following information:

1. Identification of the Study, the Test Item and Reference Item
   a) A descriptive title;
   b) A statement which reveals the nature and purpose of the study;
   c) Identification of the test item by code or name (IUPAC; CAS number, biological parameters, etc.);
   d) The reference item to be used.
2. Information Concerning the Sponsor and the Test Facility
   a) Name and address of the sponsor;
   b) Name and address of any test facilities and test sites involved;

c) Name and address of the Study Director;
d) Name and address of the Principal Investigator(s), and the phase(s) of the study delegated by the Study Director and under the responsibility of the Principal Investigator(s).
3. Dates
   a) The date of approval of the study plan by signature of the Study Director. The date of approval of the study plan by signature of the test facility management and sponsor if required by national regulation or legislation in the country where the study is being performed.
   b) The proposed experimental starting and completion dates.
4. Test Methods
   Reference to the OECD Test Guideline or other test guideline or method to be used.
5. Issues (where applicable)
   a) The justification for selection of the test system;
   b) Characterisation of the test system, such as the species, strain, substrain, source of supply, number, body weight range, sex, age and other pertinent information;
   c) The method of administration and the reason for its choice;
   d) The dose levels and/or concentration(s), frequency, and duration of administration/ application;
   e) Detailed information on the experimental design, including a description of the chronological procedure of the study, all methods, materials and conditions, type and frequency of analysis, measurements, observations and examinations to be performed, and statistical methods to be used (if any).
6. Records
   A list of records to be retained.

8.3 Conduct of the Study

1. A unique identification should be given to each study. All items concerning this study should carry this identification. Specimens from the study should be identified to confirm their origin. Such identification should enable traceability, as appropriate for the specimen and study.
2. The study should be conducted in accordance with the study plan.

3. All data generated during the conduct of the study should be recorded directly, promptly, accurately, and legibly by the individual entering the data. These entries should be signed or initialled and dated.
4. Any change in the raw data should be made so as not to obscure the previous entry, should indicate the reason for change and should be dated and signed or initialled by the individual making the change.
5. Data generated as a direct computer input should be identified at the time of data input by the individual(s) responsible for direct data entries. Computerised system design should always provide for the retention of full audit trails to show all changes to the data without obscuring the original data. It should be possible to associate all changes to data with the persons having made those changes, for example, by use of timed and dated (electronic) signatures. Reason for changes should be given.

9. Reporting of Study Results

9.1 General

1. A final report should be prepared for each study. In the case of short term studies, a standardised final report accompanied by a study specific extension may be prepared.
2. Reports of Principal Investigators or scientists involved in the study should be signed and dated by them.
3. The final report should be signed and dated by the Study Director to indicate acceptance of responsibility for the validity of the data. The extent of compliance with these Principles of Good Laboratory Practice should be indicated.
4. Corrections and additions to a final report should be in the form of amendments. Amendments should clearly specify the reason for the corrections or additions and should be signed and dated by the Study Director.
5. Reformatting of the final report to comply with the submission requirements of a national registration or Regulatory Authority does not constitute a correction, addition or amendment to the final report.

9.2 Content of the Final Report

The final report should include, but not be limited to, the following information:
1. Identification of the Study, the Test Item and Reference Item
   a) A descriptive title;
   b) Identification of the test item by code or name (IUPAC, CAS number, biological parameters, etc.);
   c) Identification of the reference item by name;
   d) Characterisation of the test item including purity, stability and homogeneity.
2. Information Concerning the Sponsor and the Test Facility
   a) Name and address of the sponsor;
   b) Name and address of any test facilities and test sites involved;
   c) Name and address of the Study Director;
   d) Name and address of the Principal Investigator(s) and the phase(s) of the study delegated, if applicable;
   e) Name and address of scientists having contributed reports to the final report.
3. Dates
   Experimental starting and completion dates.
4. Statement
   A Quality Assurance Programme statement listing the types of inspections made and their dates, including the phase(s) inspected, and the dates any inspection results were reported to management and to the Study Director and Principal Investigator(s), if applicable. This statement would also serve to confirm that the final report reflects the raw data.
5. Description of Materials and Test Methods
   a) Description of methods and materials used;
   b) Reference to OECD Test Guideline or other test guideline or method.
6. Results
   a) A summary of results;
   b) All information and data required by the study plan;
   c) A presentation of the results, including calculations and determinations of statistical significance;
   d) An evaluation and discussion of the results and, where appropriate, conclusions.

7. Storage
   The location(s) where the study plan, samples of test and reference items, specimens, raw data and the final report are to be stored.

10. **Storage and Retention of Records and Materials**

    10.1 The following should be retained in the archives for the period specified by the appropriate authorities:

    a) The study plan, raw data, samples of test and reference items, specimens, and the final report of each study;
    b) Records of all inspections performed by the Quality Assurance Programme, as well as master schedules;
    c) Records of qualifications, training, experience and job descriptions of personnel;
    d) Records and reports of the maintenance and calibration of apparatus;
    e) Validation documentation for computerised systems;
    f) The historical file of all Standard Operating Procedures;
    g) Environmental monitoring records.

    In the absence of a required retention period, the final disposition of any study materials should be documented. When samples of test and reference items and specimens are disposed of before the expiry of the required retention period for any reason, this should be justified and documented. Samples of test and reference items and specimens should be retained only as long as the quality of the preparation permits evaluation.

    10.2 Material retained in the archives should be indexed so as to facilitate orderly storage and retrieval.

    10.3 Only personnel authorised by management should have access to the archives. Movement of material in and out of the archives should be properly recorded.

    10.4 If a test facility or an archive contracting facility goes out of business and has no legal successor, the archive should be transferred to the archives of the sponsor(s) of the study(s).

Appendix II.II

Excerpts from the
United States Food and Drug Agency
21 Code of Federal Regulations, Part 58
Good Laboratory Practice for Nonclinical Laboratory Studies

A — General Provisions

§ 58.1 Scope.

(a) This part prescribes good laboratory practices for conducting nonclinical laboratory studies that support or are intended to support applications for research or marketing permits for products regulated by the Food and Drug Administration, including food and colour additives, animal food additives, human and animal drugs, medical devices for human use, biological products, and electronic products.

...

§ 58.3 Definitions.

As used in this part, the following terms shall have the meanings specified:

(a) ...
(b) Test article means any food additive, colour additive, drug, biological product, electronic product, medical device for human use, or any other article subject to regulation ...
(c) Control article means any food additive, colour additive, drug, biological product, electronic product, medical device for human use, or any article

other than a test article, feed, or water that is administered to the test system in the course of a nonclinical laboratory study for the purpose of establishing a basis for comparison with the test article.
(d) Nonclinical laboratory study means *in vivo* or *in vitro* experiments in which test articles are studied prospectively in test systems under laboratory conditions to determine their safety. The term does not include studies utilising human subjects or clinical studies or field trials in animals. The term does not include basic exploratory studies carried out to determine whether a test article has any potential utility or to determine physical or chemical characteristics of a test article.
(e) ...
(f) Sponsor means:
  (1) A person who initiates and supports, by provision of financial or other resources, a nonclinical laboratory study;
  (2) A person who submits a nonclinical study to the Food and Drug Administration in support of an application for a research or marketing permit; or
  (3) A testing facility, if it both initiates and actually conducts the study.
(g) Testing facility means a person who actually conducts a nonclinical laboratory study, i.e., actually uses the test article in a test system. ... Testing facility encompasses only those operational units that are being or have been used to conduct nonclinical laboratory studies.
(h) Person includes an individual, partnership, corporation, association, scientific or academic establishment, government agency, or organisational unit thereof, and any other legal entity.
(i) Test system means any animal, plant, microorganism, or subparts thereof to which the test or control article is administered or added for study. Test system also includes appropriate groups or components of the system not treated with the test or control articles.
(j) Specimen means any material derived from a test system for examination for analysis.
(k) Raw data means any laboratory worksheets, records, memoranda, notes, or exact copies thereof, that are the result of original observations and activities of a nonclinical laboratory study and are necessary for the reconstruction and evaluation of the report of that study. In the event that exact transcripts of raw data have been prepared (e.g., tapes which have been transcribed verbatim, dated, and verified accurate by signature), the exact copy or exact transcript may be substituted for the original source as raw data. Raw data may include photographs, microfilm or microfiche copies, computer printouts, magnetic media, including dictated observations, and recorded data from automated instruments.

(l) Quality Assurance unit means any person or organisational element, except the Study Director, designated by testing facility management to perform the duties relating to Quality Assurance of nonclinical laboratory studies.
(m) Study director means the individual responsible for the overall conduct of a nonclinical laboratory study.
(n) Batch means a specific quantity or lot of a test or control article that has been characterised according to § 58.105(a).
(o) Study initiation date means the date the protocol is signed by the Study Director.
(p) Study completion date means the date the final report is signed by the Study Director.

§ 58.10 Applicability to studies performed under grants and contracts.

When a sponsor conducting a nonclinical laboratory study intended to be submitted to or reviewed by the Food and Drug Administration utilises the services of a consulting laboratory, contractor, or grantee to perform an analysis or other service, it shall notify the consulting laboratory, contractor, or grantee that the service is part of a nonclinical laboratory study that must be conducted in compliance with the provisions of this part.

§ 58.15 Inspection of a testing facility.

(a) A testing facility shall permit an authorised employee of the Food and Drug Administration, at reasonable times and in a reasonable manner, to inspect the facility and to inspect (and in the case of records also to copy) all records and specimens required to be maintained regarding studies within the scope of this part. The records inspection and copying requirements shall not apply to Quality Assurance unit records of findings and problems, or to actions recommended and taken.
(b) The Food and Drug Administration will not consider a nonclinical laboratory study in support of an application for a research or marketing permit if the testing facility refuses to permit inspection. The determination that a nonclinical laboratory study will not be considered in support of an application for a research or marketing permit does not, however, relieve the applicant for such a permit of any obligation under any applicable statute or regulation to submit the results of the study to the Food and Drug Administration.

B — Organisation and Personnel

§ 58.29 Personnel.

(a) Each individual engaged in the conduct of or responsible for the supervision of a nonclinical laboratory study shall have education, training, and experience, or combination thereof, to enable that individual to perform the assigned functions.

(b) Each testing facility shall maintain a current summary of training and experience and job description for each individual engaged in or supervising the conduct of a nonclinical laboratory study.

(c) There shall be a sufficient number of personnel for the timely and proper conduct of the study according to the protocol.

(d) Personnel shall take necessary personal sanitation and health precautions designed to avoid contamination of test and control articles and test systems.

(e) Personnel engaged in a nonclinical laboratory study shall wear clothing appropriate for the duties they perform. Such clothing shall be changed as often as necessary to prevent microbiological, radiological, or chemical contamination of test systems and test and control articles.

(f) Any individual found at any time to have an illness that may adversely affect the quality and integrity of the nonclinical laboratory study shall be excluded from direct contact with test systems, test and control articles and any other operation or function that may adversely affect the study until the condition is corrected. All personnel shall be instructed to report to their immediate supervisors any health or medical conditions that may reasonably be considered to have an adverse effect on a nonclinical laboratory study.

§ 58.31 Testing facility management.

For each nonclinical laboratory study, testing facility management shall:

(a) Designate a Study Director as described in § 58.33, before the study is initiated.
(b) Replace the Study Director promptly if it becomes necessary to do so during the conduct of a study.
(c) Assure that there is a Quality Assurance unit as described in § 58.35.

(d) Assure that test and control articles or mixtures have been appropriately tested for identity, strength, purity, stability, and uniformity, as applicable.
(e) Assure that personnel, resources, facilities, equipment, materials, and methodologies are available as scheduled.
(f) Assure that personnel clearly understand the functions they are to perform.
(g) Assure that any deviations from these regulations reported by the Quality Assurance unit are communicated to the Study Director and corrective actions are taken and documented.

§ 58.33 Study director.

For each nonclinical laboratory study, a scientist or other professional of appropriate education, training, and experience, or combination thereof, shall be identified as the Study Director. The Study Director has overall responsibility for the technical conduct of the study, as well as for the interpretation, analysis, documentation and reporting of results, and represents the single point of study control. The Study Director shall assure that:

(a) The protocol, including any change, is approved as provided by § 58.120 and is followed.
(b) All experimental data, including observations of unanticipated responses of the test system are accurately recorded and verified.
(c) Unforeseen circumstances that may affect the quality and integrity of the nonclinical laboratory study are noted when they occur, and corrective action is taken and documented.
(d) Test systems are as specified in the protocol.
(e) All applicable good laboratory practice regulations are followed.
(f) All raw data, documentation, protocols, specimens, and final reports are transferred to the archives during or at the close of the study.

§ 58.35 Quality Assurance unit.

(a) A testing facility shall have a Quality Assurance unit which shall be responsible for monitoring each study to assure management that the facilities, equipment, personnel, methods, practices, records, and controls are in conformance with the regulations in this part. For any given study, the Quality Assurance unit shall be entirely separate from and independent of the personnel engaged in the direction and conduct of that study.
(b) The Quality Assurance unit shall:

(1) Maintain a copy of a master schedule sheet of all nonclinical laboratory studies conducted at the testing facility indexed by test article and containing the test system, nature of study, date study was initiated, current status of each study, identity of the sponsor, and name of the Study Director.
(2) Maintain copies of all protocols pertaining to all nonclinical laboratory studies for which the unit is responsible.
(3) Inspect each nonclinical laboratory study at intervals adequate to assure the integrity of the study and maintain written and properly signed records of each periodic inspection showing the date of the inspection, the study inspected, the phase or segment of the study inspected, the person performing the inspection, findings and problems, action recommended and taken to resolve existing problems, and any scheduled date for reinspection. Any problems found during the course of an inspection which are likely to affect study integrity shall be brought to the attention of the Study Director and management immediately.
(4) Periodically submit to management and the Study Director written status reports on each study, noting any problems and the corrective actions taken.
(5) Determine that no deviations from approved protocols or standard operating procedures were made without proper authorization and documentation.
(6) Review the final study report to assure that such report accurately describes the methods and standard operating procedures, and that the reported results accurately reflect the raw data of the nonclinical laboratory study.
(7) Prepare and sign a statement to be included with the final study report which shall specify the dates inspections were made and findings reported to management and to the Study Director.

(c) The responsibilities and procedures applicable to the Quality Assurance unit, the records maintained by the Quality Assurance unit, and the method of indexing such records shall be in writing and shall be maintained. These items including inspection dates, the study inspected, the phase or segment of the study inspected, and the name of the individual performing the inspection shall be made available for inspection to authorised employees of the Food and Drug Administration.

(d) ...

C — Facilities

§ 58.41 General.

Each testing facility shall be of suitable size and construction to facilitate the proper conduct of nonclinical laboratory studies. It shall be designed so that there is a degree of separation that will prevent any function or activity from having an adverse effect on the study.

§ 58.43 Animal care facilities.

(a) A testing facility shall have a sufficient number of animal rooms or areas, as needed, to assure proper:

  (1) Separation of species or test systems,
  (2) isolation of individual projects,
  (3) quarantine of animals, and
  (4) routine or specialised housing of animals.

(b) A testing facility shall have a number of animal rooms or areas separate from those described in paragraph (a) of this section to ensure isolation of studies being done with test systems or test and control articles known to be biohazardous, including volatile substances, aerosols, radioactive materials, and infectious agents.

(c) Separate areas shall be provided, as appropriate, for the diagnosis, treatment, and control of laboratory animal diseases. These areas shall provide effective isolation for the housing of animals either known or suspected of being diseased, or of being carriers of disease, from other animals.

(d) When animals are housed, facilities shall exist for the collection and disposal of all animal waste and refuse or for safe sanitary storage of waste before removal from the testing facility. Disposal facilities shall be so provided and operated as to minimise vermin infestation, odors, disease hazards, and environmental contamination.

§ 58.45 Animal supply facilities.

There shall be storage areas, as needed, for feed, bedding, supplies, and equipment. Storage areas for feed and bedding shall be separated from areas housing the test systems and shall be protected against infestation or contamination. Perishable supplies shall be preserved by appropriate means.

§ 58.47 Facilities for handling test and control articles.

(a) As necessary to prevent contamination or mixups, there shall be separate areas for:
   (1) Receipt and storage of the test and control articles.
   (2) Mixing of the test and control articles with a carrier, e.g., feed.
   (3) Storage of the test and control article mixtures.
(b) Storage areas for the test and/or control article and test and control mixtures shall be separate from areas housing the test systems and shall be adequate to preserve the identity, strength, purity, and stability of the articles and mixtures.

§ 58.49 Laboratory operation areas.

Separate laboratory space shall be provided, as needed, for the performance of the routine and specialised procedures required by nonclinical laboratory studies.

§ 58.51 Specimen and data storage facilities.

Space shall be provided for archives, limited to access by authorised personnel only, for the storage and retrieval of all raw data and specimens from completed studies.

D — Equipment

§ 58.61 Equipment design.

Equipment used in the generation, measurement, or assessment of data and equipment used for facility environmental control shall be of appropriate design and adequate capacity to function according to the protocol and shall be suitably located for operation, inspection, cleaning, and maintenance.

§ 58.63 Maintenance and calibration of equipment.

(a) Equipment shall be adequately inspected, cleaned, and maintained. Equipment used for the generation, measurement, or assessment of data shall be adequately tested, calibrated and/or standardised.

(b) The written standard operating procedures required under § 58.81(b)(11) shall set forth in sufficient detail the methods, materials, and schedules to be used in the routine inspection, cleaning, maintenance, testing, calibration, and/or standardization of equipment, and shall specify, when appropriate, remedial action to be taken in the event of failure or malfunction of equipment. The written standard operating procedures shall designate the person responsible for the performance of each operation.

(c) Written records shall be maintained of all inspection, maintenance, testing, calibrating and/or standardizing operations. These records, containing the date of the operation, shall describe whether the maintenance operations were routine and followed the written standard operating procedures. Written records shall be kept of nonroutine repairs performed on equipment as a result of failure and malfunction. Such records shall document the nature of the defect, how and when the defect was discovered, and any remedial action taken in response to the defect.

E — Testing Facilities Operation

§ 58.81 Standard operating procedures.

(a) A testing facility shall have standard operating procedures in writing setting forth nonclinical laboratory study methods that management is satisfied are adequate to insure the quality and integrity of the data generated in the course of a study. All deviations in a study from standard operating procedures shall be authorised by the Study Director and shall be documented in the raw data. Significant changes in established standard operating procedures shall be properly authorised in writing by management.

(b) Standard operating procedures shall be established for, but not limited to, the following:

    (1) Animal room preparation.
    (2) Animal care.
    (3) Receipt, identification, storage, handling, mixing, and method of sampling of the test and control articles.

Appendix II: US-FDA GLP Regulations

(4) Test system observations.
(5) Laboratory tests.
(6) Handling of animals found moribund or dead during study.
(7) Necropsy of animals or postmortem examination of animals.
(8) Collection and identification of specimens.
(9) Histopathology.
(10) Data handling, storage, and retrieval.
(11) Maintenance and calibration of equipment.
(12) Transfer, proper placement, and identification of animals.

(c) Each laboratory area shall have immediately available laboratory manuals and standard operating procedures relative to the laboratory procedures being performed. Published literature may be used as a supplement to standard operating procedures.

(d) A historical file of standard operating procedures, and all revisions thereof, including the dates of such revisions, shall be maintained.

§ 58.83 Reagents and solutions.

All reagents and solutions in the laboratory areas shall be labelled to indicate identity, titer or concentration, storage requirements, and expiration date. Deteriorated or outdated reagents and solutions shall not be used.

§ 58.90 Animal care.

(a) There shall be standard operating procedures for the housing, feeding, handling, and care of animals.

(b) All newly received animals from outside sources shall be isolated and their health status shall be evaluated in accordance with acceptable veterinary medical practice.

(c) At the initiation of a nonclinical laboratory study, animals shall be free of any disease or condition that might interfere with the purpose or conduct of the study. If, during the course of the study, the animals contract such a disease or condition, the diseased animals shall be isolated, if necessary. These animals may be treated for disease or signs of disease provided that such treatment does not interfere with the study. The diagnosis, authorizations of treatment, description of treatment, and each date of treatment shall be documented and shall be retained.

(d) Warmblooded animals, excluding suckling rodents, used in laboratory procedures that require manipulations and observations over an extended period of time or in studies that require the animals to be removed from and returned to their home cages for any reason (e.g., cage cleaning, treatment, etc.), shall receive appropriate identification. All information needed to specifically identify each animal within an animal housing unit shall appear on the outside of that unit.

(e) Animals of different species shall be housed in separate rooms when necessary. Animals of the same species, but used in different studies, should not ordinarily be housed in the same room when inadvertent exposure to control or test articles or animal mixup could affect the outcome of either study. If such mixed housing is necessary, adequate differentiation by space and identification shall be made.

(f) Animal cages, racks and accessory equipment shall be cleaned and sanitised at appropriate intervals.

(g) Feed and water used for the animals shall be analysed periodically to ensure that contaminants known to be capable of interfering with the study and reasonably expected to be present in such feed or water are not present at levels above those specified in the protocol. Documentation of such analyses shall be maintained as raw data.

(h) Bedding used in animal cages or pens shall not interfere with the purpose or conduct of the study and shall be changed as often as necessary to keep the animals dry and clean.

(i) If any pest control materials are used, the use shall be documented. Cleaning and pest control materials that interfere with the study shall not be used.

F — Test and Control Articles

§ 58.105 Test and control article characterisation.

(a) The identity, strength, purity, and composition or other characteristics which will appropriately define the test or control article shall be determined for each batch and shall be documented. Methods of synthesis, fabrication, or derivation of the test and control articles shall be documented by the sponsor or the testing facility. In those cases where marketed prod-

ucts are used as control articles, such products will be characterised by their labelling.
(b) The stability of each test or control article shall be determined by the testing facility or by the sponsor either: (1) Before study initiation, or (2) concomitantly according to written standard operating procedures, which provide for periodic analysis of each batch.
(c) Each storage container for a test or control article shall be labelled by name, chemical abstract number or code number, batch number, expiration date, if any, and, where appropriate, storage conditions necessary to maintain the identity, strength, purity, and composition of the test or control article. Storage containers shall be assigned to a particular test article for the duration of the study.
(d) For studies of more than 4 weeks' duration, reserve samples from each batch of test and control articles shall be retained for the period of time provided by § 58.195.

§ 58.107 Test and control article handling.

Procedures shall be established for a system for the handling of the test and control articles to ensure that:

(a) There is proper storage.
(b) Distribution is made in a manner designed to preclude the possibility of contamination, deterioration, or damage.
(c) Proper identification is maintained throughout the distribution process.
(d) The receipt and distribution of each batch is documented. Such documentation shall include the date and quantity of each batch distributed or returned.

§ 58.113 Mixtures of articles with carriers.

(a) For each test or control article that is mixed with a carrier, tests by appropriate analytical methods shall be conducted:
   (1) To determine the uniformity of the mixture and to determine, periodically, the concentration of the test or control article in the mixture.
   (2) To determine the stability of the test and control articles in the mixture as required by the conditions of the study either:
      (i) Before study initiation, or

(ii) Concomitantly according to written standard operating procedures which provide for periodic analysis of the test and control articles in the mixture.

(b) [Reserved]

(c) Where any of the components of the test or control article carrier mixture has an expiration date, that date shall be clearly shown on the container. If more than one component has an expiration date, the earliest date shall be shown.

G — Protocol for and Conduct of a Nonclinical Laboratory Study

§ 58.120 Protocol.

(a) Each study shall have an approved written protocol that clearly indicates the objectives and all methods for the conduct of the study. The protocol shall contain, as applicable, the following information:

(1) A descriptive title and statement of the purpose of the study.
(2) Identification of the test and control articles by name, chemical abstract number, or code number.
(3) The name of the sponsor and the name and address of the testing facility at which the study is being conducted.
(4) The number, body weight range, sex, source of supply, species, strain, substrain, and age of the test system.
(5) The procedure for identification of the test system.
(6) A description of the experimental design, including the methods for the control of bias.
(7) A description and/or identification of the diet used in the study as well as solvents, emulsifiers, and/or other materials used to solubilize or suspend the test or control articles before mixing with the carrier. The description shall include specifications for acceptable levels of contaminants that are reasonably expected to be present in the dietary materials and are known to be capable of interfering with the purpose or conduct of the study if present at levels greater than established by the specifications.
(8) Each dosage level, expressed in milligrams per kilogram of body weight or other appropriate units, of the test or control article to be administered and the method and frequency of administration.

(9) The type and frequency of tests, analyses, and measurements to be made.
(10) The records to be maintained.
(11) The date of approval of the protocol by the sponsor and the dated signature of the Study Director.
(12) A statement of the proposed statistical methods to be used.

(b) All changes in or revisions of an approved protocol and the reasons therefor shall be documented, signed by the Study Director, dated, and maintained with the protocol.

§ 58.130 Conduct of a nonclinical laboratory study.

(a) The nonclinical laboratory study shall be conducted in accordance with the protocol.

(b) The test systems shall be monitored in conformity with the protocol.

(c) Specimens shall be identified by test system, study, nature, and date of collection. This information shall be located on the specimen container or shall accompany the specimen in a manner that precludes error in the recording and storage of data.

(d) Records of gross findings for a specimen from postmortem observations should be available to a pathologist when examining that specimen histopathologically.

(e) All data generated during the conduct of a nonclinical laboratory study, except those that are generated by automated data collection systems, shall be recorded directly, promptly, and legibly in ink. All data entries shall be dated on the date of entry and signed or initialled by the person entering the data. Any change in entries shall be made so as not to obscure the original entry, shall indicate the reason for such change, and shall be dated and signed or identified at the time of the change. In automated data collection systems, the individual responsible for direct data input shall be identified at the time of data input. Any change in automated data entries shall be made so as not to obscure the original entry, shall indicate the reason for change, shall be dated, and the responsible individual shall be identified.

J — Records and Reports

§ 58.185 Reporting of nonclinical laboratory study results.

(a) A final report shall be prepared for each nonclinical laboratory study and shall include, but not necessarily be limited to, the following:
   (1) Name and address of the facility performing the study and the dates on which the study was initiated and completed.
   (2) Objectives and procedures stated in the approved protocol, including any changes in the original protocol.
   (3) Statistical methods employed for analysing the data.
   (4) The test and control articles identified by name, chemical abstracts number or code number, strength, purity, and composition or other appropriate characteristics.
   (5) Stability of the test and control articles under the conditions of administration.
   (6) A description of the methods used.
   (7) A description of the test system used. Where applicable, the final report shall include the number of animals used, sex, body weight range, source of supply, species, strain and substrain, age, and procedure used for identification.
   (8) A description of the dosage, dosage regimen, route of administration, and duration.
   (9) A description of all cirmcumstances that may have affected the quality or integrity of the data.
   (10) The name of the Study Director, the names of other scientists or professionals, and the names of all supervisory personnel, involved in the study.
   (11) A description of the transformations, calculations, or operations performed on the data, a summary and analysis of the data, and a statement of the conclusions drawn from the analysis.
   (12) The signed and dated reports of each of the individual scientists or other professionals involved in the study.
   (13) The locations where all specimens, raw data, and the final report are to be stored.
   (14) The statement prepared and signed by the Quality Assurance unit as described in § 58.35(b)(7).

(b) The final report shall be signed and dated by the Study Director.

(c) Corrections or additions to a final report shall be in the form of an amendment by the Study Director. The amendment shall clearly identify that part of the final report that is being added to or corrected and the reasons for the correction or addition, and shall be signed and dated by the person responsible.

§ 58.190 Storage and retrieval of records and data.

(a) All raw data, documentation, protocols, final reports, and specimens (except those specimens obtained from mutagenicity tests and wet specimens of blood, urine, feces, and biological fluids) generated as a result of a nonclinical laboratory study shall be retained.

(b) There shall be archives for orderly storage and expedient retrieval of all raw data, documentation, protocols, specimens, and interim and final reports. Conditions of storage shall minimise deterioration of the documents or specimens in accordance with the requirements for the time period of their retention and the nature of the documents or specimens. A testing facility may contract with commercial archives to provide a repository for all material to be retained. Raw data and specimens may be retained elsewhere provided that the archives have specific reference to those other locations.

(c) An individual shall be identified as responsible for the archives.

(d) Only authorised personnel shall enter the archives.

(e) Material retained or referred to in the archives shall be indexed to permit expedient retrieval.

§ 58.195 Retention of records.

(a) ...

(b) ...

(c) Wet specimens (except those specimens obtained from mutagenicity tests and wet specimens of blood, urine, feces, and biological fluids), samples of test or control articles, and specially prepared material, which are relatively fragile and differ markedly in stability and quality during storage, shall be retained only as long as the quality of the preparation affords evaluation. ....

(d) The master schedule sheet, copies of protocols, and records of Quality Assurance inspections, as required by § 58.35(c) shall be maintained by the Quality Assurance unit as an easily accessible system of records ....

(e) Summaries of training and experience and job descriptions required to be maintained by § 58.29(b) may be retained along with all other testing facility employment records ....

(f) Records and reports of the maintenance and calibration and inspection of equipment, as required by § 58.63(b) and (c), shall be retained ....

(g) Records required by this part may be retained either as original records or as true copies such as photocopies, microfilm, microfiche, or other accurate reproductions of the original records.

(h) If a facility conducting nonclinical testing goes out of business, all raw data, documentation, and other material specified in this section shall be transferred to the archives of the sponsor of the study. The Food and Drug Administration shall be notified in writing of such a transfer.

Appendix II.III

Excerpts from the Draft
United States Environmental Protection Agency
Code of Federal Regulations, Part 806
Good Laboratory Practice Standards

A — General Provisions

806.1 Scope.

(1) This part prescribes good laboratory practices for conducting studies that support or are intended to support applications for research or marketing permits for pesticide products regulated by the EPA. This part is intended to assure the quality and integrity of data submitted ...

(2) This part also prescribes good laboratory practices for conducting studies relating to health effects, environmental effects, and chemical fate testing pursuant to the Toxic Substances Control Act (TSCA) ...

(3) It is EPA's policy that all data developed for submission ... be in accordance with provisions of this part. If data are not developed in accordance with the provisions of this part, EPA will consider such data insufficient to evaluate the health and environmental effects of the chemical substances unless the submitter provides additional information demonstrating that the data are reliable and adequate.

806.3 Definitions.

As used in this part, the following terms shall have the meanings specified:

...

Batch means a specific quantity or lot of a test, control, or reference substance that has been characterised ...

Carrier means any material, including but not limited to feed, water, soil, air, or nutrient media, with which the test substance is combined for administration to a test system.

Control substance means any chemical substance or mixture, or any other material other than a test substance, feed, or water, that is administered to the test system in the course of a study for the purpose of establishing a basis for comparison with the test substance for known chemical or biological measurements.

EPA means the U.S. Environmental Protection Agency.

Experimental start date means the first date the test substance is applied to the test system.

Experimental termination date means the last date on which data are collected directly from the study.

FDA means the U.S. Food and Drug Administration.

Person includes an individual, partnership, corporation, association, scientific or academic establishment, government agency, or organisational unit thereof, and any other legal entity.

Quality Assurance unit means any person or organisational element (except individual(s) directly involved in the conduct of the study, including the Study Director), designated by testing facility management to perform the duties relating to Quality Assurance of the studies.

Raw data means any laboratory worksheets, records, memoranda, notes, or exact copies thereof, that are the result of original observations and activities of a study and are necessary for the reconstruction and evaluation of the report of that study. In the event that exact transcripts of raw data have been prepared (e.g., tapes which have been transcribed verbatim, dated, and verified accurate by signature), the exact copy or exact transcript may be substituted for the original source as raw data. Raw data may include photographs, microfilm or microfiche copies, computer printouts, any original data captured electronically or by some other medium, dictated observations, and recorded data from automated instruments.

Reference substance means any chemical substance or mixture, or analytical standard, or material other than a test substance, feed, or water, that is administered to or used in analysing the test system in the course of a study for the purposes of establishing a basis for comparison with the test substance for known chemical or biological measurements.

Specimens means any material or sample derived from a test system for examination or analysis.

Sponsor means:
  (1) A person who initiates and supports, by provision of financial or other resources, a study;
  (2) A person who submits a study to the EPA...; or
  (3) A testing facility, if it both initiates and actually conducts the study.

Study means any experiment at one or more test sites, in which a test substance is studied in a test system under laboratory conditions or in the environment to determine or help predict its effects, metabolism, product performance (pesticide efficacy studies only as required by 40 CFR 158.640) environmental and chemical fate, persistence, or residue, or other characteristics in humans, other living organisms, or media. The term ``study" does not include basic exploratory studies carried out to determine whether a test substance or a test method has any potential utility.

Study completion date means the date the final report is signed by the Study Director.

Study director means the individual responsible for the overall conduct of a study.

Study initiation date means the date the protocol is signed by the Study Director.

Test substance means a substance or mixture administered or added to a test system in a study, which substance or mixture:
  (1) Is the subject of an application for a research or marketing permit supported by the study, or is the contemplated subject of such an application; or
  (2) Is an ingredient, impurity, degradation product, metabolite, or radioactive isotope of a substance described by paragraph (1) of this definition, or some other substance related to a substance described by that paragraph, which is used in the study to assist in characterizing the toxicity, metabolism, or other characteristics of a substance described by that paragraph; or
  (3) Is used to develop data to meet the requirements of a TSCA section 4 test rule and/or is developed under a TSCA section 4 testing consent agreement/order or TSCA section 5 consent order to the extent the agreement, rule, or order references this part.

Test system means any animal, plant, microorganism, chemical or physical matrix, including but not limited to soil, water or air, or subparts thereof, to which the test, control, or reference substance is administered or added for study. "Test system" also includes appropriate groups or components of the system not treated with the test, control, or reference substance.

Testing facility means a person who actually conducts a study, i.e., actually uses the test substance in a test system. Testing facility encompasses only those operational units that are being or have been used to conduct studies.

Vehicle means any agent which facilitates the mixture, dispersion, or solubilization of a test substance with a carrier (e.g., water, mineral oil, animal feed).

806.10 Applicability to studies performed under grant and contracts.

When a sponsor or other person utilises the services of a consulting laboratory, contractor, or grantee to perform all or a part of a study to which this part applies, that sponsor or person shall notify the consulting laboratory, contractor, or grantee, in writing, that the service is, or is part of, a study that must be conducted in compliance with the provisions of this part, prior to initiation of the study.

806.12 Statement of compliance or non-compliance.

Any person who submits to EPA either an application for a research or marketing permit and who, in connection with the application, submits data from a study to which this part applies, or a test required by a test rule or testing consent agreement/order ... , shall include in the application or submission a true and correct statement, signed by the applicant, the sponsor, and the Study Director, of one of the following types:

(a) A statement that the study was conducted in accordance with this part.
(b) A statement describing in detail all differences between the practices used in the study and those required by this part.
(c) A statement that the person was not a sponsor of the study, did not conduct the study, and does not know whether the study was conducted in accordance with this part.

806.15 Inspection of a testing facility.

(a) Testing facility management shall permit an authorised employee or duly designated representative of EPA or FDA, at reasonable times and in a reasonable manner, to inspect the facility and to inspect (and in the case of records also to copy) all records and specimens required to be maintained regarding studies to which this part applies. The records inspection and copying requirements shall not apply to Quality Assurance unit records of findings and problems, or to actions recommended and taken, except that EPA may seek production of these records in litigation or formal adjudicatory hearings.

(b) EPA will not consider reliable for purposes of supporting an application for a research or marketing permit, or showing that a chemical substance or mixture does not present a risk of injury to health or the environment, any data developed by a testing facility or sponsor that refuses to permit inspection in accordance with this part. The determination that a study will not be considered in support of an application for a research or marketing permit or reliable for other purposes does not, however, relieve the applicant for such a permit or the sponsor of a required test of any obligation under any applicable statute or regulation to submit the results of the study to EPA.

(c) ....

806.17 Effects of non-compliance.

(1) EPA may refuse to consider reliable for purposes of supporting an application for a research or marketing permit any data from a study which was not conducted in accordance with this part.

(2) ...

(3) ...

(4) EPA, at its discretion, may not consider reliable for purposes of showing that a chemical substance or mixture does not present a risk of injury to health or the environment any study which was not conducted in accordance with this part. EPA, at its discretion, may rely upon such studies for purposes of showing adverse effects. The determination that a study will not be considered reliable does not, however, relieve the sponsor of a required test of the obligation under any applicable statute or regulation to submit the results of the study to EPA.

(3) If data submitted to fulfil a requirement of a test rule or testing consent agreement/order issued ... are not developed in accordance with this part, EPA may determine that the sponsor has not fulfilled its obligations ... and may require the sponsor to develop data in accordance with the requirements of this part in order to satisfy such obligations.

B — Organisation and Personnel

806.29 Personnel.

(a) Each individual engaged in the conduct of or responsible for the supervision of a study shall have the appropriate education, training, and experience, or a combination thereof, to enable that individual to perform the assigned functions.

(b) Each testing facility shall maintain a current summary of training and experience and job description for each individual engaged in or supervising the conduct of a study.

(c) There shall be a sufficient number of personnel for the timely and proper conduct of the study according to the protocol.

(d) Personnel shall take necessary personal sanitation and health precautions designed to avoid contamination of test systems and test, control, and reference substances.

(e) Personnel engaged in a study shall wear clothing appropriate for the duties they perform. Such clothing shall be changed as often as necessary to prevent microbiological, radiological, or chemical contamination of test systems and test, control, and reference substances.

(f) Any individual found at any time to have an illness that may adversely affect the quality and integrity of the study shall be excluded from direct contact with test systems, test, control, and reference substances, and any other operation or function that may adversely affect the study until the health or medical condition is corrected. All personnel shall be instructed to report to their immediate supervisors any health or medical conditions that may reasonably be considered to have an adverse effect on a study.

806.31 Testing facility management.

For each study, testing facility management shall:

(a) Designate a Study Director ... before the study is initiated.
(b) Replace the Study Director promptly if it becomes necessary to do so during the conduct of a study.
(c) Assure that there is a Quality Assurance unit ...
(d) Assure that test, control, and reference substances or mixtures have been appropriately tested for identity, strength, purity, stability, and uniformity, as applicable.
(e) Assure that personnel, resources, facilities, equipment, materials and methodologies are available as scheduled.
(f) Assure that personnel clearly understand the functions they are to perform.
(g) Assure that any deviations from these regulations reported by the Quality Assurance unit are communicated to the Study Director and corrective actions are taken and documented.

806.33 Study director.

For each study, a scientist or other professional of appropriate education, training, and experience, or combination thereof, shall be identified as the Study Director. The Study Director has overall responsibility for the technical conduct of the study, as well as for the interpretation, analysis, documentation, and reporting of results, and represents the single point of study control. The Study Director shall assure that:

(a) The protocol, including any change, is approved ... and is followed.
(b) All experimental data, including observations of unanticipated responses of the test system are accurately recorded and verified.
(c) Unforeseen circumstances that may affect the quality and integrity of the study are noted when they occur, and corrective action is taken and documented.
(d) Test systems are as specified in the protocol.
(e) All applicable GLP regulations are followed.
(f) All raw data, documentation, protocols, specimens, and final reports are transferred to the archives during or at the close or termination of the study.

806.35 Quality Assurance unit.

(a) A testing facility shall have a Quality Assurance unit which shall be responsible for monitoring each study to assure management that the facilities, equipment, personnel, methods, practices, records, and controls are in conformance with the regulations in this part. For any given study, the Quality Assurance unit shall be entirely separate from and independent of the personnel engaged in the direction and conduct of that study. The Quality Assurance unit shall conduct inspections and maintain records appropriate to the study.

(b) The Quality Assurance unit shall:

   (1) Maintain a copy of a master schedule sheet of all studies conducted at the testing facility indexed to permit expedient retrieval, which identifies the test substance, the test system, nature of study, date study was initiated, current status of each study, date of completion or termination if study is not ongoing, identity of the sponsor, and name of the Study Director.
   (2) Maintain copies of all protocols until study completion pertaining to all studies for which the unit is responsible.
   (3) Inspect each study at intervals adequate to ensure the integrity of the study and maintain written and properly signed records of each periodic inspection showing the date of the inspection, the study inspected, the phase or segment of the study inspected, the person performing the inspection, findings and problems, action recommended and taken to resolve existing problems, and any scheduled date for reinspection. Any problems which are likely to affect study integrity found during the course of an inspection shall be brought to the attention of the Study Director and management immediately.
   (4) Periodically submit to management and the Study Director written status reports on each study, noting any problems and the corrective actions taken.
   (5) Determine that no deviations from approved protocols or standard operating procedures were made without proper authorization and documentation.
   (6) Review the final study report to assure that such report accurately describes the methods and standard operating procedures, and that the reported results accurately reflect the raw data of the study.
   (7) Prepare and sign a statement to be included with the final study report which shall specify the dates inspections were made and findings reported to management and to the Study Director.

(c) The responsibilities and procedures applicable to the Quality Assurance unit, the records maintained by the Quality Assurance unit, and the method of indexing such records shall be in writing and shall be maintained. These items including inspection dates, the study inspected, the phase or segment of the study inspected, and the name of the individual performing the inspection shall be made available for inspection to authorised employees or duly designated representatives of EPA or FDA.

(d) An authorised employee or a duly designated representative of EPA or FDA shall have access to the written procedures established for the inspection and may request testing facility management to certify that inspections are being implemented, performed, documented, and followed-up in accordance with this paragraph.

C — Facilities

Sec. 806.41 General.

Each testing facility shall be of suitable size and construction to facilitate the proper conduct of studies. Testing facilities which are not located within an indoor controlled environment shall be of suitable location to facilitate the proper conduct of studies. Testing facilities shall be designed so that there is a degree of separation that will prevent any function or activity from having an adverse effect on the study.

806.43 Test system care facilities.

(a) A testing facility shall have a sufficient number of animal rooms or other test system areas, as needed, to ensure: proper separation of species or test systems, isolation of individual projects, quarantine or isolation of animals or other test systems, and routine or specialised housing of animals or other test systems.

   (1) In tests with plants or aquatic animals, proper separation of species can be accomplished within a room or area by housing them separately in different chambers or aquaria. Separation of species is unnecessary where the protocol specifies the simultaneous exposure of two or more species in the same chamber, aquarium, or housing unit.

(2) Aquatic toxicity tests for individual projects shall be isolated to the extent necessary to prevent cross-contamination of different chemicals used in different tests.

(b) A testing facility shall have a number of animal rooms or other test system areas separate from those described in paragraph (a) of this section to ensure isolation of studies being done with test systems or test, control, and reference substances known to be biohazardous, including volatile substances, aerosols, radioactive materials, and infectious agents.

(c) Separate areas shall be provided, as appropriate, for the diagnosis, treatment, and control of laboratory test system diseases. These areas shall provide effective isolation for the housing of test systems either known or suspected of being diseased, or of being carriers of disease, from other test systems.

(d) Facilities shall have proper provisions for collection and disposal of contaminated water, soil, or other spent materials. When animals are housed, facilities shall exist for the collection and disposal of all animal waste and refuse or for safe sanitary storage of waste before removal from the testing facility. Disposal facilities shall be so provided and operated as to minimise vermin infestation, odors, disease hazards, and environmental contamination.

(e) Facilities shall have provisions to regulate environmental conditions (e.g., temperature, humidity, photoperiod) as specified in the protocol.

(f) For marine test organisms, an adequate supply of clean sea water or artificial sea water (prepared from deionized or distilled water and sea salt mixture) shall be available. The ranges of composition shall be as specified in the protocol.

(g) For freshwater organisms, an adequate supply of clean water of the appropriate hardness, pH, and temperature, and which is free of contaminants capable of interfering with the study, shall be available as specified in the protocol.

(h) For plants, an adequate supply of soil of the appropriate composition, as specified in the protocol, shall be available as needed.

806.45 Test system supply facilities.

(a) There shall be storage areas, as needed, for feed, nutrients, soils, bedding, supplies, and equipment. Storage areas for feed nutrients, soils, and bedding shall be separated from areas where the test systems are located and shall be protected against infestation or contamination. Perishable supplies shall be preserved by appropriate means.

(b) When appropriate, plant supply facilities shall be provided. As specified in the protocol, these include:
   (1) Facilities for holding, culturing, and maintaining algae and aquatic plants.
   (2) Facilities for plant growth, including, but not limited to, greenhouses, growth chambers, light banks, and fields.

(c) When appropriate, facilities for aquatic animal tests shall be provided. These include, but are not limited to, aquaria, holding tanks, ponds, and ancillary equipment, as specified in the protocol.

806.47 Facilities for handling test, control, and reference substances.

(a) As necessary to prevent contamination or mixups, there shall be separate areas for:
(1) Receipt and storage of the test, control, and reference substances.
(2) Mixing of the test, control, and reference substances with a carrier, e.g., feed.
(3) Storage of the test, control, and reference substance mixtures.
(b) Storage areas for test, control, and/or reference substance and for test, control, and/or reference mixtures shall be separate from areas housing the test systems and shall be adequate to preserve the identity, strength, purity, and stability of the substances and mixtures.

806.49 Laboratory operation areas.

Separate laboratory space and other space shall be provided, as needed, for the performance of the routine and specialised procedures required by studies.

806.51 Specimen and data storage facilities.

Space shall be provided for archives, limited to access by authorised personnel only, for the storage and retrieval of all raw data and specimens from completed or terminated studies.

D — Equipment

806.61 Equipment design.

Equipment used in the generation, measurement, or assessment of data and equipment used for facility environmental control shall be of appropriate design and adequate capacity to function according to the protocol and shall be suitably located for operation, inspection, cleaning, and maintenance.

806.63 Maintenance and calibration of equipment.

(a) Equipment shall be adequately inspected, cleaned, and maintained. Equipment used for the generation, measurement, or assessment of data shall be adequately tested, calibrated, and/or standardised.

(b) The written standard operating procedures ... shall set forth in sufficient detail the methods, materials, and schedules to be used in the routine inspection, cleaning, maintenance, testing, calibration, and/or standardization of equipment, and shall specify, when appropriate, remedial action to be taken in the event of failure or malfunction of equipment. The written standard operating procedures shall designate the person(s) responsible for the performance of each operation.

(c) Written records shall be maintained of all inspection, maintenance, testing, calibrating, and/or standardizing operations. These records, containing the date of the operations, shall describe whether the maintenance operations were routine and followed the written standard operating procedures. Written records shall be kept of nonroutine repairs performed on equipment as a result of failure and malfunction. Such records shall document the nature of the defect, how and when the defect was discovered, and any remedial action taken in response to the defect.

(d) The integrity of data from computers, data processors, and automated laboratory procedures involved in the collection, generation, or measurement of data shall be ensured through appropriate validation processes, maintenance procedures, disaster recovery, and security measures.

E — Testing Facilities Operation

806.81 Standard operating procedures.

(a) A testing facility shall have standard operating procedures in writing setting forth study methods that management is satisfied are adequate to ensure the quality and integrity of the data generated in the course of a study. All deviations in a study from standard operating procedures shall be authorised by the Study Director and shall be documented in the raw data. Significant changes in established standard operating procedures shall be properly authorised in writing by management.

(b) Standard operating procedures shall be established for, but not limited to, the following:

(1) Test system area preparation.
(2) Test system care.
(3) Receipt, identification, storage, handling, mixing, and method of sampling of the test, control, and reference substances.
(4) Test system observations.
(5) Laboratory or other tests.
(6) Handling of test systems found moribund or dead during study.
(7) Necropsy of test systems or postmortem examination of test systems.
(8) Collection and identification of specimens.
(9) Histopathology.
(10) Data handling, storage, and retrieval.
(11) Maintenance and calibration of equipment.
(12) Transfer, proper placement, and identification of test systems.

(c) Each laboratory or other study area shall have immediately available manuals and standard operating procedures relative to the laboratory or field procedures being performed. Published literature may be used as a supplement to standard operating procedures.

(d) A historical file of standard operating procedures, and all revisions thereof, including the dates of such revisions, shall be maintained.

806.83 Reagents and solutions.

All reagents and solutions in the laboratory areas shall be labelled to indicate identity, titer or concentration, storage requirements, and expiration date. Deteriorated or outdated reagents and solutions shall not be used. As an alternative to labelling wash bottles and transfer bottles with the expiration date, the testing facility may develop a well-documented performance standard to ensure that the reagents or solutions have not deteriorated or are outdated.

806.90 Animal and other test system care.

(a) There shall be standard operating procedures for the housing, feeding, handling, and care of animals and other test systems.

(b) All newly received test systems from outside sources shall be isolated and their health status or appropriateness for the study shall be evaluated. This evaluation shall be in accordance with acceptable veterinary medical practice or scientific methods.

(c) At the initiation of a study, test systems shall be free of any disease or condition that might interfere with the purpose or conduct of the study. If during the course of the study, the test systems contract such a disease or condition, the diseased test systems should be isolated, if necessary. These test systems may be treated for disease or signs of disease provided that such treatment does not interfere with the study. The diagnosis, authorization of treatment, description of treatment, and each date of treatment shall be documented and shall be retained.

(d) Warm-blooded animals, adult reptiles, and adult terrestrial amphibians used in laboratory procedures that require manipulations and observations over an extended period of time or in studies that require these test systems to be removed from and returned to their test system-housing units for any reason (e.g., cage cleaning, treatment, etc.), shall receive appropriate identification (e.g., tattoo, colour code, ear tag, ear punch, etc.). All information needed to specifically identify each test system within the test system-housing unit shall appear on the outside of that unit. Suckling mammals and juvenile birds are excluded from the requirement of individual identification unless otherwise specified in the protocol.

(e) Except as specified in paragraph (e)(1) of this section, test systems of different species shall be housed in separate rooms when necessary. Test systems of the same species, but used in different studies, should not ordinarily be housed in the same room when inadvertent exposure to test, control, or reference substances or test system mixup could affect the outcome of either study. If such mixed housing is necessary, adequate differentiation by space and identification shall be made.

  (1) Plants, invertebrate animals, aquatic vertebrate animals, and organisms that may be used in multispecies tests need not be housed in separate rooms, provided that they are adequately segregated to avoid mixup and cross contamination.

(f) Cages, racks, pens, enclosures, aquaria, holding tanks, ponds, growth chambers, and other holding, rearing and breeding areas, and accessory equipment, shall be cleaned and sanitised at appropriate intervals.

(g) Feed, soil, and water used for the test systems shall be analysed periodically to ensure that contaminants known to be capable of interfering with the study and reasonably expected to be present in such feed, soil, or water are not present at levels above those specified in the protocol. Documentation of such analyses shall be maintained as raw data.

(h) Bedding used in animal cages or pens shall not interfere with the purpose or conduct of the study and shall be changed as often as necessary to keep the animals dry and clean.

(i) If any pest control or cleaning materials are used, the use shall be documented. Cleaning and pest control materials that interfere with the study shall not be used.

(j) All plant and animal test systems shall be acclimatized to the environmental conditions of the test, prior to their use in a study.

F — Test, Control, and Reference Substances

806.105 Test, control, and reference substance characterisation.

(a) The identity, strength, purity, and composition, or other characteristics which will appropriately define the test, control, or reference substance shall be determined for each batch and shall be documented before its use in a study. Methods of synthesis, fabrication, or derivation of the test, control, or reference substance shall be documented by the sponsor or the testing facility, and the location of such documentation shall be specified.

(b) When relevant to the conduct of the study, the solubility of each test, control, or reference substance shall be determined by the testing facility or the sponsor before the experimental start date or concurrently according to written standard operating procedures, which provide for periodic analysis of each batch. The stability of the test, control, or reference substance shall be determined before the experimental start date or concurrently according to written standard operating procedures, which provide for periodic analysis of each batch.

(c) Each storage container for a test, control, or reference substance shall be labelled by name, Chemical Abstracts Service (CAS) registry number or code number, batch number, expiration date, if any, and storage conditions necessary to maintain the identity, strength, purity, and composition of the test, control, or reference substance. Storage containers shall be assigned to a particular test substance for the duration of the study. With the Study Director's written approval, test substance storage containers need not be retained after use, provided that full documentation of the disposition of the containers is maintained as raw data for the study. This documentation shall include:

(1) (i) Information of shipments pertaining to each container leaving the storage site (examples of such records are shipping request records, bills of lading, carrier bills, and monthly inventories of warehouse activity).
(ii) Test substance receipt records at each testing facility.
(iii) Complete use logs of material taken from containers.
(iv) A record of the final destination of the container, including the place and date of disposal or reclaiming, and any appropriate receipts.

(2) An inventory record of empty containers before disposal, including sufficient information to uniquely identify containers, maintained in an up-to-date manner recording all arrivals of empty containers and their disposal. This record shall be maintained as raw data for this study.

(3) Locations of facilities; where test substance is stored; where empty containers are stored prior to disposal; where records of use, shipment, and disposal of containers are maintained; and where the test substance is used in studies (i.e., testing facility).

(d) For studies of more than 4 weeks from the experimental start to completion dates, reserve samples from each batch of test, control, and reference substances shall be retained f...

(e) The stability of test, control, and reference substances under storage conditions at the test site shall be known for all studies.

806.107 Test, control, and reference substance handling.

Procedures shall be established for a system for the handling of the test, control, and reference substances to ensure that:

(a) There is proper storage.

(b) Distribution is made in a manner designed to preclude the possibility of contamination, deterioration, or damage.

(c) Proper identification is maintained throughout the distribution process.

(d) The receipt and distribution of each batch is documented. Such documentation shall include the date and quantity of each batch distributed or returned.

806.113 Mixtures of substances with carriers.

(a) For each test, control, or reference substance that is mixed with a carrier, tests by appropriate analytical methods shall be conducted:

(1) To determine the uniformity of the mixture and to determine, periodically, the concentration of the test, control, or reference substance in the mixture.

(2) When relevant to the conduct of the study, to determine the solubility of each test, control, or reference substance in the mixture; or if the solubility of the substance is difficult to determine, appropriate homogeneity data, by the testing facility or the sponsor before the experimental start date.

(3) To determine the stability of the test, control, or reference substance in the mixture before the experimental start date or concomitantly according to written standard operating procedures, which provide for periodic analysis of each batch.

(b) Tank mixes prepared for application to soil or plants by typical agricultural practices within a 12-hour period between preparation and application, and solutions prepared for immediate administration in mammalian acute toxicology studies, metabolism studies, or mutagenicity studies, are exempt from requirements for concentration determinations (but not from uniformity determinations) under paragraph (a)(1) of this section and are exempt from requirements for solubility determinations under paragraph (a)(2) of this section.

(c) Where any of the components of the test, control, or reference substance carrier mixture has an expiration date, that date shall be clearly shown on the container. If more than one component has an expiration date, the earliest date shall be shown.

(d) If a vehicle is used to facilitate the mixing of a test substance with a carrier, assurance shall be provided that the vehicle does not interfere with the integrity of the test.

G — Protocol for and Conduct of a Study

806.120 Protocol.

(a) Each study shall have an approved written protocol that clearly indicates the objectives and all methods for the conduct of the study. The protocol shall contain but shall not necessarily be limited to the following information:

(1) A descriptive title and statement of the purpose of the study.
(2) Identification of the test, control, and reference substance by name, Chemical Abstracts Service (CAS) registry number or code number. When a reference substance for a metabolite cannot be identified prior to the beginning of a study (only in the case of metabolism studies), it is not necessary to identify the substance in the protocol. However, a statement must be included that the identity of the reference substance will be determined during the course of the study and maintained as raw data.
(3) The name and address of the sponsor and the name and address of the testing facility at which the study is being conducted.
(4) The proposed experimental start and termination dates.
(5) Justification for selection of the test system.
(6) Where applicable, the number, body weight range, sex, source of supply, species, strain, substrain, and age of the test system.

(7) The procedure for identification of the test system.

(8) A description of the experimental design, including methods for the control of bias.

(9) Where applicable, a description and/or identification of the diet used in the study as well as solvents, emulsifiers and/or other materials used to solubilize or suspend the test, control, or reference substances before mixing with the carrier. The description shall include specifications for acceptable levels of contaminants that are reasonably expected to be present in the dietary materials and are known to be capable of interfering with the purpose or conduct of the study if present at levels greater than established by the specifications.

(10) The route of administration and the reason for its choice.

(11) Each dosage level, expressed in milligrams per kilogram of body or test system weight or other appropriate units, of the test, control, or reference substance to be administered and the method and frequency of administration.

(12) The type and frequency of tests, analyses, and measurements to be made.

(13) The records to be maintained.

(14) The date of approval of the protocol by the sponsor and the dated signature of the Study Director.

(15) A statement of the statistical method to be used.

(b) All changes in or revisions of an approved protocol and the reasons therefore shall be documented, signed by the Study Director, dated, and maintained with the protocol.

(c) Discontinued studies or studies otherwise terminated before completion shall be finalised by writing a protocol amendment providing the reason(s) for termination. All documentation for terminated studies including the protocol, protocol amendment(s), and raw data, if collected, shall be retained as provided at Sec. 806.195.

806.130 Conduct of a study.

(a) The study shall be conducted in accordance with the protocol.

(b) The test systems shall be monitored in conformity with the protocol.

(c) Specimens shall be identified by test system, study, nature, and date of collection. This information shall be located on the specimen container or shall accompany the specimen in a manner that precludes error in the recording and storage of data.

(d) In animal studies where histopathology is required, records of gross findings for a specimen from postmortem observations shall be available to a pathologist when examining that specimen histopathologically.

(e) All data generated during the conduct of a study, except those that are generated by automated data collection systems, shall be recorded directly, promptly, and legibly in ink. All data entries shall be dated on the day of entry and signed or initialled by the person entering the data. Any change in entries shall be made so as not to obscure the original entry, shall indicate the reason for such change, and shall be dated and signed or identified at the time of the change. In automated data collection systems, the individual responsible for direct data input shall be identified at the time of data input. Any change in automated data entries shall be made so as not to obscure the original entry, shall indicate the reason for change, shall be dated, and the responsible individual shall be identified.

806.135 Physical and chemical characterisation studies.

(a) All provisions of the GLPS shall apply to physical and chemical characterisation studies designed to determine stability, solubility, octanol water partition coefficient, volatility, and persistence (such as biodegradation, photodegradation, and chemical degradation studies) of test, control, or reference substances.

(b) ...

H and I — [Reserved]

J — Records and Reports

806.185 Reporting of study results.

(a) With the exception of discontinued or otherwise terminated studies, as provided at Sec. 806.120(c), a final report shall be prepared for each study and shall include, but not necessarily be limited to, the following:

Appendix III: US-EPA Draft GLP Regulations

(1) Name and address of the facility performing the study and the dates on which the study was initiated and was completed.
(2) Objectives and procedures stated in the approved protocol, including any changes in the original protocol.
(3) Statistical methods employed for analysing the data.
(4) The test, control, and reference substances identified by name, Chemical Abstracts Service (CAS) registry number or code number, strength, purity, and composition, or other appropriate characteristics.
(5) Stability and, when relevant to the conduct of the study, solubility of the test, control, and reference substances under the conditions of administration.
(6) A description of the methods used.
(7) A description of the test system used. Where applicable, the final report shall include the number of animals used, sex, body weight range, source of supply, species, strain and substrain, age, and procedure used for identification. For other test organisms (plants, bacteria), similarly detailed descriptions of the test system are required.
(8) A description of the dosage, dosage regimen, route of administration, and duration.
(9) A description of all circumstances that may have affected the quality or integrity of the data.
(10) The name of the Study Director, the names of other scientists or professionals, and the names of all supervisory personnel, involved in the study.
(11) A description of the transformations, calculations, or operations performed on the data, a summary and analysis of the data, and a statement of the conclusions drawn from the analysis.
(12) The signed and dated reports of each of the individual scientists or other professionals involved in the study, including each person who, at the request or direction of the testing facility or sponsor, conducted an analysis or evaluation of data or specimens from the study after data generation was completed.
(13) The locations where all specimens, raw data, and the final report are to be stored.
(14) The statement prepared and signed by the Quality Assurance unit as described in Sec. 806.35(b)(7).

(b) The final report shall be signed and dated by the Study Director.
(c) Corrections or additions to a final report shall be in the form of an amendment by the Study Director. The amendment shall clearly identify that part of the final report that is being added to or corrected and the reasons for

the correction or addition, and shall be signed and dated by the person responsible. Modification of a final report to comply with the submission requirements of EPA does not constitute a correction, addition, or amendment to a final report.

(d) A copy of the final report and of any amendment to it shall be maintained by the sponsor and the test facility.

806.190 Storage and retrieval of records and data.

(a) All raw data, documentation, records, protocols, specimens, and final reports generated as a result of a study shall be retained. Specimens obtained from mutagenicity tests, specimens of soil, water, and plants, and wet specimens of blood, urine, feces, and biological fluids, do not need to be retained after Quality Assurance verification. Correspondence and other documents relating to interpretation and evaluation of data, other than those documents contained in the final report, also shall be retained.

(b) There shall be archives for orderly storage and expedient retrieval of all raw data, documentation, protocols, specimens, and interim and final reports. Conditions of storage shall minimise deterioration of the documents or specimens in accordance with the requirements for the time period of their retention and the nature of the documents of specimens. A testing facility may contract with commercial archives to provide a repository for all material to be retained. Raw data and specimens may be retained elsewhere provided that the archives have specific reference to those other locations.

(c) An individual shall be identified as responsible for the archives.

(d) Only authorised personnel shall enter the archives.

(e) Material retained or referred to in the archives shall be indexed to permit expedient retrieval.

806.195 Retention of records.

(a) ...
(b) ...
(c) Wet specimens, samples of test, control, or reference substances, and specially prepared material which are relatively fragile and differ markedly in sta-

bility and quality during storage, shall be retained only as long as the quality of the preparation affords evaluation. Specimens obtained from mutagenicity tests, specimens of soil, water, and plants, and wet specimens of blood, urine, feces, and biological fluids, do not need to be retained after Quality Assurance verification. ...

(d) The master schedule sheet, copies of protocols, and records of Quality Assurance inspections ... shall be maintained by the Quality Assurance unit as an easily accessible system of records ...

(e) Summaries of training and experience and job descriptions ... may be retained along with all other testing facility employment records ...

(f) Records and reports of the maintenance and calibration and inspection of equipment ... shall be retained ...

(g) If a facility conducting testing or an archive contracting facility goes out of business, all raw data, documentation, and other material specified in this section shall be transferred to the archives of the sponsor of the study. EPA shall be notified in writing of such a transfer.

(h) Specimens, samples, or other non-documentary materials need not be retained after EPA has notified in writing the sponsor or testing facility holding the materials that retention is no longer required by EPA. Such notification normally will be furnished upon request after EPA or FDA has completed an audit of the particular study to which the materials relate and EPA has concluded that the study was conducted in accordance with this part.

(i) Records required by this part may be retained either as original records or as true copies such as photocopies, microfilm, microfiche, or other accurate reproductions of the original records.

Appendix II.IV

Excerpts from the
United States Food and Drug Administration
21 Code of Federal Regulations, Part 11
Electronic Records; Electronic Signatures

A — General Provisions

§ 11.1 Scope.

(a) The regulations in this part set forth the criteria under which the agency considers electronic records, electronic signatures, and handwritten signatures executed to electronic records to be trustworthy, reliable, and generally equivalent to paper records and handwritten signatures executed on paper.

(b) This part applies to records in electronic form that are created, modified, maintained, archived, retrieved, or transmitted, under any records requirements set forth in agency regulations. ... However, this part does not apply to paper records that are, or have been, transmitted by electronic means.

(c) ...

(d) ...

(e) Computer systems (including hardware and software), controls, and attendant documentation maintained under this part shall be readily available for, and subject to, FDA inspection.

Appendix IV: US-FDA "Rule 11"

§ 11.2 Implementation.

(a) ...

(b) ...

§ 11.3 Definitions.

(a) The definitions and interpretations of terms contained in section 201 of the act apply to those terms when used in this part.

(b) The following definitions of terms also apply to this part:
   (1) Act means the Federal Food, Drug, and Cosmetic Act.
   (2) Agency means the Food and Drug Administration.
   (3) Biometrics means a method of verifying an individual's identity based on measurement of the individual's physical feature(s) or repeatable action(s) where those features and/or actions are both unique to that individual and measurable.
   (4) Closed system means an environment in which system access is controlled by persons who are responsible for the content of electronic records that are on the system.
   (5) Digital signature means an electronic signature based upon cryptographic methods of originator authentication, computed by using a set of rules and a set of parameters such that the identity of the signer and the integrity of the data can be verified.
   (6) Electronic record means any combination of text, graphics, data, audio, pictorial, or other information representation in digital form that is created, modified, maintained, archived, retrieved, or distributed by a computer system.
   (7) Electronic signature means a computer data compilation of any symbol or series of symbols executed, adopted, or authorised by an individual to be the legally binding equivalent of the individual's handwritten signature.
   (8) Handwritten signature means the scripted name or legal mark of an individual handwritten by that individual and executed or adopted with the present intention to authenticate a writing in a permanent

form. The act of signing with a writing or marking instrument such as a pen or stylus is preserved. The scripted name or legal mark, while conventionally applied to paper, may also be applied to other devices that capture the name or mark.

(9) Open system means an environment in which system access is not controlled by persons who are responsible for the content of electronic records that are on the system.

B — Electronic Records

§ 11.10 Controls for closed systems.

Persons who use closed systems to create, modify, maintain, or transmit electronic records shall employ procedures and controls designed to ensure the authenticity, integrity, and, when appropriate, the confidentiality of electronic records, and to ensure that the signer cannot readily repudiate the signed record as not genuine. Such procedures and controls shall include the following:

(a) Validation of systems to ensure accuracy, reliability, consistent intended performance, and the ability to discern invalid or altered records.

(b) The ability to generate accurate and complete copies of records in both human readable and electronic form suitable for inspection, review, and copying by the agency. Persons should contact the agency if there are any questions regarding the ability of the agency to perform such review and copying of the electronic records.

(c) Protection of records to enable their accurate and ready retrieval throughout the records retention period.

(d) Limiting system access to authorised individuals.

(e) Use of secure, computer-generated, time-stamped audit trails to independently record the date and time of operator entries and actions that create, modify, or delete electronic records. Record changes shall not obscure previously recorded information. Such audit trail documentation shall be retained for a period at least as long as that required for the subject electronic records and shall be available for agency review and copying.

(f) Use of operational system checks to enforce permitted sequencing of steps and events, as appropriate.

(g) Use of authority checks to ensure that only authorised individuals can use the system, electronically sign a record, access the operation or computer system input or output device, alter a record, or perform the operation at hand.

(h) Use of device (e.g., terminal) checks to determine, as appropriate, the validity of the source of data input or operational instruction.

(i) Determination that persons who develop, maintain, or use electronic record/electronic signature systems have the education, training, and experience to perform their assigned tasks.

(j) The establishment of, and adherence to, written policies that hold individuals accountable and responsible for actions initiated under their electronic signatures, in order to deter record and signature falsification.

(k) Use of appropriate controls over systems documentation including:

(1) Adequate controls over the distribution of, access to, and use of documentation for system operation and maintenance.

(2) Revision and change control procedures to maintain an audit trail that documents time-sequenced development and modification of systems documentation.

§ 11.30 Controls for open systems.

Persons who use open systems to create, modify, maintain, or transmit electronic records shall employ procedures and controls designed to ensure the authenticity, integrity, and, as appropriate, the confidentiality of electronic records from the point of their creation to the point of their receipt. Such procedures and controls shall include those identified in § 11.10, as appropriate, and additional measures such as document encryption and use of appropriate digital signature standards to ensure, as necessary under the circumstances, record authenticity, integrity, and confidentiality.

§ 11.50 Signature manifestations.

(a) Signed electronic records shall contain information associated with the signing that clearly indicates all of the following:

    (1) The printed name of the signer;

    (2) The date and time when the signature was executed; and

(3) The meaning (such as review, approval, responsibility, or authorship) associated with the signature.

(b) The items identified in paragraphs (a)(1), (a)(2), and (a)(3) of this section shall be subject to the same controls as for electronic records and shall be included as part of any human readable form of the electronic record (such as electronic display or printout).

§ 11.70 Signature/record linking.

Electronic signatures and handwritten signatures executed to electronic records shall be linked to their respective electronic records to ensure that the signatures cannot be excised, copied, or otherwise transferred to falsify an electronic record by ordinary means.

C—Electronic Signatures

§ 11.100 General requirements.

(a) Each electronic signature shall be unique to one individual and shall not be reused by, or reassigned to, anyone else.

(b) Before an organisation establishes, assigns, certifies, or otherwise sanctions an individual's electronic signature, or any element of such electronic signature, the organisation shall verify the identity of the individual.

(c) Persons using electronic signatures shall, prior to or at the time of such use, certify to the agency that the electronic signatures in their system, used on or after August 20, 1997, are intended to be the legally binding equivalent of traditional handwritten signatures.

(1) The certification shall be submitted in paper form and signed with a traditional handwritten signature, to the Office of Regional Operations (HFC-100), 5600 Fishers Lane, Rockville, MD 20857.

(2) Persons using electronic signatures shall, upon agency request, provide additional certification or testimony that a specific electronic signature is the legally binding equivalent of the signer's handwritten signature.

§ 11.200 Electronic signature components and controls.

(a) Electronic signatures that are not based upon biometrics shall:
   (1) Employ at least two distinct identification components such as an identification code and password.
      (i) When an individual executes a series of signings during a single, continuous period of controlled system access, the first signing shall be executed using all electronic signature components; subsequent signings shall be executed using at least one electronic signature component that is only executable by, and designed to be used only by, the individual.
      (ii) When an individual executes one or more signings not performed during a single, continuous period of controlled system access, each signing shall be executed using all of the electronic signature components.
   (2) Be used only by their genuine owners; and
   (3) Be administered and executed to ensure that attempted use of an individual's electronic signature by anyone other than its genuine owner requires collaboration of two or more individuals.
(b) Electronic signatures based upon biometrics shall be designed to ensure that they cannot be used by anyone other than their genuine owners.

§ 11.300 Controls for identification codes/passwords.

Persons who use electronic signatures based upon use of identification codes in combination with passwords shall employ controls to ensure their security and integrity. Such controls shall include:

(a) Maintaining the uniqueness of each combined identification code and password, such that no two individuals have the same combination of identification code and password.
(b) Ensuring that identification code and password issuances are periodically checked, recalled, or revised (e.g., to cover such events as password aging).
(c) Following loss management procedures to electronically deauthorize lost, stolen, missing, or otherwise potentially compromised tokens, cards, and other devices that bear or generate identification code or password information, and to issue temporary or permanent replacements using suitable, rigorous controls.

(d) Use of transaction safeguards to prevent unauthorized use of passwords and/or identification codes, and to detect and report in an immediate and urgent manner any attempts at their unauthorized use to the system security unit, and, as appropriate, to organisational management.

(e) Initial and periodic testing of devices, such as tokens or cards, that bear or generate identification code or password information to ensure that they function properly and have not been altered in an unauthorized manner.

III. How can Good Laboratory Practice be Introduced in a Test Facility?

1. Introduction

When the necessity arises for a test facility to introduce a formal quality system, a number of questions have to be answered before a reasonable decision about the most appropriate system can be made. One of the foremost considerations in this respect will be the nature of the studies which are to be conducted under this quality system, coupled with the question of the quality target. If this target consists of convincing the future sponsors of the precision, reproducibility and general quality of the data generated by the test facility, then the implementation of an ISO- or accreditation-based system could be better suited to the needs of this facility. If, on the other hand, the studies do require the conduct under the rules of GLP because they may be considered as safety-related and apt to be submitted to a Regulatory Authority, then this test facility would have no other choice than to adopt the GLP Principles as its quality system. The decision to implement the GLP Principles in the test facility may also be influenced by other considerations, of which the wishes of, or requests from, the prospective sponsors form an economically very important part. Additionally, the attitude of the national compliance monitoring authority with regard to the applicability of GLP in borderline cases might also be taken into account, and a discussion with the relevant authority about the possibility of being entered in the national monitoring program is certainly to be advised.

These aspects have already be discussed exhaustively in section 4 of the first part (see page 19) and need therefore not to be repeated here. In the following sections it is thus assumed that these primary questions have been answered in the affirmative, i.e. that it has been determined that it is Good Laboratory Practice which is needed by the test facility.

2. General Aspects

The successful implementation of GLP in a test facility "from scratch" is a labour-intensive, time and resource consuming activity. Therefore, the careful and well considered preparation of the necessary steps to be taken will be extremely important. An absolute prerequisite is hereby the total commitment of the test facility management to bring this work to a successful end, and only through its full engagement, combined with the absolute conviction about the value of GLP, can this goal be achieved. It should be obvious that experts should be consulted early in this process, as they will be instrumental in planning the introduction and implementation of GLP in a logical way. Thus, the setting-up of the Quality Assurance, although possibly in an early, skeletal form only, will be one of the very first activities. Management has to remind itself at all times that the effort necessary to introduce GLP does not come cheap and more or less by itself. Adequate resources in terms of finances and personnel have to be provided, on the one hand for dealing with the many different tasks with respect to structures, documents and facilities, as well as on the other for the instruction and training of the test facility personnel. Implementation of GLP calls thus in the first instance for a concentrated management effort, addressing these very questions and points, and management might therefore be well advised to formulate an implementation plan in a detailed policy document.

It will be of great advantage to the orderly and as smooth as possible introduction of GLP, if management would, at this early stage already, acquaint itself thoroughly with GLP through printed information, through attendance at meetings, seminars and workshops dealing with this theme, as well as through personal relations and discussions with management from GLP-compliant test facilities. Also those individuals selected for specific responsibilities in the implementation proces and under the future regime of GLP should be encouraged to undertake such educational efforts.

According to the situation two different approaches may be possible. A "big bang" approach might be necessary in certain instances, e.g. when it is planned to establish a completely new test facility, which should be able to conduct studies in a GLP compliant way from the very start of its operations. On the other hand, a step-wise introduction of GLP could be preferable in many other situations, especially when an existing test facility would feel the need to comply with GLP. In the former instance, extensive planning is necessary anyway for the establishment of the test facility, and the simultaneous introduction of GLP would just call for an additional planning segment. The situation will be different, however, for a facility that is already active and the operations of which should not be disturbed too much by the introduction of GLP concomitantly with the still on-going daily activities. The latter approach of a step-wise implementation could in such cases certainly relieve some of the pressure exerted on management and the test facility as a whole, especially if pre-existing documents and models could be adapted to suit the purposes of GLP. To this end it is important that, at the very beginning, a test facility take stock of what is already available, what could be used in an adapted way, and what needs to be generated *de novo*. This inventory would have to involve not only the question of documents to be adapted or generated, but would as well have to address the necessary adaptations in facility allocations and use.

The least difficult task would probably be the description of the organisational structure of the test facility together with the definition of the test facility management. CVs and job descriptions of employees will probably be already available in the personnel files, and they would possibly need only slight adaptations to the GLP requirements, e.g. their conversion to a general format, or a constant up-dating with training records. Also the required apparatus calibration and maintenance records could be developed from existing log-books. The most time consuming effort will certainly be the development of the necessary SOPs. Only a part of them, especially the ones related to test systems and study conduct, might be adapted from pre-existing documents like descriptions of experimental methods used in the test facility. Therefore, it should be advisable to define firstly the experimental activities performed at the test facility, to identify subsequently the areas where SOPs should be necessary, and to draw up finally a tentative list of these SOPs. This list will then enable the identification of individuals best suited to tackle the writing task. The expenditures in time and manpower necessary to achieve a more or less coherent set of SOPs should not, however, be underestimated, nor should the intellectual efforts to realise this objective be belittled.

In another activity lane the facilities, rooms and areas available at the test facility should be investigated with regard to their suitability for the GLP compliant conduct of studies and for the various ancillary purposes in the context of GLP. The allocation of rooms and areas for the various activities may need to be changed from their actual use depending on the possibilities of fulfilling the requirements of GLP which, in turn, may again influence the ways in which the different activities are organised with respect to each other. Therefore it is not only the facilities themselves which need to be considered at this point, but in a concomitant way the processes running at the test facility will have to be scrutinised for their need to be redefined and adapted to the changed conditions.

While the previously described tasks can be considered as rather straightforward, the issue of the suitability determination of apparatus, equipment and computerised systems may involve complex investigations, inquiries at the manufacturer or vendor, extensive acceptance testing and validations and/or vendor audits in order to render these systems GLP compliant. Most certainly the amount of work involved in these aspects will necessitate the prioritisation of the various systems in use at the test facility. In the assumed case of a test facility having been in operation for some time already, albeit not under GLP, these apparatus and systems may be credited to a certain extent with the assumption of suitability for their purposes. In a first round, therefore, only the relevant documentation already available on their performance need be collected, while a formal retrospective evaluation and acceptance testing may be deferred to a later time point. The policy document of test facility management dealing with the time plan for the introduction of GLP should include therefore also a timetable for such further activities to be performed after the successful implementation of GLP.

Last but not least, one of the most important, but sometimes a little bit neglected, aspects in the implementation of GLP concerns the instruction, education and training of personnel. This does not only involve technicians, laboratory workers, animal caretakers or field hands, but also – and this has to be especially emphasised – the prospective Study Directors. Only if all individuals in the test facility can be considered to be on an equivalent level of theoretical and practical knowledge with regard to the application of the GLP rules can the test facility be expected to work in a perfectly compliant way.

When all these issues have been addressed, all these documents have been produced, all these processes have been defined and all these activities have been concluded, then GLP is by no means already and finally implemented! As the last step in the introduction of GLP into the operations of a test facility, there has to be a run-in period, in which two to four studies will have to be conducted to the full extent of the GLP requirements. It certainly may be a good idea, already in the preliminary stages of GLP introduction, to perform studies according to the GLP rules available at these time points in order to acquaint and familiarise the test facility personnel with the new working conditions. The "proof of the pudding" lies, however, in the execution of a number of studies in a practically faultless, GLP compliant way. Then, and only then, will the test facility be able to claim GLP compliance, and indeed only then, a national compliance monitoring authority will consider the request for inclusion of the test facility in the national monitoring programme.

3. A General Way to Implementation

In this last section, a step-by-step approach will be described that may be utilised by any test facility wishing to introduce the Principles of Good Laboratory Practice. Although presented in a somewhat general way it should be applicable to test facilities of all denominations.

3.1 The preliminaries

Test facility management has decided that Good Laboratory Practice is really the quality system it needs to introduce for increasing or at least continuing the economic success of its operations, or to improve the quality of its scientific work. This decision will form the basis of the next steps which have to be taken by management. First of all, in deciding to introduce GLP, test facility management will have to define itself in order to fulfil the very first requirement of the GLP Principles, namely to "*ensure that a statement exists which identifies the individual(s) within a test facility who fulfil the responsibilities of management as defined by these Principles of Good Laboratory Practice*". At this stage, a single individual should be nominated to be responsible for the whole process, and a Quality Assurance expert should be called in.

> The decision to introduce GLP will be formulated in a policy document which identifies the individuals who will act as test facility management under the GLP regulations. The advice of a Quality Assurance expert will also have to be sought already at this stage of GLP implementation.

3.2 *The organisation*

The first activity will then consist of scrutinising the present organisation of the test facility in order to determine whether it would have to be adapted or changed to satisfy the spirit of GLP and its requirements. It has to be kept in mind that a clear separation of GLP- and non-GLP-activities throughout the test facility will greatly facilitate the adherence to the GLP Principles. Therefore, if the test facility is at present organised in a way as shown in figure 32, where it is not readily discernible in which parts GLP will have to be followed, or where areas and sites that have to be compliant with GLP are interspersed with those that don't, then a re-organisation along the lines suggested in figure 33 would be advisable, if not outright necessary. At this point those individuals among the personnel who will work under GLP may be already designated, with special emphasis on the designation of the future Study Directors.

> The organisation of the test facility has to be adapted so as to clearly separate the organisational units under GLP from those which do not need to comply with these Principles. Additionally the designation of the future Study Director(s) will be an important step, since this (these) individual(s) will be instrumental in the further implementation steps.

3.3 *Separation and distribution of facilities and equipment*

Now the time has come to take stock of the activities, study types and test systems which are envisaged to come under GLP. The organisational separation which has taken place has now to be translated into the physical separation of sites, areas, rooms, laboratories, greenhouses or field plots, and the concomitant allocation of the necessary equipment to these GLP sites. The foremost consideration in this respect should not be the organisational ease with which the separation may be pulled through, but the GLP requirement of "suitability". For each activity, test system and study type the most suitable places, areas or rooms have to be singled out. Of course, if the maximal suitability concept should yield a distribution of rooms and areas that would be impractical to a large extent, e.g. if the GLP areas were isolated from each

3.3 A Way to Implementation

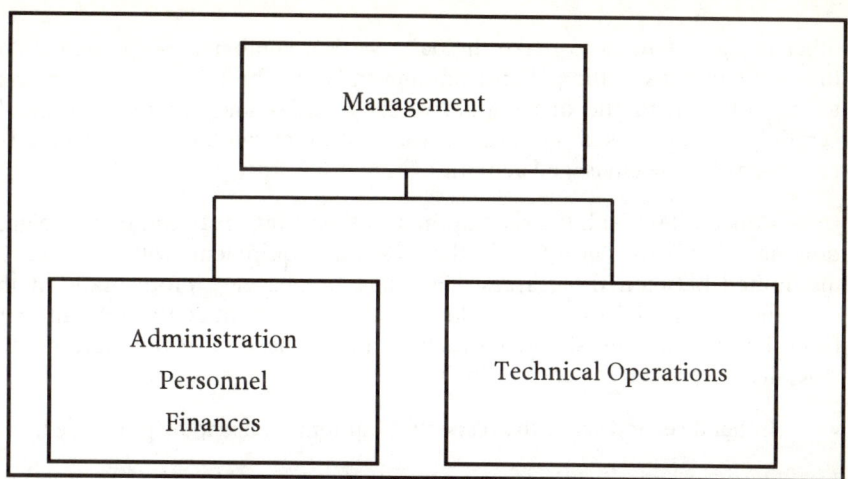

Figure 32: Model of the organisation chart of a company without GLP; included in the department "Technical Operations" may be any research, development, validation and testing activities.

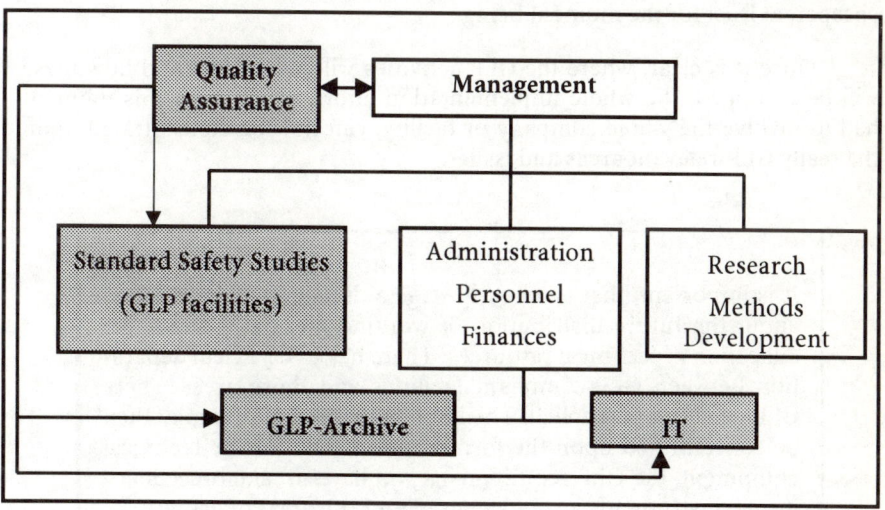

Figure 33: The same company after reorganisation to achieve separation of GLP- and Non-GLP areas

other to the extent as to make the daily work a cumbersome, perpetual and time-consuming shuttling affair from one end of the building or company plot to the other, then the distribution of these areas and laboratories should certainly be optimised in such a way as to allow the most favourable conditions for the envisaged activities.

Concomitant with the separation of the working areas and their designation as "GLP" or "non-GLP", the relevant equipment will have to be distributed between these areas. This may not be very problematic, if the activities under GLP were to employ a set of equipment completely different from the one the non-GLP activities would be in need of. If this were not the case, then two solutions are possible:

- Either a second set of the respective equipment has to be purchased, or
- the respective instrument or apparatus has to be placed under the regime of GLP, irrespective of the ratio of GLP *versus* non-GLP work conducted with it.

The former solution might have the advantage that the looming problem of having to validate retrospectively the respective instruments might be defused or at least alleviated, while of course the latter option might be cheaper, at least for the moment being.

Once it is clear, where the GLP activities will take place, and how these will be equipped, the whole implementation effort, which until this moment had to involve the whole company or facility, can now be concentrated upon the really GLP-relevant areas and issues.

> It is important, that at an early stage a decision will be reached about the future distribution of working areas and about the allocation of technical resources. There has to be a clear separation between GLP-compliant facilities and those areas where GLP need not be applied. The implementation efforts may then be concentrated upon the former ones, while for the technical equipment, the GLP requirements will have to take precedence if such instruments were to be used for both areas of activity.

3.4 Interlude: Personnel documentation

At the same time when the facilities and the technical resources are being distributed between GLP-compliant facilities and "common" areas, the need to attribute also the personnel to these two different sectors becomes apparent. With the assumption of work under GLP the respective personnel documentation will have to suffice also the GLP requirements. This will, as has been already mentioned, not pose major problems, since the relevant information will certainly be available to a great extent in the files of the personnel administration. It will probably be just a question of whether all the personnel documentation should be brought to the format required by GLP, or not; as this would not involve any major additional efforts, the price would possibly be well worth to pay, regarding the future possibilities of shifting persons from non-GLP to GLP areas. It should be easy to devise a common template for the curricula vitae, as well as for the relevant job descriptions. These templates may then be filled in with information that most probably will already be available at the personnel administration office. The main effort in this respect would probably be to bring the training records of the various people up-to-date. GLP requires that personnel should be qualified for performing the respective functions within a GLP study, and it requires from management that records of "*qualifications, training, experience and job description for each professional and technical individual*" should be present at the test facility. What is frequently forgotten to include in these records is an indication of the training the individual has received with regard to the GLP Principles and their application to the actual working environment.

> One of the various templates that will have to be created in the course of the implementation of GLP in a test facility are the templates for the personnel. While curricula vitae may differ in format (and anyway in content), the job descriptions should obtain a common format. Furthermore, training record forms have to be created, which will provide evidence for the fact that training of the individual has been up-to-date, not only in respect of his or her technical abilities, but also with regard to the training in, and understanding of, GLP matters-in general and especially those relevant to the actual work to be performed.

3.5 Distributing Responsibilities

Under GLP a number of special positions with special responsibilities have to be created. Although some of these positions may already be present in a test facility, they merit very thorough reflection and consideration at this point. The responsibility of test facility management to assume the leading role in the implementation of GLP has already been described, as has the early appointment of a Quality Assurance manager or expert. There are the prospective Study Directors to be chosen, since they will have their central responsibility in the conduct of the studies, and thus they will have to have time to acquaint themselves with their new roles and responsibilities.

It should not be forgotten, however, that GLP requires management to name a person as the responsible for the archives. Archiving under non-GLP conditions may have been a haphazard affair, with every laboratory head having his or her own cherished system for archiving, maybe even retaining data and records in the laboratory or in the personal files at whim. Therefore the organisation of a GLP archive has to be started also at an early stage of GLP implementation. But even when an archiving system has been in operation it must be checked for GLP compliance and for the future operability under GLP. The individual nominated to become responsible for the GLP archives would most certainly be the very person to do this job.

Another organisational task will be the development of the test and reference item handling and accounting system. Here, too, it should become the task of the individual chosen to become responsible for this activity under the future GLP conditions of developing this system in a coherent way. Therefore, it should be advantageous also to nominate this person at a relatively early stage.

> The implementation of GLP has to be connected very early on with the distribution of the various responsibilities. Certain tasks require somebody to be in absolute control over their initiation, development, implementation and conclusion. Where the (external or part-time) Quality Assurance expert may not provide for sufficient co-ordination between the various issues to be tackled, then somebody should be nominated for this function, and should be made responsible for the orderly development of GLP implementation.

3.6   The Major Task: Standard Operating Procedures

Well prepared with lists of study types, test systems, and apparatus to be used in the future GLP compliant test facility, the task of writing the various SOPs can now be tackled in earnest.

The first effort in this will certainly be the development of a standard format for the SOPs to be written. The Quality Assurance expert, the consulting of whom has already been described as instrumental for the success of the whole operation, will certainly provide some proven ideas about this matter, and an "SOP on SOPs" will consequently be developed in the first instance. This general template for SOPs will consequently be utilised in the generation of the further SOPs, which will have to be written by those individuals who are most familiar with the respective subjects. As a guidance for the topics to be covered, the OECD Principles present a general list of areas where SOPs will be needed. This list will have to be adapted to the actual necessities of the test facility, but it will be a valuable guide in the multiple decisions for the preparation of the actual SOPs.

It might be easiest to start with the SOPs in the area of apparatus, instruments and equipment. On the one hand, the manuals and directions for use which are normally provided by the manufacturer or the vendor can either be utilised as templates for the description of the standard way to use them, or they may even be just appended to the cover page of the respective SOP and serve as such. Also for the maintenance, cleaning and calibration the respective procedures may already be described, or even prescribed, in these manuals. On the other hand the limited number of apparatus present in a test facility will make the task a relatively easily overseeable one. A further advantage can be seen in the fact that - at least for the bigger and more expensive instruments - some kind of list or register would possibly already be available, thus forestalling the need to draw up such a list anew. In preparation for the task of writing the set of "apparatus SOPs" it should furthermore be considered, whether every single piece of equipment would need an individual SOP, or whether similar, or identical, instruments could be covered by one single SOP, valid for the entire test facility, even if used in different departments or laboratories.

The computerised systems will certainly occupy a major part of the "brain power" for the implementation of GLP also in the SOP area. Since it may not be assumed that in the "era before GLP" very extensive thoughts had been spent on the validation of computerised systems, especially those "hidden" in purchased "standard" apparatus, the compilation of a list of apparatus containing electronic devices and functionalities, and their prioritisation in terms of validation necessity should also be contemplated at this point.

Turning back to the creation of SOPs, a major effort would probably also be required for the compilation and writing of study-related SOPs, those governing the conduct of the studies which are to be performed under GLP at the test facility. In this regard, it is very important to have an absolutely clear idea or concept of how the test facility's studies are being conducted. To this end, the drawing-up of a flow chart of every single study type the test facility is going to place under GLP is certainly of advantage. In doing this, a number of areas, activities and processes in need of Standard Operating Procedures will become apparent. In looking at the flow chart of a field test, it can be easily seen that a number of areas, which the analytical scientist - living in the laboratory under, and being used to, controlled conditions - might not think of, will have to be regulated in order to attain full GLP compliance. As an example, the general flow chart of a field test is presented in figure 34 in which the different areas in need of SOPs can then be entered.

There are different ways to come to grips with the problem of complete - or as complete as "complete" can be - SOP coverage. The solitary thinker might lock himself up and devise logically the way certain studies are being conducted and the areas where specific SOPs would be needed. More sociable people might resort to the brain-storming exercise and try in this way to arrive at a compilation of SOPs which should deal with each and every activity imaginable. Another possibility is to contact colleagues from test facilities and trying to obtain a copy of their SOP list, thus enabling the "rookies" to check their own ideas against solutions which have already been proven useful.

While the organisational aspects in the compilation of SOP themes and SOP lists are important for the full extent of topics to be covered, the actual writing of these documents poses another challenge. The individuals writing the SOPs will certainly, because of their expertise in the fields and areas to be described, make sure that their SOPs will be, as required by the GLP Principles "technically valid". Whether they are intelligible and utilisable, however, remains to be determined in practice. Therefore, any SOP should be tested in the daily life of the laboratory before approval. The test should involve the use of the respective SOP as a working guideline by persons other than the author, and it will thus provide the opportunity for a comparison of the descriptions presented with the possibly engrained, customary way of performing the so

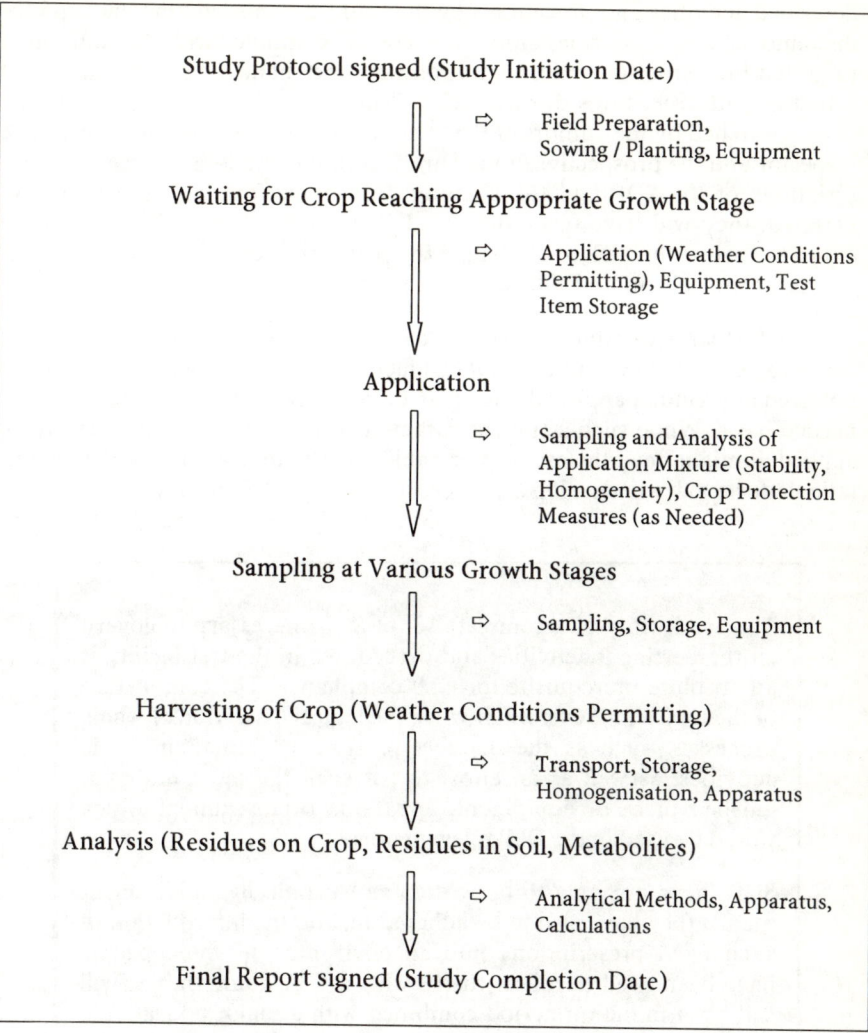

Figure 34: General flow chart of a field study, showing the various phases of the experimental and "maintenance" work to be performed.

described activities and procedures by the "old lab hands" on the one, and for the removal of redundancies or too much detail, combined with the addition of forgotten but important information on the other hand. The performance of activities and procedures during such a "run-in" period for SOPs should be closely watched by the authors of the SOPs as well as by the Quality Assurance inspector and the prospective Study Director(s). If deviations between the prescriptions of the SOP and the common way of performing an activity are detected, they will have to be discussed immediately, since this is the time either to correct the SOP or to change the performance of the activity (whichever is "technically more valid" or easier to achieve or enforce).

A further area which has to be dealt is the generation of all sorts of forms to be used in various areas of test facility activities. Since these forms do not need any kind of approval, they may be composed and produced on an as-needed basis. While such a (maybe rather chaotic) approach can certainly be applied, it might nevertheless be preferable for the "first round" of implementation efforts to try to develop also these forms in a coherent way.

> The availability of a complete set of SOPs necessary to govern all the pertinent activities and procedures in the test facility is an absolute prerequisite for GLP compliance. The compilation of the topics for SOPs will involve the logical dissection of whole processes, such as the conduct of studies, into their single activities, as well as an effort to list even the most undistinguished piece of equipment, apparatus or instrument which would be used in the GLP relevant areas.
>
> SOPs have not only to be written as "technically valid" documents, they have also to be adhered to, and the introduction of such fixed prescriptions into an environment, where people have been used to follow their own ways of doing things, will call for a initiation period combined with a major educational initiative.

3.7 Second Interlude: Quality Assurance and IT

It may wondered at the lack of detailed treatment of the Quality Assurance and IT issues here. This is not to say that they lack importance, but it has already been mentionned before that for these two ares, specialists should be employed, who may be found either externally, or even within the test facility. Since it has to be avoided that "the blind would be leading the blind" these individuals should have extensive knowledge and expertise in these two areas, they would not need much guidance from this text.

Suffice it to say here, that the compiling of the Quality Assurance Programme with the respective SOPs, and the validation policies and validation efforts for the IT applications should certainly be one, preferably two, steps ahead of the other implementation efforts.

3.8 The Personnel: Education and Training

Already in the above section the importance of education and training has been mentioned, in this case with regard to respecting the prescriptions of SOPs. It will be important, however, that all other aspects of how to work in compliance with GLP are looked at early on, since to effect a change in "cherished habits" may be rather difficult for some people.

General training sessions or small workshops should in the first instance provide the personnel with the reasons for the planned implementation of GLP in the test facility, followed by an introduction to the Principles themselves. These theoretical introductions will then have to be followed by the "hands-on" exercises with the practical application of the GLP Principles and the getting used to the important "new ways" of working. The latter point concerns mainly the recording and documentation practices, which will have to be learned and to be applied correctly, faithfully and – especially important ! – invariably all the time. The immediate, legible and indelible recording of all events, data and other occurrences which by definition need recording, together with their dating and initialling and correctly introducing changes into the records, will necessitate a more or less prolonged training period.

There are a number of other tasks, prescribed by the respective SOPs, which will have to be executed with regularity. While it may not have been usual to control the function of the refrigerators / freezers in any regular way or at regular intervals in the opinion that any malfunction would anyway be detected, it will become mandatory under GLP to document their continuous correct functioning, e.g. by reading off at daily or weekly intervals (whatever the SOP commands) the temperature of the interior, recording it on a "Fridge Control Form" and dating and initialling this entry. In the same way, the environmental records of temperature and humidity in animal rooms, which had probably already been measured till the advent of GLP, but which may not have been kept for any required time, and which most certainly had not been dated and initialled, will now have to suffice these requirements of GLP. For such tasks, which had not been done in this regular fashion, at regular intervals and with the strict identification requirements, it may be a favourable or advisable idea to list all these regular activities, ordered according to frequency, and to post this list on the door of the laboratory, the refrigerator, the animal room, the various storage lockers in order to make the personnel constantly aware of these requirements.

A very important aspect in this phase of GLP implementation will be the efficient coaching and supervision with regard to these changes in "personnel behaviour". In the first time it may be necessary to install a regimen of more or less constant coaching, in order to enable the immediate detection, admonition and correction of slips, errors, omissions or neglect. This coaching task will certainly involve, to a very great degree, what has been mentioned already in part II (section 4, see page 123 top paragraph): These interactions between the coach and the coached will necessitate tact and very good communication skills on the part of the coach / supervisor.

In order to facilitate this interaction the progress of the test facility and its personnel in achieving adherence to the GLP Principles has to be communicated in a "positive re-enforcement paradigm". This involves the measuring of certain parameters, like the number of sloppy, not justified or otherwise incorrect changes in the raw data, over a fixed period of time or number of raw data pages, say every two weeks, and posting the (hopefully) improved figures on the test facility message board or in a e-mail to the test facility personnel. Such measures of improving the GLP-relevant quality of work will also have their value in judging the work of the Quality Assurance unit (Berny and Long, 1989).

> Education in the ways of GLP is a necessity from the very beginning of the implementation of GLP in a test facility. Nowadays, many laboratory technicians may have worked already in a test facility under the requirements of the GLP Principles; to these the introduction of GLP will pose no major problems short of having possibly to acquaint themselves to new SOPs, new study types and generally a new environment, which, however, occurs anyway at each change of position. In areas, where GLP is not in as widespread use as it is in the safety testing of chemicals, e.g. in pharmaceutical toxicology laboratories, the notions of GLP will be new to many employees, and these individuals will need very careful and extensive education, coaching and training. Providing good motivation to perform the relevant tasks in a correct way should be a foremost consideration in guiding these individuals towards full GLP compliance.

3.9 Study Plans

The study plan is the central document for the orderly, planned conduct of a study. The generation of study plans has thus to occupy a prominent part of the preparations for performing the first GLP compliant studies. As it has been described above for the compilation and writing of SOPs it is extremely important that this task should be tackled with forethought. It might be advisable to generate model study plans based on a list of the various types of studies to be conducted at the test facility. These model study plans, written probably by the respective future Study Directors, will first have to be scrutinised very carefully by the Quality Assurance expert for compliance with the GLP rules. It might be also a favourable idea to have the other Study Directors (if there are any) look at these plans and to discuss them together with Quality Assurance.

While the formulation of the scientific rationale for the studies should pose no problems to the Study Directors - if there were problems with this topic, then the respective individual should not be allowed to act as Study Director! - and while the observations, measurements and data to be collected are prescribed either by international guidelines (as in toxicology studies), or by the declared purpose of the study, the whole timing and organisation of the process will have to be looked at. A flow chart of study conduct, like the one already shown in figure 34, will certainly help very much in delineating the chronological order of activities to be performed.

> Writing study plans may not be a very big problem for Study Directors used to planning ahead. Since in a test facility under GLP, the study plan has to contain information enabling a later reconstruction of the study, some training in the correct drawing up of study plans. Quality Assurance will have, in this phase even more than later, the responsibility for a timely review of these plans for GLP compliance.

3.10 Test and Reference Item Issues

Along with the organisation of the various rooms and areas which will be used for GLP-relevant activities, one specific area should be set aside from the beginning: An area or room for the handling and storing of test and reference items. Together with the provision of adequate conditions for the respective tasks, and together with the nomination of an individual as responsible in the future for the test and reference item handling, it will be necessary also to organise the process of test item receipt, storage, release for use, taking back and return to the sponsor or disposal in a clear-cut way, which should ensure the full traceability of test and reference item fates.

It is especially in this area of activities, where a well devised set of forms should be available from the beginning, in order to ascertain the possibility of a coherent accounting of test item utilisation.

3.11 Study Conduct

When all these things are done, the first study under GLP-like conditions may be started. It is not to be expected that this first study will comply completely to GLP, and neither management, nor personnel, Study Directors or the

Quality Assurance should be surprised by this "fact of life". One of the main points to be monitored, and possibly a somewhat difficult one to bend into the right direction, will be the observance of the prescriptions regarding the question of deviations from, and amendments to, the study plan by the Study Director. A scientist who has until now been used to simply look into some activity performed by his assistant and telling him to change some parameter in the study in the sense of "let's try it, if it doesn't work, we can turn back to the original setting at any time", may possibly have a hard time in adjusting to faithfully observing the requirements of writing an amendment, dating and signing it and adding it to the study plan before the change is effected.

During the conduct of the first studies, it will become apparent, which ones of the various parts that have to interplay with each other for the full GLP compliance, are indeed able to provide this interplay, and in which ones some need is apparent for changes and improvements.

Conducting studies according to the rules GLP will be the final effort needed to place a test facility under the realm of the GLP Principles. Indeed it is only after some studies have been planned, conducted, reported and archived that the GLP compliance of a test facility can conclusively be judged. Therefore, GLP Compliance Monitoring Authorities will in most cases insist on the presentation of some GLP-studies before a final judgement on the compliance of the respective test facility can be made.

It will finally become evident in the execution of studies, their documented transparency, their reconstructability and thus, in summary, in their quality and integrity that GLP has indeed been successfully implemented. The final "proof of the pudding" lies then not in the eating, but in the successful survival of an official Compliance Monitoring inspection (see the next part).

Good Luck For That !

IV. How is Compliance with Good Laboratory Practice Monitored ?

1. Introduction

The best rules can be worthless if there is no control over the adherence to them. This is true for every legal regulation, and GLP certainly is no exception. A roadside board in an Egyptian town reads:

> Respecting the traffic lights
>
> is a sign of civilisation
>
> and is good practice.

"Good Traffic Light Respecting Practice" may be another one of the "Good Practices" that has not been mentioned in the first part of this book, but is certainly one worth of consideration. Even if there might be disagreement about whether it is indeed a sign of civilisation, there should be absolutely no disagreement about its life-saving properties. It illustrates in a very lucid way, however, that not even the most obvious advantages in the long run can induce individuals to respect regulations, when in the short run an immediate advantage may be gained, although for the price of some risk. It would seem unimportant whether this involves risking one's own life in the traffic light situation, the risk of being caught and fined in the case of some petty tax evasion, or the risk of having a study rejected by a Regulatory Authority because of a serious violation of the GLP Principles.

In the same way as the traffic rules are known in principle to every Egyptian driver, the GLP rules are known to test facilities and their personnel all over the world, but if adherence to these rules would depend on the goodwill of the single individuals only, inequalities in the level of respecting them would certainly become observable. In the same way as the respect for traffic lights can only be asserted through monitoring by police and fining the offenders, the correct implementation and respect of the GLP Principles needs assertion by a monitoring authority. This necessity is not only based on the need for monitoring compliance as such, but also on the need to obtain a similar level of compliance in all test facilities. Even if test facilities were to proclaim voluntary adherence to the rules of GLP, it could by no means be assumed that the implementation of these principles would follow the same standards in each country or indeed in each test facility. Especially since the GLP Principles are worded in a general way and may thus need interpretation, there could be widely divergent ways of "adherence to GLP", if no supervisory bodies would provide for a more or less uniform application and implementation of these guidelines.

Therefore, the OECD Council, in its Decision-Recommendation on Compliance with Principles of Good Laboratory Practice, adopted on October 2, 1989 [C(89)87(Final)], decided that the member countries should establish national procedures for monitoring compliance with GLP Principles and that they should designate authorities to discharge the functions required by these compliance monitoring procedures. In deciding so, the Council also recommended that the member countries of the OECD should, for the development, implementation and establishment of these authorities and procedures, apply the two documents appended as an integral part to the Decision-Recommendation itself. The two documents were also published separately in the OECD Series on GLP as numbers 2 and 3, and they are reprinted as Appendix I and II to this part, too.

2. National Monitoring Authorities

While the necessity of monitoring the compliance of test facilities with the Principles of GLP may be considered obvious, no general way for the formation and placing of the national authorities with respect to other governmental structures can be derived from the Principles themselves or the recommendations of the OECD. Therefore, different countries have taken different approaches in the creation and administrative placement of their respective GLP Compliance Monitoring Authorities, which may have been influenced by factors such as the historical development of GLP in the respective country, the relative importance of the various branches of industry to be subjected to the GLP regulations, or simply the number of test facilities to be monitored.

The spectrum of approaches taken by various countries thus ranges from governmental inspectorates dealing exclusively with monitoring GLP compliance, to offices charged with the supervision of all quality systems, from ISO standards and accreditation to the GXPs. In relation to the respective governmental, ministerial structures, different official bodies may be declared responsible for dealing with different parts of industry subject to GLP. One example of such a distribution of responsibilities are the United States, where the FDA regulates and monitors GLP compliance with respect to safety studies on foods, drugs and cosmetics, while the EPA forms the counterpart responsible for the analogous studies on chemicals, pesticides and general environmental issues.

Although the affiliation of GLP monitoring authorities may thus vary between countries according to their administrative structures, their specific situation with regard to industrial activities and their needs for the economical management of GLP-related monitoring, there is one aspect which in all cases deserves well considered attention. This aspect concerns the relationship between the GLP compliance monitoring authority on one side and the receiving, registration authority on the other. Since the GLP compliant conduct of "human health and environmental safety" studies is a prerequisite to their acceptability for the authority receiving the respective submissions, it should be clear that the two authorities will have to co-operate closely with regard to establishing the GLP status of submitted studies. The closeness of these ties will, however, be again dictated by the administrative structures and other factors, and may range from "silent" working agreements up to situations where the same staff is performing both tasks. In any case, the communication system between the two parties should be so well developed as to allow on the one hand a good knowledge about the working modalities of the respective authorities and on the other hand to avoid contradictory decisions between the two authorities with respect to the acceptability of a study.

With the guidance for inspections and study audits provided by the OECD documents, the national monitoring authority assumes the role of a "Super-Quality Assurance", in the sense that it will inspect the adherence of test facilities to the GLP Principles in an analogous way as the individual Quality Assurance units do it for the single test facility. Indeed the OECD document presented in Appendix IV.II ("Revised Guidance for the Conduct of

Laboratory Inspections and Study Audits") may be regarded to serve as template for the respective Quality Assurance check-lists, as a comparison of the relevant parts of this document with the examples given in figures 9 and 10 (see pages 139 and 140) will make obvious.

There is one difference evident between the activities, and especially the conclusions drawn from them, of the two. The Quality Assurance unit of a test facility will monitor on a continuous basis the procedures and activities at the test facility, while the inspections of the monitoring authority, although also conducted at regular intervals, are only snapshot pictures of the test facility's compliance valid only for a very limited time point. This makes it all the more important that test facilities will regard the admonitions and critiques provided by the inspectors of the monitoring authorities not as singular events but as starting points to check the whole facility through with respect to this and similar faults. The agency inspector may, e.g., point to a non-compliantly executed correction in the raw data by one employee. The test facility and its Quality Assurance should not, however, consider the situation as settled ("The employee has been instructed to pay better attention to the way in which corrections have to be made"), but Quality Assurance and test facility management should subsequently consider whether this single incident could reflect some deficiency in the way employees are instructed about GLP.

Study audits can be conducted with two purposes in mind. On the one hand, they provide the Regulatory Authority with the assurance that the study indeed has been conducted according to the rules of GLP and that therefore the integrity and validity of data and study can be assumed. This is the reason why certain Regulatory Authorities routinely require from GLP monitoring authorities to have specific, pivotal studies audited, since the assessment of the whole submission may hinge on the reliability of these studies. On the other hand, study audits may be considered as suitable means to obtain a better picture about the continuous observation of the GLP Principles during the course of a study. In this sense study audits may be used by the compliance monitoring authority as a substitute for longer-term inspections. It has been stated above, that an authority inspection of a test facility corresponds to a snapshot, registering only the momentary situation, which may, moreover, be influenced in one way or the other by the inspection itself and the test facility's preparation thereto. Not all activities and critical phases of all kinds of studies conducted at the test facility will therefore be inspected, and a systematic, but rare, violation of the GLP Principles might thus go undetected. A study audit, however, will present more of the actual ways in which the study had been conducted, in which data have been recorded, in which the Study Director has exerted his control, in which the study plan had been followed, and in which the study report had been written.

In comparing the tasks of the compliance monitoring inspection teams to the work of test facility Quality Assurances, an additional parallel may be drawn. The observations made during inspections and audits, and the conclusions drawn therefrom, are finally the subject of inspection reports. Since these reports are instrumental for the official decision about the GLP compliance of the inspected test facility or the audited studies, there is potentially some interest by other compliance monitoring or Regulatory Authorities to receive them, should some question arise in another country with regard to a submission. Furthermore, inspections and study audits may have been conducted in response to a request of another authority, and the respective report should then be submitted to the requesting authority. In order to improve the comparability of such inspection and audit reports the OECD has again resorted to the instrument of a Consensus Document, which details the way in which inspectional procedures, results and conclusions should be presented in an internationally harmonised way, so that no important information about the inspection or audit might get lost.

3. MOUs, MRAs, and MJVs

The acronyms of this section's title stand for "Memorandum of Understanding", "Mutual Recognition Agreement", and "Mutual Joint Visit", respectively. The first two terms denote formal agreements between two or more countries, in the present context related to mutual acceptance of decisions by the partner country and its (or their) monitoring authorities. The question might now be asked, why such agreements would at all be necessary, because the mutual acceptance of data and recognition of monitoring results should be provided through the OECD Council Decisions related to safety testing in general and GLP in particular.

The OECD Council, in its "Decision concerning the Mutual Acceptance of Data in the Assessment of Chemicals [C(81)30(Final)]", decided that data generated in one member country in accordance with OECD Test Guidelines and the OECD Principles of GLP should be accepted also in other member countries, with the intention that such studies would not need to be repeated, nor the GLP compliance of such studies be questioned. In an analogous way, in its "Decision-Recommendation on Compliance with Principles of Good Laboratory Practice [C(89)87(Final)]" the OECD Council decided in Part II of this document that OECD member countries should recognise and accept the assurance of any other member country that data had been generated and safety studies had been conducted in accordance with the GLP Principles, thus obviating the need to conduct test facility inspections and study audits in other member countries, as long as these were complying with the requirements of the Council Decision with regard to the designation of a responsible compliance monitoring authority.

It can be easily seen, however, that, as well as there can be differences in the interpretation and implementation of the GLP Principles by different test facilities, there can exist also differences in the interpretation of the Principles by monitoring authorities as well, and therefore the standards achieved in test facilities recognised as GLP-compliant by the monitoring authority in one country might be different from the ones in another country, which would make the comparative assessment of the true meaning of a "GLP Compliance Certificate" or "GLP Compliance Statement" rather difficult and would therefore tend to jeopardise the acceptability of the "assurance by another member country" as foreseen in the Council Decision.

To resolve this obvious problem, the foremost and unanimously accepted consideration had been to increase, among countries and their respective monitoring and Regulatory Authorities, the mutual trust into the capabilities of the respective inspectorate, and into the similarity of views with regard to the interpretation of the Principles, to inspectional procedures and to the final judgements about what constitutes GLP compliance. Any two countries which felt compelled to work more closely together could thus meet at the negotiation table, work out a way in which to recognise the other one's system as equivalent, test this in practice by observing the other partner's monitoring activities and procedures, and then to conclude an agreement on the mutual acceptability of the respective GLP compliance monitoring systems. The agreements enclosed in the terms MOU and MRA are thus the expression of the will of two countries to accept in mutual trust the data generated in the respective partner country and to accept the affirmation of one country's authority that the generation of these data represents GLP compliance. There may be legal niceties that distinguish an MRA from an MOU; for both cases, however, a process of reciprocal observation of the implementation of "Good Inspectional Practices", of the adherence to similar standards in the judgement of GLP compliance in test facilities, and of a mutual understanding of each other's philosophy in the conduct of the monitoring functions will have been instrumental in the final conclusion of such agreements.

The problems regarding the possibility of differing inspectional standards between the monitoring authorities of different countries, or at least the fear that such differences might exist, had been the trigger for concluding MRAs and MOUs. These "contracts" between the monitoring authorities were primarily concluded between countries which expected the respective partner to observe similar standards, and such agreements between countries with widely differing standards would certainly have been impossible to conclude. As long as such agreements had to be concluded only in cases of obvious need or of mutual interest, the number of these special bi-lateral connections could be kept low and uncomplicated. In the European Union, the issue became a rather complex one, since with the development of the common market, and with the centralised procedure for the registration of pharmaceuticals, the mutual acceptance of national statements of GLP compliance would have be a prerequisite. To conclude, however, bi-lateral agreements between any possible combination of the 15 member states of the EU, or between them and the European Commission as the central authority, would certainly have led to such a complex network of relations that the situation would have become worse instead of better. In this situation the European countries engaged in an endeavour, termed "Mutual Joint Visits" or MJV. Briefly, its intentions were to improve trust into the capabilities, the organisation and the output of each national GLP monitoring authority in a similar way as it would have been done between two individual countries. For such an MJV experts from the monitoring authorities of three member states visited the authority of a fourth country, inspecting the legal basis and the organisational aspects of the authority itself, and observing a test facility inspection conducted by the personnel of the visited authority. In this way a good picture could be obtained on the compliance of the visited authority with the respective requirements as set forth in the OECD Council Decision-Recommendation of 1989 which had been transferred also into a European Commission Directive (99/12/EC, adapting to technical progress for the second time the annex to Council directive 88/320/EEC). These visits were not conceived with the objective to grade national authorities with respect to their compliance or experience, but to draw attention to those points in the procedures of an authority where possibilities for improvements towards a common European standard could be identified.

Meanwhile, another development had taken place which made this system for monitoring "official compliance" attractive also to the situation within the OECD. While the membership of this organisation consists of industrialised countries, it was recognised that in view of the opening of world trade and the removal of non-tariff barriers to trade in the World Trade Organisation agreements, the mutual acceptance of data should be widened and not rest restricted to the member countries only. Consequently, in 1997 the OECD Council reached another decision with regard to the mutual data acceptance, concerned with the "Adherence of Non-member Countries to the Council Acts related to the Mutual Acceptance of Data in the Assessment of Chemicals" [C(97)114/Final], in which the provisions for the mutual acceptance of data were expanded to such non-member countries expressing their willingness and demonstrating their ability to participate in the implementation of the related OECD Council Acts. Many countries expressed their interest in the offered possibility, but the broadening of the circle of countries asserting their adherence to the GLP Principles, increased again the call for a mechanism within OECD which would serve to ensure that similar standards were to be adopted in all countries and by all monitoring authorities. The excellent experiences with the MJV programme in the EU made this an attractive choice for the requested monitoring mechanism of OECD, and in 1997 the OECD Working Group on GLP proposed to conduct such MJVs as a pilot project, and on a voluntary basis, in the first instance.

The first such "visits" took place in 1998, and it remains to be seen what impact these mutual inspections will have in the future on the world-wide acceptance of GLP compliance monitoring and data exchange.

Appendix IV.I

OECD Series on GLP, No. 2:

Revised Guides for Compliance Monitoring Procedures for Good Laboratory Practice

(reprinted by permission of OECD)*

To facilitate the mutual acceptance of test data generated for submission to Regulatory Authorities of OECD Member countries, harmonization of the procedures adopted to monitor good laboratory practice compliance, as well as comparability of their quality and rigour, are essential. The aim of this document is to provide detailed practical guidance to OECD Member countries on the structure, mechanisms and procedures they should adopt when establishing national Good Laboratory Practice compliance monitoring programmes so that these programmes may be internationally acceptable.

It is recognised that Member countries will adopt GLP Principles and establish compliance monitoring procedures according to national legal and administrative practices, and according to priorities they give to, e.g., the scope of initial and subsequent coverage concerning categories of chemicals and types of testing. Since Member countries may establish more than one Good Laboratory Practice Monitoring Authority due to their legal framework for chemicals control, more than one Good Laboratory Practice Compliance Programme may be established. The guidance set forth in the following paragraphs concerns each of these Authorities and Compliance Programmes, as appropriate.

Definitions of Terms

The definitions of terms in the "OECD Principles of Good Laboratory Practice" [Annex 2 to Council Decision C(81)30(Final)] are applicable to this document. In addition, the following definitions apply:

---

* Revised Guides for Compliance Monitoring Procedures for Good Laboratory Practice. Copyright OECD Paris, 1995. Material available on OECD website at http:\\www.oecd.org/ehs/ehsmono/index.htm#GLP

GLP Principles: Principles of good laboratory practice that are consistent with the OECD Principles of Good Laboratory Practice as set out in Annex 2 of Council Decision C(81)30(Final)4.

GLP Compliance Monitoring: The periodic inspection of test facilities and/or auditing of studies for the purpose of verifying adherence to GLP Principles.

(National) GLP Compliance Programme: The particular scheme established by a Member country to monitor good laboratory practice compliance by test facilities within its territories, by means of inspections and study audits.

(National) GLP Monitoring Authority: A body established within a Member country with responsibility for monitoring the good laboratory practice compliance of test facilities within its territories and for discharging other such functions related to good laboratory practice as may be nationally determined. It is understood that more than one such body may be established in a Member country.

Test Facility Inspection: An on-site examination of the test facility's procedures and practices to assess the degree of compliance with GLP Principles. During inspections, the management structures and operational procedures of the test facility are examined, key technical personnel are interviewed, and the quality and integrity of data generated by the facility are assessed and reported.

Study Audit: A comparison of raw data and associated records with the interim or final report in order to determine whether the raw data have been accurately reported, to determine whether testing was carried out in accordance with the study plan and Standard Operating Procedures, to obtain additional information not provided in the report, and to establish whether practices were employed in the development of data that would impair their validity.

Inspector: A person who performs the test facility inspections and study audits on behalf of the (National) GLP Monitoring Authority.

GLP Compliance Status: The level of adherence of a test facility to the GLP Principles as assessed by the (National) GLP Monitoring Authority.

Regulatory Authority: A national body with legal responsibility for aspects of the control of chemicals.

Components of Good Laboratory Practice Compliance Monitoring Procedures

Administration

A (National) GLP Compliance Programme should be the responsibility of a properly constituted, legally identifiable body adequately staffed and working within a defined administrative framework.

Member countries should:

— ensure that the (National) GLP Monitoring Authority is directly responsible for an adequate "team" of inspectors having the necessary technical/scientific expertise or is ultimately responsible for such a "team";

— publish documents relating to the adoption of GLP Principles within their territories;

— publish documents providing details of the (National) GLP Compliance Programme, including information on the legal or administrative framework within which the programme operates and references to published acts, normative documents (e.g., regulations, codes of practice), inspection manuals, guidance notes, periodicity of inspections and/or criteria for inspection schedules, etc.;

— maintain records of test facilities inspected (and their GLP Compliance Status) and of studies audited for both national and international purposes.

Confidentiality

(National) GLP Monitoring Authorities will have access to commercially valuable information and, on occasion, may even need to remove commercially sensitive documents from a test facility or refer to them in detail in their reports.

Member countries should:

— make provision for the maintenance of confidentiality, not only by Inspectors but also by any other persons who gain access to confidential information as a result of GLP Compliance Monitoring activities;

— ensure that, unless all commercially sensitive and confidential information has been excised, reports of Test Facility Inspections and Study Audits are made available only to Regulatory Authorities and, where appropriate, to the test facilities inspected or concerned with Study Audits and/or to study sponsors.

Personnel and Training

(National) GLP Monitoring Authorities should:

— *ensure that an adequate number of Inspectors is available*

The number of Inspectors required will depend upon:

i) the number of test facilities involved in the (National) GLP Compliance Programme;

ii) the frequency with which the GLP Compliance Status of the test facilities is to be assessed;

iii) the number and complexity of the studies undertaken by those test facilities;

iv) the number of special inspections or audits requested by Regulatory Authorities.

— *ensure that Inspectors are adequately qualified and trained*

Inspectors should have qualifications and practical experience in the range of scientific disciplines relevant to the testing of chemicals. **(National) GLP Monitoring Authorities should:**

i) ensure that arrangements are made for the appropriate training of GLP Inspectors, having regard to their individual qualifications and experience;

ii) encourage consultations, including joint training activities where necessary, with the staff of (National) GLP

Monitoring Authorities in other Member countries in order to promote international harmonization in the interpretation and application of GLP Principles, and in the monitoring of compliance with such Principles.

— *ensure that inspectorate personnel, including experts under contract, have no financial or other interests in the test facilities inspected, the studies audited or the firms sponsoring such studies*

— *provide Inspectors with a suitable means of identification (e.g., an identity card).*

Inspectors may be:

— on the permanent staff of the (National) GLP Monitoring Authority;

— on the permanent staff of a body separate from the (National) GLP Monitoring Authority; or

— employed on contract, or in another way, by the (National) GLP Monitoring Authority to perform Test Facility Inspections or Study Audits. In the latter two cases, the (National) GLP Monitoring Authority should have ultimate responsibility for determining the GLP Compliance Status of test facilities and the quality/acceptability of a Study Audit, and for taking any action based on the results of Test Facility Inspections or Study Audits which may be necessary.

(National) GLP Compliance Programmes

GLP Compliance Monitoring is intended to ascertain whether test facilities have implemented GLP Principles for the conduct of studies and are capable of assuring that the resulting data are of adequate quality. As indicated above, Member countries should publish the details of their (National) GLP Compliance Programmes. Such information should, *inter alia*:

— *define the scope and extent of the Programme*

A (National) GLP Compliance Programme may cover only a limited range of chemicals, e.g., industrial chemicals, pesticides, pharmaceuticals, etc., or may include all chemicals. The scope of the monitoring for compliance should be defined, both with respect to the categories of chemicals and to the types of tests subject to it, e.g., physical, chemical, toxicological and/or ecotoxicological.

— *provide an indication as to the mechanism whereby test facilities enter the Programme*

The application of GLP Principles to health and environmental safety data generated for regulatory purposes may be mandatory. A mechanism should be available whereby test facilities may have their compliance with GLP Principles monitored by the appropriate (National) GLP Monitoring Authority.

— *provide information on categories of Test Facility Inspections/Study Audits.*

A (National) GLP Compliance Programme should include:

i) provision for Test Facility Inspections. These inspections include both a general Test Facility Inspection and a Study Audit of one or more on-going or completed studies;

ii) provision for special Test Facility Inspections/Study Audits at the request of a Regulatory Authority — e.g., prompted by a query arising from the submission of data to a Regulatory Authority.

iii) define the powers of Inspectors for entry into test facilities and their access to data held by test facilities (including specimens, SOP's, other documentation, etc.)

While Inspectors will not normally wish to enter test facilities against the will of the facility's management, circumstances may arise where test facility entry and access to data are essential to protect public health or the environment. The powers available to the (National) GLP Monitoring Authority in such cases should be defined.

— *describe the Test Facility Inspection and Study Audit procedures for verification of GLP compliance.*

The documentation should indicate the procedures which will be used to examine both the organisational processes and the conditions under which studies are planned, performed, monitored and recorded. Guidance for such procedures is available in Guidance for the Conduct of Test Facility Inspections and Study Audits (No. 3 in the OECD series on Principles of GLP and Compliance Monitoring).

— *describe actions that may be taken as follow-up to Test Facility Inspections and Study Audits.*

Follow-up to Test Facility Inspections and Study Audits

When a Test Facility Inspection or Study Audit has been completed, the Inspector should prepare a written report of the findings.

Member countries should take action where deviations from GLP Principles are found during or after a Test Facility Inspection or Study Audit. The appropriate actions should be described in documents from the (National) GLP Monitoring Authority.

If a Test Facility Inspection or Study Audit reveals only minor deviations from GLP Principles, the facility should be required to correct such minor deviations. The Inspector may need, at an appropriate time, to return to the facility to verify that corrections have been introduced.

Where no or where only minor deviations have been found, the (National) GLP Monitoring Authority may:

— issue a statement that the test facility has been inspected and found to be operating in compliance with GLP Principles. The date of the inspections and, if appropriate, the categories of test inspected in the test facility at that time should be included. Such statements may be used to provide information to (National) GLP Monitoring Authorities in other Member countries; and/or

— provide the Regulatory Authority which requested a Study Audit with a detailed report of the findings.

Where serious deviations are found, the action taken by (National) GLP Monitoring Authorities will depend upon the particular circumstances of each case and the legal or administrative provisions under which GLP Compliance Monitoring has been established within their countries. Actions which may be taken include, but are not limited to, the following:

— issuance of a statement, giving details of the inadequacies or faults found which might affect the validity of studies conducted in the test facility;

— issuance of a recommendation to a Regulatory Authority that a study be rejected;

— suspension of Test Facility Inspections or Study Audits of a test facility and, for example and where administratively possible, removal of the test facility from the (National) GLP Compliance Programme or from any existing list or register of test facilities subject to GLP Test Facility Inspections;

— requiring that a statement detailing the deviations be attached to specific study reports;

— action through the courts, where warranted by circumstances and where legal/ administrative procedures so permit.

Appeals Procedures

Problems, or differences of opinion, between Inspectors and test facility management will normally be resolved during the course of a Test Facility Inspection or Study Audit. However, it may not always be possible for agreement to be reached. A procedure should exist whereby a test facility may make representations relating to the outcome of a Test Facility Inspection or Study Audit for GLP Compliance Monitoring and/or relating to the action the GLP Monitoring Authority proposes to take thereon.

Appendix IV.II

Revised Guidance for the Conduct of Laboratory Inspections and Study Audits

(reprinted by permission of OECD)*

Introduction

The purpose of this document is to provide guidance for the conduct of Test Facility Inspections and Study Audits which would be mutually acceptable to OECD Member countries. It is principally concerned with Test Facility Inspections, an activity which occupies much of the time of GLP Inspectors. A Test Facility Inspection will usually include a Study Audit or "review" as a part of the inspection, but Study Audits will also have to be conducted from time to time at the request, for example, of a Regulatory Authority. General guidance for the conduct of Study Audits will be found at the end of this document. Test Facility Inspections are conducted to determine the degree of conformity of test facilities and studies with GLP Principles and to determine the integrity of data to assure that resulting data are of adequate quality for assessment and decision-making by national Regulatory Authorities. They result in reports which describe the degree of adherence of a test facility to the GLP Principles. Test Facility Inspections should be conducted on a regular, routine basis to establish and maintain records of the GLP compliance status of test facilities. Further clarification of many of the points in this document may be obtained by referring to the OECD Consensus Documents on GLP (on, e.g., the Role and Responsibilities of the Study Director).

Definitions of Terms

The definitions of terms in the "OECD Principles of Good Laboratory Practice" [Annex II to Council Decision C(81)30(Final)] and in the "Guides for Compliance Monitoring Procedures for Good Laboratory Practice" [Annex I to Council Decision-Recommendation C(89)87(Final)/revised in C(95)8(Final)] are applicable to this document.

---

* Revised Guidance for the Conduct of Laboratory Inspections and Study Audits. Copyright OECD Paris, 1995. Material available on OECD website at
http:\\www.oecd.org/ehs/ehsmono/index.htm#GLP

Test Facility Inspections

Inspections for compliance with GLP Principles may take place in any test facility generating health or environmental safety data for regulatory purposes. Inspectors may be required to audit data relating to the physical, chemical, toxicological or ecotoxicological properties of a substance or preparation. In some cases, Inspectors may need assistance from experts in particular disciplines.

The wide diversity of facilities (in terms both of physical layout and management structure), together with the variety of types of studies encountered by Inspectors, means that the Inspectors must use their own judgement to assess the degree and extent of compliance with GLP Principles. Nevertheless, Inspectors should strive for a consistent approach in evaluating whether, in the case of a particular test facility or study, an adequate level of compliance with each GLP Principle has been achieved.

In the following sections, guidance is provided on the various aspects of the testing facility, including its personnel and procedures, which are likely to be examined by Inspectors. In each section, there is a statement of purpose, as well as an illustrative list of specific items which could be considered during the course of a Test Facility Inspection. These lists are not meant to be comprehensive and should not be taken as such.

Inspectors should not concern themselves with the scientific design of the study or the interpretation of the findings of studies with respect to risks for human health or the environment. These aspects are the responsibility of those Regulatory Authorities to which the data are submitted for regulatory purposes.

Test Facility Inspections and Study Audits inevitably disturb the normal work in a facility. Inspectors should therefore carry out their work in a carefully planned way and, so far as practicable, respect the wishes of the management of the test facility as to the timing of visits to certain sections of the facility.

Inspectors will, while conducting Test Facility Inspections and Study Audits, have access to confidential, commercially valuable information. It is essential that they ensure that such information is seen by authorised personnel only. Their responsibilities in this respect will have been established within their (National) GLP Compliance Monitoring Programme.

Inspection Procedures

Pre-Inspection

PURPOSE: To familiarise the Inspector with the facility which is about to be inspected in respect of management structure, physical layout of buildings and range of studies.

Prior to conducting a Test Facility Inspection or Study Audit, Inspectors should familiarise themselves with the facility which is to be visited. Any existing information on the facility should be reviewed. This may include previous inspection reports, the layout of the facility, organisation charts, study reports, protocols and curricula vitae (CVs) of personnel. Such documents would provide information on:

— the type, size and layout of the facility;

— the range of studies likely to be encountered during the inspection;

— the management structure of the facility.

Inspectors should note, in particular, any deficiencies from previous Test Facility Inspections. Where no previous Test Facility Inspections have been conducted, a pre-inspection visit can be made to obtain relevant information.

Test Facilities may be informed of the date and time of Inspector's arrival, the objective of their visit and the length of time they expect to be on the premises. This could allow the test facility to ensure that the appropriate personnel and documentation are available. In cases where particular documents or records are to be examined, it may be useful to identify these to the test facility in advance of the visit so that they will be immediately available during the Test Facility Inspection.

Starting Conference

PURPOSE: To inform the management and staff of the facility of the reason for the Test Facility Inspection or Study Audit that is about to take place, and to identify the facility areas, study(ies) selected for audit, documents and personnel likely to be involved.

The administrative and practical details of a Test Facility Inspection or Study Audit should be discussed with the management of the facility at the start of the visit. At the starting conference, Inspectors should:

outline the purpose and scope of the visit;

— describe the documentation which will be required for the Test Facility Inspection, such as lists of on-going and completed studies, study plans, standard operating procedures, study reports, etc. Access to and, if necessary, arrangements for the copying of relevant documents should be agreed upon at this time;

— clarify or request information as to the management structure (organisation) and personnel of the facility;

— request information as to the conduct of studies not subject to GLP Principles in the areas of the test facility where GLP studies are being conducted;

— make an initial determination as to the parts of the facility to be covered during the Test Facility Inspection;

— describe the documents and specimens that will be needed for on-going or completed study(ies) selected for Study Audit;

— indicate that a closing conference will be held at the completion of the inspection.

Before proceeding further with a Test Facility Inspection, it is advisable for the Inspector(s) to establish contact with the facility's Quality Assurance (QA) Unit.

As a general rule, when inspecting a facility, Inspectors will find it helpful to be accompanied by a member of the QA unit..

Inspectors may wish to request that a room be set aside for examination of documents and other activities.

Organisation and Personnel

PURPOSE: To determine whether: the test facility has sufficient qualified personnel, staff resources and support services for the variety and number of studies undertaken; the organisational structure is appropriate; and management has established a policy regarding training and staff health surveillance appropriate to the studies undertaken in the facility.

The management should be asked to produce certain documents, such as:

— floor plans;
— facility management and scientific organisation charts;
— CVs of personnel involved in the type(s) of studies selected for the Study Audit;
— list(s) of on-going and completed studies with information on the type of study, initiation/completion dates, test system, method of application of test substance and name of Study Director;
— staff health surveillance policies;
— staff job descriptions and staff training programmes and records;
— an index to the facility's Standard Operating Procedures (SOPs);
— specific SOPs as related to the studies or procedures being inspected or audited;
— list(s) of the Study Directors and sponsors associated with the study(ies) being audited. The Inspector should check, in particular:
— lists of on-going and completed studies to ascertain the level of work being undertaken by the test facility;
— the identity and qualifications of the Study Director(s), the head of the Quality Assurance unit and other personnel;
— existence of SOPs for all relevant areas of testing.

Quality Assurance Programme

PURPOSE: To determine whether the mechanisms used to assure management that studies are conducted in accordance with GLP Principles are adequate.

The head of the Quality Assurance (QA) Unit should be asked to demonstrate the systems and methods for QA inspection and monitoring of studies, and the system for recording observations made during QA monitoring. Inspectors should check:

— the qualifications of the head of QA, and of all QA staff;
— that the QA unit functions independently from the staff involved in the studies;
— how the QA unit schedules and conducts inspections, how it monitors identified critical phases in a study, and what resources are available for QA inspections and monitoring activities;
— that where studies are of such short duration that monitoring of each study is impracticable, arrangements exist for monitoring on a sample basis;
— the extent and depth of QA monitoring during the practical phases of the study;
— the extent and depth of QA monitoring of routine test facility operation;
— the QA procedures for checking the final report to ensure its agreement with the raw data;
— that management receives reports from QA concerning problems likely to affect the quality or integrity of a study;
— the actions taken by QA when deviations are found;
— the QA role, if any, if studies or parts of studies are done in contract laboratories;
— the part played, if any, by QA in the review, revision and updating of SOPs.

National GLP monitoring authorities may request information relating to the nature and dates of Quality Assurance inspections. However, Quality Assurance inspection reports should not normally be examined for their contents by national monitoring authorities as this may inhibit Quality Assurance when preparing inspection reports. Nevertheless, national monitoring authorities may occasionally require access to the contents of inspection reports in order to verify the adequate functioning of Quality Assurance. They should not inspect such reports merely as an easy way to identify inadequacies in the studies carried out.

Facilities

PURPOSE: To determine if the test facility, whether indoor or outdoor, is of suitable size, design and location to meet the demands of the studies being undertaken. The Inspector should check that:

— the design enables an adequate degree of separation so that, e.g., test substances, animals, diets, pathological specimens, etc. of one study cannot be confused with those of another;

— environmental control and monitoring procedures exist and function adequately in critical areas, e.g., animal and other biological test systems rooms, test substance storage areas, laboratory areas;

— the general housekeeping is adequate for the various facilities and that there are, if necessary, pest control procedures.

Care, Housing and Containment of Biological Test Systems

PURPOSE: To determine whether the test facility, if engaged in studies using animals or other biological test systems, has support facilities and conditions for their care, housing and containment, adequate to prevent stress and other problems which could affect the test system and hence the quality of data.

A test facility may be carrying out studies which require a diversity of animal or plant species as well as microbial or other cellular or sub-cellular systems. The type of test systems being used will determine the aspects relating to care, housing or containment that the Inspector will monitor. Using his judgement, the Inspector will check, according to the test systems, that:

— there are facilities adequate for the test systems used and for testing needs;

- there are arrangements to quarantine animals and plants being introduced into the facility and that these arrangements are working satisfactorily;
- there are arrangements to isolate animals (or other elements of a test system, if necessary) known to be, or suspected of being, diseased or carriers of disease;
- there is adequate monitoring and record-keeping of health, behaviour or other aspects, as appropriate to the test system;
- the equipment for maintaining the environmental conditions required for each test system is adequate, well maintained, and effective;
- animal cages, racks, tanks and other containers, as well as accessory equipment, are kept sufficiently clean;
- analyses to check environmental conditions and support systems are carried out as required;
- facilities exist for removal and disposal of animal waste and refuse from the test systems and that these are operated so as to minimise vermin infestation, odours, disease hazards and environmental contamination;
- storage areas are provided for animal feed or equivalent materials for all test systems; that these areas are not used for the storage of other materials such as test substances, pest control chemicals or disinfectants, and that they are separate from areas in which animals are housed or other biological test systems are kept;
- stored feed and bedding are protected from deterioration by adverse environmental conditions, infestation or contamination.

Apparatus, Materials, Reagents and Specimens

PURPOSE: To determine whether the test facility has suitably located, operational apparatus in sufficient quantity and of adequate capacity to meet the requirements of the tests being conducted in the facility and that the materials, reagents and specimens are properly labelled, used and stored.

The Inspector should check that:
- apparatus is clean and in good working order;

- records have been kept of operation, maintenance, verification, calibration and validation of measuring equipment and apparatus (including computerised systems);
- materials and chemical reagents are properly labelled and stored at appropriate temperatures and that expiry dates are not being ignored. Labels for reagents should indicate their source, identity and concentration and/or other pertinent information;
- specimens are well identified by test system, study, nature and date of collection;
- apparatus and materials used do not alter to any appreciable extent the test systems.

Test Systems

PURPOSE: To determine whether adequate procedures exist for the handling and control of the variety of test systems required by the studies undertaken in the facility, e.g., chemical and physical systems, cellular and microbic systems, plants or animals.

Physical and Chemical Systems

The Inspector should check that:

- where required by study plans, the stability of test and reference substances was determined and that the reference substances specified in test plans were used;
- in automated systems, data generated as graphs, recorder traces or computer print-outs are documented as raw data and archived.

Biological Test Systems

Taking account of the relevant aspects referred to above relating to care, housing or containment of biological test systems, the Inspector should check that:

- test systems are as specified in study plans;

- test systems are adequately and, if necessary and appropriate, uniquely identified throughout the study; and that records exist regarding receipt of the test systems and document fully the number of test systems received, used, replaced or discarded;
- housing or containers of test systems are properly identified with all the necessary information;
- there is an adequate separation of studies being conducted on the same animal species (or the same biological test systems) but with different substances;
- there is an adequate separation of animal species (and other biological test systems) either in space or in time;
- the biological test system environment is as specified in the study plan or in SOPs for aspects such as temperature, or light/dark cycles;
- the recording of the receipt, handling, housing or containment, care and health evaluation is appropriate to the test systems;
- written records are kept of examination, quarantine, morbidity, mortality, behaviour, diagnosis and treatment of animal and plant test systems or other similar aspects as appropriate to each biological test system;
- there are provisions for the appropriate disposal of test systems at the end of tests.

Test and Reference Substances

PURPOSE: To determine whether the test facility has procedures designed (i) to ensure that the identify, potency, quantity and composition of test and reference substances are in accordance with their specifications, and (ii) to properly receive and store test and reference substances.

The Inspector should check that:

- there are written records on the receipt (including identification of the person responsible), and for the handling, sampling, usage and storage of tests and reference substances;
- test and reference substances containers are properly labelled;

- storage conditions are appropriate to preserve the concentration, purity and stability of the test and reference substances;
- there are written records on the determination of identity, purity, composition, stability, and for the prevention of contamination of test and reference substances, where applicable;
- there are procedures for the determination of the homogeneity and stability of mixtures containing test and reference substances, where applicable;
- containers holding mixtures (or dilutions) of the test and reference substances are labelled and that records are kept of the homogeneity and stability of their contents, where applicable;
- when the test is of longer than four weeks' duration, samples from each batch of test and reference substances have been taken for analytical purposes and that they have been retained for an appropriate time;
- procedures for mixing substances are designed to prevent errors in identification or cross-contamination.

Standard Operating Procedures

PURPOSE: To determine whether the test facility has written SOPs relating to all the important aspects of the its operations, considering that one of the most important management techniques for 16 controlling facility operations is the use of written SOPs. These relate directly to the routine elements of tests conducted by the test facility.

The Inspector should check that:
- each test facility area has immediately available relevant, authorised copies of SOPs;
- procedures exist for revision and updating of SOPs;
- any amendments or changes to SOPs have been authorised and dated;
- historical files of SOPs are maintained;
- SOPs are available for, but not necessarily limited to, the following activities:

i) receipt; determination of identity, purity, composition and stability; labelling; handling; sampling; usage; and storage of test and reference substances;

ii) use, maintenance, cleaning, calibration and validation of measuring apparatus, computerised systems and environmental control equipment;

iii) preparation of reagents and dosing formulations;

iv) record-keeping, reporting, storage and retrieval of records and reports;

v) preparation and environmental control of areas containing the test systems;

vi) receipt, transfer, location, characterisation, identification and care of test systems;

vii) handling of the test systems before, during and at the termination of the study;

viii) disposal of test systems;

xi) use of pest control and cleaning agents;

x) Quality Assurance Programme operations.

Performance of the Study

PURPOSE: To verify that written study plans exist and that the plans and the conduct of the study are in accordance with GLP Principles.

The Inspector should check that:

— the study plan was signed by the Study Director;
— any amendments to the study plan were signed and dated by the Study Director;
— the date of the agreement to the study plan by the sponsor was recorded (where applicable);
— measurements, observations and examinations were in accordance with the study plan and relevant SOPs;

- the results of these measurements, observations and examinations were recorded directly, promptly, accurately and legibly and were signed (or initialled) and dated;
- any changes in the raw data, including data stored in computers, did not obscure previous entries, included the reason for the change and identified the person responsible for the change and the date it was made;
- computer-generated or stored data have been identified and that the procedures to protect them against unauthorised amendments or loss are adequate;
- the computerised systems used within the study are reliable, accurate and have been validated;
- any unforeseen events recorded in the raw data have been investigated and evaluated;
- the results presented in the reports of the study (interim or final) are consistent and complete and that they correctly reflect the raw data.

Reporting of Study Results

PURPOSE: To determine whether final reports are prepared in accordance with GLP Principles.

When examining a final report, the Inspector should check that:

- it is signed and dated by the Study Director to indicate acceptance of responsibility for the validity of the study and confirming that the study was conducted in accordance with GLP Principles;
- it is signed and dated by other principal scientists, if reports from co-operating disciplines are included;
- a Quality Assurance statement is included in the report and that it is signed and dated;
- any amendments were made by the responsible personnel;
- it lists the archive location of all samples, specimens and raw data.

Storage and Retention of Records

PURPOSE: To determine whether the facility has generated adequate records and reports and whether adequate provision has been made for the safe storage and retention of records and materials;

The Inspector should check:

— that a person has been identified as responsible for the archive;
— the archive facilities for the storage of study plans, raw data (including that from discontinued GLP Studies), final reports, samples and specimens and records of education and training of personnel;
— the procedures for retrieval of archived materials;
— the procedures whereby access to the archives is limited to authorised personnel and records are kept of personnel given access to raw data, slides, etc.;
— that an inventory is maintained of materials removed from, and returned to, the archives;
— that records and materials are retained for the required or appropriate period of time and are protected from loss or damage by fire, adverse environmental conditions, etc.

Study Audits

Test Facility inspections will generally include, inter alia, Study Audits, which review on-going or completed studies. Specific Study Audits are also often requested by Regulatory Authorities, and can be conducted independently of Test Facility Inspections. Because of the wide variation in the types of studies which might be audited, only general guidance is appropriate, and Inspectors and others taking part in Study Audits will always need to exercise judgement as to the nature and extent of their examinations. The objective should be to reconstruct the study by comparing the final report with the study plan, relevant SOPs, raw data and other archived material.

In some cases, Inspectors may need assistance from other experts in order to conduct an effective Study Audit, e.g., where there is a need to examine tissue sections under the microscope.

When conducting a Study Audit, the Inspector should:

— obtain names, job descriptions and summaries of training and experience for selected personnel engaged in the study(ies) such as the Study Director and principal scientists;

— check that there is sufficient staff trained in relevant areas for the study(ies) undertaken;

— identify individual items of apparatus or special equipment used in the study and examine the calibration, maintenance and service records for the equipment;

— review the records relating to the stability of the test substances, analyses of test substance and formulations, analyses of feed, etc.;

— attempt to determine, through the interview process if possible, the work assignments of selected individuals participating in the study to ascertain if these individuals had the time to accomplish the tasks specified in the study plan or report;

— obtain copies of all documentation concerning control procedures or forming integral parts of the study, including:

i) the study plan;

ii) SOPs in use at the time the study was done;

iii) log books, laboratory notebooks, files, worksheets, print-outs of computer-stored data, etc.; check calculations, where appropriate;

iv) the final report.

In studies in which animals (i.e., rodents and other mammals) are used, the Inspectors should follow a certain percentage of individual animals from their arrival at the test facility to autopsy. They should pay particular attention to the records relating to:

— animal body weight, food/water intake, dose formulation and administration, etc.;

— clinical observations and autopsy findings;

— clinical chemistry;

— pathology.

Completion of Inspection or Study Audit

When a Test Facility Inspection or Study Audit has been completed, the Inspector should be prepared to discuss his findings with representatives of the test facility at a Closing Conference and should prepare a written report, i.e., the Inspection Report..

A Test Facility Inspection of any large facility is likely to reveal a number of minor deviations from GLP Principles but, normally, these will not be sufficiently serious to affect the validity of studies emanating from that test facility. In such cases, it is reasonable for an Inspector to report that the facility is operating in compliance with GLP Principles according to the criteria established by the (National) GLP Monitoring Authority. Nevertheless, details of the inadequacies or faults detected should be provided to the test facility and assurances sought from its senior management that action will be taken to remedy them. The Inspector may need to revisit the facility after a period of time to verify that necessary action has been taken.

If a serious deviation from the GLP Principles is identified during a Test Facility Inspection or Study Audit which, in the opinion of the Inspector, may have affected the validity of that study, or of other studies performed at the facility, the Inspector should report back to the (National) GLP Monitoring Authority. The action taken by that Authority and/or the Regulatory Authority, as appropriate, will depend upon the nature and extent of the non-compliance and the legal and/or administrative provisions within the GLP Compliance Programme.

Where a Study Audit has been conducted at the request of a Regulatory Authority, a full report of the findings should be prepared and sent via the relevant (National) GLP Monitoring Authority to the Regulatory Authority concerned.

References

Beernaert, H., Caroli, S., Clausing, P., Doherty, B., Edelmann, A., Hede, R., Helder, Th., Lange, J., Lehn, H., McCormack, J., Seiler, J.P., Turnheim, D. (2000), The Revised Principles of Good Laboratory Practice of the Organisation for Economic Cooperation and Development—Changes, Chances, and Controversies, Drug Information Journal 34 (1), pp. 33 – 45

Berny, C., and Long, D. (1989), Quality Assurance Impact: Can It Be Measured?, J. Amer. Coll. Toxicol. 8 (2), pp. 419 - 428.

Clausen, T., and Riius, P. (1997) Simple guidelines for experimental reports, documentation and storage of data. A new initiative to further research integrity, Dan. Med. Bull. 44 (1) pp. 85-87.

Committee on Public Labor and Welfare (1975), Preclinical and Clinical Testing by the Pharmaceutical Industry 1975, Joint Hearings before the Subcommittee on Health of the Committee on Labor and Public Welfare and the Subcommittee on Administrative Practice and Procedure of the Committee on the Judiciary. United States Senate, Ninetyfourth Congress. First Session on Examination of the process of drug testing and FDA's role in the regulation and conditions under which such testing is carried out, July 10 and 11, 1975.

Coombes, P. (2000), The Future of Electronic Signatures, Pharm. Tech. Europe 12 (4), pp.24 – 26.

Cooper-Hannan, R., Harbell, J.W., Coecke, S., Balls, M., Bowe, G., Cervinka, M., Clothier, R., Hermann, F., Klahm, L.K., de Lange, J., Liebsch, M., and Vanparys, Ph. (1999), The Principles of Good Laboratory Practice: Application to *In Vitro* Toxicology Studies, ATLA 27, pp.539 – 577.

Dalton, R. (1999), Roche's *Taq* patent "obtained by deceit", rules US court, Nature (news) 402, 16 December, p. 709.

Dent, N.J. (1994), GLP Compliance in the European Clinical Laboratory, Appl. Clin. Trials 3(4), 44 - 54.

DHHS (1992), Guidelines for the Conduct of Research Within the Public Health Service, US Department of Health and Human Services, Public Health Service, Office of the Assistant Secretary for Health, January 1 1992.

EPA (1995), Good Automated Laboratory Practices, Office of Information Resources Management, Research Triangle Park, NC 27711

Finney, D.J. (2000), Symbols and Terminology in Biometry, Biometrical Journal 42 (1), pp. 5 - 16.

Fox, M., Bloom, J.C., Conolly, M.M., and DeRisi, M.F. (1995), Picking a Centralized Laboratory, Appl. Clin. Trials, Supplement 1, p. 13.

Goodman, B. (1996), Decision In Imanishi-Kari Appeal Spurs Call for Changes In System, The Scientist, 10 (16), pp. 1 and 6 - 7.

Homberger, F.R., Boot, R., Feinstein, R., Kornerup-Hansen, A., and van der Logt, J. (1999), FELASA guidance paper for the accreditation of animal diagnostic laboratories, Lab. Animals 33 (Suppl. 1), pp. 19 – 38.

Humphrey, G.F. (1994), Scientific Fraud: The McBride Case - Judgment, Med. Sci. Law 34 (4), pp. 299 - 306.

Lane, R. (1997), Software Validation in the Pharmaceutical Industry – A Supplier's View, Pharm. Tech. Europe 9 (2), pp. 36 – 37.

Law, J. (1999), Fighting fraud in clinical trials, Scrip Magazine, September 1999, pp. 33 - 37.

OECD Series on Principles of Good Laboratory Practice and Compliance Monitoring

— Number 1: OECD Principles on Good Laboratory Practice (as Revised in 1997), Paris 1998

— Number 2 (Revised): Guidance for GLP Monitoring Authorities, Revised Guides for Compliance Monitoring Procedures for Good Laboratory Practice, Paris 1995

— Number 3 (Revised): Guidance for GLP Monitoring Authorities, Revised Guidance for the Conduct of Laboratory Inspections and Study Audits, Paris 1995

— Number 4 (Revised): Consensus Document, Quality Assurance and GLP, Paris 1999

— Number 5 (Revised): Consensus Document, Compliance of Laboratory Suppliers with GLP Principles, Paris 1999

- Number 6 (Revised): Consensus Document, The Application of the GLP Principles to Field Studies, Paris 1999

- Number 7 (Revised): Consensus Document, The Application of the GLP Principles to Short Term Studies, Paris 1999

- Number 8 (Revised(: Consensus Document, The Role and Responsibilities of the Dtudy Director in GLP Studies, Paris 1999

- Number9: Guidance for GLP Monitoring Authorities, Guidance for the Preparation of GLP Inspection Reports, Paris 1995

- Number 10: Consensus Document, The Application of the Principles of GLP to Computerised Systems, Paris 1995

- Number 11: Advisory Document of the Panel on Good Laboratory Practice, The Role and Responsibilities of the Sponsor in the Application of the Principles of GLP, Paris 1998

- Position of the OECD Panel on Good Laboratory Practice, The Use of Laboratory Accreditation with Reference to GLP Compliance Monitoring, Paris 1994

Rossiter, E.J.R. (1992), Reflections of a whistle-blower, Nature 357 (June 11), pp. 434 - 436.Schneider, K. (1983), Faking It - The Case Against Industrial Bio-Test Laboratories, The Amicus Journal, Spring 1983, pp. 14 - 26.

Segalstad, S.H. (1996), Supplier Auditing and Software, Eur. Pharm. Review, September, pp.37 - 44.

Stokes, T. (1999), Computer Validation in the Regulated Laboratory, Pharm. Tech. Europe 11 (12), pp. 16 - 24.

UK (1989), The aplication of GLP Principles to computer systems (Advisory Leaflet number 1), UK GLP Compliance Programme, Department of Health, London.

Weiss, R.B., Rifkin, R.M., Stewart, F.M., Theriault, R.L., Williams, L.A., Herman, A.A., and Beveridge, R.A. (2000), High-dose chemotherapy for high-risk primary breast cancer: an on-site review of the Bezwoda study, The Lancet, 355 (9208), pp. 999 - 1003.

Useful Internet Addresses

OECD	www.oecd.org/ehs/glp.htm
FDA (Homepage)	www.fda.gov
EPA	www.epa.gov/oeca/polyguid/glp/glp.html
Code of Federal Regulations (USA)	www.access.gpo.gov/nara/cfr/index.htm
Federal Register (USA)	www.access.gpo.gov/su_docs/aces/aces140.htm
European Union, GLP site	http:\\europa.eu.int/comm/enterprise/chemicals/glp/glp.htm
UK GLP	www.doh.gov.uk/practice.htm
British Association for Research Quality Assurance (BARQA)	www.barqa.com/GLP/body_glp.html
International Organisation for Standardisation (ISO)	www.iso.ch/
Society for Quality Assurance (SQA, USA)	www.sqa.org
GLPOnline (internet site of P. Lepore, former head of FDA's GLP Compliance Monitoring)	www.glpguru.com/

... with further links to be found on these sites

Subject Index

Access restriction	162, 164, 267	CRO	97, 110, 120, 210
Accountability	16, 203	Data retention	41
Accreditation	48, 337, 359	Deputy	60, 97, 126, 254
Amendment	80, 241, 257	Deviation	80, 241
Animal drug	23	Disposal	263
Animal protection	199	Efficacy study	24
Apparatus	167, 347	Electronic records	86
Appointment (SD/PI)	96	Environmental conditions	157, 179
Archive	98, 165, 259, 345	EPA	9, 30, 51, 77, 99, 34, 143, 151, 173
- indexing	264		
- retrieval	264	EU council directive	12, 13, 363
- suitability	269	Evaluation, retrospective	188
Audit	146, 360		
Audit trail	180	Experiment	65
Back-up	192	Exploratory study	25
Calibration	16, 168, 171, 234	FDA	7, 9, 22, 30, 51, 17, 134, 143, 151, 173, 182, 251
Change control	190		
Check list	5		
Code of Federal Regulations	9, 30	Field study	58, 82, 135, 216, 262, 348
Communication	99, 105, 135, 174, 245	Flexibility	83
		GCP	20
Compliance statement	103, 251, 362	GLP Monitoring	29, 358
		GLP Introduction	338
Computerised system	101, 172	Guideline	40, 244, 348, 361
- complexity	194	Hard-copy	182

Information technology (IT)	175, 268	Quality assurance	63, 89, 90, 93, 98, 121, 266, 350
Inspection report	141	- statement	150, 154, 249
Isolation	160	- independence	123
Kennedy Hearings	7	- inspections	128
LIMS	101, 173, 187, 194	- SOP	128, 130, 138, 236
Maintanance	168, 190, 227	Quality system(s)	45, 48, 53, 176, 337
Management	54, 92, 94, 141, 177, 228, 238, 266, 338	Raw data	85, 115, 180, 262, 352
Master schedule	91, 136	- verified copies	87
Materials	167	Reagents	167
Metabolism	26	Reconstructability	15, 54, 75, 167, 202, 227, 355
Misconduct	33		
Multi-site study	62, 100, 135, 147, 154, 245	Recording, (immediate)	35, 178, 180, 245
OECD advisory document	117	Reference item	202, 354
		Regulatory study	65
OECD consensus documents	12, 52, 57, 71, 74, 89, 130, 134, 154, 173, 200, 217, 361	Replacement (SD/PI)	96, 104
		Report amendment	155, 255
		Report, re-formatting	256
OECD Council Decision	11, 13, 29, 30, 51, 358, 361, 363	Reproducibility	46
		Responsibility	94, 102, 249, 266, 346
Personnel	95, 114, 351		
Precision	46, 337	Sample retention	71, 220, 260
Principal Investigator	57, 69, 89, 100, 102, 107, 243, 249	Security	165, 192, 265
		Separation	158, 160, 340
		Short-term study	70, 153, 245, 252
Process-based inspections	130	SOP	63, 99, 100, 122, 178, 223, 339, 347
Quality	20, 47, 121, 127, 168	- approval	228

- content	232	- logistics	203, 346
- distribution	228	- preparation	163
- format	225	- stability	211, 216
- on-line	231	Test site	61, 89
Sponsor	110, 116, 210, 241, 270	Test systems	196
		- biological	198
Storage	159, 163	- characterisation	208
- period	259	- identification	199
Study conduct	246, 245, 354	- physical/chemical	197
- control	113	Tidiness	159
- dates	76, 243	Traceability	16, 170, 195, 247
- director	57, 67, 100, 102, 145, 178, 210, 238, 249, 270, 350	Tranparency	254
		user requirements	184
- identification	245	- specifications	184
- phase	60, 88, 151, 243	Validation	177, 182, 340
- phase, critical	89, 128	- plan	183
- plan	37, 73, 76, 80, 98, 119, 240, 353	Vendor audit	187, 340
- plan, approval	241		
- reconstruction	43		
- report	73, 103, 249		
- report, reformatting	118		
- splitting	66		
Suitability	167, 176, 178, 187, 199, 340, 342		
Test facility	61, 156, 339		
Test item	67, 162, 202, 354		
- characterisation	119, 208		
- expiry date	218		
- labelling	206		

Printing: Weihert-Druck GmbH, Darmstadt
Binding: Buchbinderei Schäffer, Grünstadt